Exterior Billiards

Alexander Plakhov

Exterior Billiards

Systems with Impacts Outside Bounded Domains

Springer

Alexander Plakhov
University of Aveiro
Aveiro, Portugal
and
Institute for Information
Transmission Problems
Moscow, Russia

ISBN 978-1-4614-4480-0 ISBN 978-1-4614-4481-7 (eBook)
DOI 10.1007/978-1-4614-4481-7
Springer New York Heidelberg Dordrecht London

Library of Congress Control Number: 2012944740

Mathematics Subject Classification (2010): 37D50, 49Q10, 49Q20, 49K30, 37N05, 76G25, 76M28

Printed on acid-free paper

Springer is part of Springer Science+Business Media (www.springer.com)

Introduction

Imagine that we are going to design a spaceship for a long voyage in open space. During the voyage, the ship will cross huge rarefied clouds of interstellar gas. Our goal is to make the ship's shape as streamlined as possible, so as to minimize the loss of velocity when moving in the clouds.

To specify this task, we need to make a number of assumptions concerning the state of the cloud, its interaction with the spaceship surface, the nature of the ship's motion, and a description of shapes the ship could have. (In what follows the spaceship will be called the *body*, and the cloud will be called the *medium*.) Throughout the book it is assumed that the medium is homogeneous and consists of point particles, and in addition the following conditions are fulfilled:

1. The particles of the medium do not interact with each other.
2. When hitting the body surface, the particles are reflected in a perfectly elastic manner.

Condition 1 is ensured by the fact that the space cloud is highly rarefied, so that the mutual interaction of particles can be neglected. Condition 2 means that the interaction of particles with the body is billiardlike.

Different settings of the problem correspond to cases where the medium temperature equals zero and where it is positive. The zero-temperature assumption is justified in the case where the velocity of thermal motion of the particles is much smaller than the spaceship's velocity; in general, this significantly simplifies the task. Further, the problem settings and methods of study are completely different in the case of translational motion and in the case where the body performs both translational and rotational motion. Finally, the nature of the problem and approaches to its solution vary greatly depending on the class of admissible bodies.

In particular, in the case of translational motion of *convex* bodies, the drag force (usually called the *resistance*) can be represented analytically as a function of the body's shape, and variational methods can be used to solve the minimum resistance problem. This kind of problem has a long history originating from the publication by I. Newton in his *Principia* of the famous problem on minimal resistance of convex axisymmetric bodies and continuing today in a series of papers in the 1990s and 2000s related to the minimal resistance of convex (not necessarily symmetric) bodies [8, 11, 12, 29, 30]. If we consider *nonconvex* bodies, an explicit analytical expression for the resistance becomes impossible and one needs to use billiard techniques to minimize the resistance. If, additionally, the body rotates in the course of forward motion, one must appeal to methods of optimal mass transportation.

This book provides a review of these problems and describes methods for solving them. The main part of the book is dedicated to results obtained by the author and his collaborators. Most of the attention is given to the case where the body is nonconvex, and therefore reflections of particles from its surface are generally multiple. Chapter 3, related to the motion of convex bodies in media with nonzero temperatures, is an exception.

These problems originating from classical mechanics also allow a natural interpretation from the viewpoint of geometrical optics, where particles incident on a body are replaced with light rays falling on the specular surface of the body and reflected according to the rule that says the angle of incidence equals the angle of reflection. In some cases, the optical setting is in better agreement with empirical reality than the mechanical one. Indeed, light rays practically do not mutually interact, and the elastic reflection law approximation is usually much more precise for them than for gas particles.

Optical problems on light scattering by a reflecting surface have their own specific character. We consider, in particular, problems on invisible bodies and retroreflectors. Invisibility in a certain direction means that any light ray falling on a body in this direction and its extension behind the point of the last reflection lie on the same straight line. A retroreflector is a body that reverses the direction of any incident light ray. A well-known example of a "partial" retroreflector is the inner part of a cube corner: a portion of incident light rays makes three successive reflections from the cube's faces and then moves in the direction opposite to the direction of incidence. From a mechanical point of view, an invisible body has *zero resistance* when moving through a medium *in a fixed direction*, and a retroreflector has the *greatest possible resistance* when moving *in any direction*.

The next important problem is related to a description of elastic scattering of particles by a rough surface. We consider a surface that looks smooth to the naked eye but contains "microscopic" unevennesses invisible to the naked eye, e.g., dimples, grooves, cracks. A point particle falling on a body and entering into a dimple or groove makes one or several reflections there and eventually escapes in a direction that does not obey the law of reflection, according to which the angle of incidence equals the angle of reflection. Moreover, one cannot predict the direction of escape; instead, the statistical distribution for this direction can be determined. That is, the *billiard scattering law* at a given point of a surface and for a given velocity of incidence should describe the probability distribution over the velocities of escape. We will see subsequently that it is natural to define the scattering law at a point as a joint distribution of a pair of vectors (*velocity of incidence, velocity of reflection*), and the law of scattering by a whole rough surface is naturally defined as a joint distribution of a vector triple (*velocity of incidence, velocity of reflection, normal to the surface at the point of impact*).

There is a vast body of literature in the natural sciences dedicated to rough surfaces. A variety of models of *real* rough surfaces utilizing, for example, periodic, fractal, and random functions have been developed. In contrast, we provide a unique description of all *geometrically* possible rough surfaces (where the molecular structure of *real* bodies is ignored).

There is a huge variety of shapes of roughness, and it seems probable that the variety of the corresponding scattering laws is also very large. One of the book's chapters is dedicated to the characterization of scattering laws. In very general terms, the solution is as follows: a joint distribution of two or three vectors is a law of scattering by a rough surface if, first, it is symmetric with respect to a certain vector exchange and, second, two natural projections of this distribution coincide with some predetermined measures.

We believe that studying billiard scattering by rough surfaces is of potential interest for space aerodynamics. Consider again an illustrative example of a spaceship moving through an interstellar cloud. Imagine that as a result of the movement of astronauts in inner compartments, the ship very slowly turns around its center of mass in a random uncontrollable fashion, that is, does somersaults. Originally, the ship is a convex body. Our goal is to apply a roughening on its surface so that the (time-averaged) resulting resistance force is minimal. This problem reduces to minimizing a certain functional defined on the set of scattering laws and can be reformulated in terms of optimal mass transportation, where the initial and final mass distributions are concentrated on the unit sphere and correspond to the distributions over velocities of the incident and reflected particle flows. The mass transfer is identified with the scattering law, and the cost of the transfer with the resistance force. A separate chapter is devoted to solving special problems of mass transportation related to the problems of minimal resistance in which we are interested.

We will see that the force of resistance of a slowly somersaulting body can be decreased by means of roughening by 3.05 % at most. The very fact that the resistance can be decreased by roughening is quite surprising and nonintuitive; on the other hand, the insignificance of the decrease is disappointing. [Notice that a "wrong" roughening can result in an (at most twofold) increase of the resistance—this fact does not seem strange at all.] In the case of *fast* rotation, the relation between the roughness and the body dynamics is much more complicated and diverse; we study here the simplest example of a spinning rough two-dimensional disc.

In Chap. 1, basic mathematical notions used throughout the book are defined, and a brief review of the main results is given. Our intention is that the reader who reads only this chapter should get a clear idea of the main results of the book (but not their proofs). In Chap. 2, problems of minimal resistance as applied to *translational* motion of bodies in a medium are considered. In Chap. 3 a generalization of Newton's problem to convex axisymmetric bodies moving in media with positive temperature is studied. Auxiliary results on billiard scattering by nonconvex and rough bodies are presented in Chap. 4. In Chap. 5, some special problems on optimal mass transportation are solved explicitly. We believe they are of independent interest since they extend the (quite short at present) list of explicitly solvable optimal transportation problems. The results of Chaps. 4 and 5 are used in Chap. 6, where the problems of minimum and maximum resistance for translating and, at the same time, slowly rotating (somersaulting) bodies are considered. Chapter 7 examines the Magnus effect. This effect means that there exists a nonzero transversal component of the force acting on a spinning body in a flow of particles. In Chaps. 8 and 9, billiards possessing extremal properties of best and worst streamlining are studied. That is, we design bodies of zero resistance and bodies invisible in one and two directions, on the one hand, and bodies reversing the direction of particle flows, on the other hand.

I am grateful to G. Buttazzo, A. Stepin, and E. Lakshtanov for fruitful discussions on the subject. It was G. Buttazzo who persuaded me to write this book. The work was prompted by a reading of the book by V. Tikhomirov *Stories About Maxima and Minima* as I was preparing my classes for undergraduate students. Many results of the book are coauthored with A. Aleksenko, P. Bachurin, P. Gouveia, K. Khanin, J. Marklof, G. Mishuris, V. Roshchina, T. Tchemisova, and D. Torres, and some results are based on personal communications by V. Protasov and J. Zilinskas. Many of these results have been reported at the Sinai seminar (Institute for Information Transmission Problems, Moscow), Anosov – Stepin seminar (Moscow State University), Kozlov – Treshchev seminar (Moscow State University) and discussed with the participants of these seminars. I am very grateful to all of them. Last but not least, I would like to thank my wife Alla for her patience and continued support of my work.

This work was partly supported by *FEDER* funds through *COMPETE*—Competitiveness Factors Operational Programme—and by Portuguese sources of funds through the Center for Research and Development in Mathematics and Applications (CIDMA) and the Portuguese Foundation for Science and Technology (FCT), within Project PEst-C/MAT/UI4106/2011 with COMPETE No. FCOMP-01-0124-FEDER-022690; by the FCT Research Projects PTDC/MAT/72840/2006 and PTDC/MAT/113470/2009; and by Grants of the President of Russia for Leading Scientific Schools NSh-8508.2010.1 and NSh-5998.2012.1.

Preface

This book contains an account of results obtained by the author and his collaborators on billiards in the complement of bounded domains and their applications in aerodynamics and geometrical optics.

We consider several problems related to the aerodynamics of bodies in highly rarefied media. It is assumed that the medium particles do not interact with each other and are elastically reflected when colliding with the body boundary; these assumptions drastically simplify the aerodynamics and make it possible to reduce it to a number of purely mathematical problems.

First we examine problems of minimal resistance in the case of translational motion of bodies. These problems generalize the Newton problem of least resistance; the difference is that the bodies are generally nonconvex in our case, and therefore the particles can make multiple reflections from the body surface. It is proved that typically the infimum of resistance equals zero; thus, there exist "almost perfectly streamlined" bodies.

Next we consider the generalization of Newton's problem on the minimal resistance of convex axisymmetric bodies to the case of media with a thermal motion of particles. Two kinds of solution are found: first, Newton-like bodies and, second, shapes obtained by gluing together two Newton-like bodies along their rear ends.

Further, we present results on the characterization of billiard scattering by nonconvex and rough bodies; next we solve some special problems of optimal mass transportation. These two groups of results are applied to problems of minimal and maximal resistance for bodies that move forward and at the same time slowly rotate. It is found, in particular, that the resistance of a three-dimensional convex body can be increased at most twice and decreased at most by 3.05 % by roughening its surface.

Next, we consider a rapidly rotating rough disc moving in a rarefied medium on a plane. It is shown that the force acting on the disc is not generally parallel to the direction of the disc motion, that is, it has a nonzero transversal component. This phenomenon is called the Magnus effect (proper or inverse, depending on the direction of the transversal component). We show that the kind of Magnus effect depends on the kind of disc roughness, and we study this dependence. The problem of finding all admissible values of the force acting on the disc is formulated in terms of a vector-valued problem of optimal mass transportation.

Finally, we describe bodies that have zero resistance when being translated through a medium and present results on the existence or nonexistence of bodies with a mirror surface invisible in one or several directions. We also consider the problem of constructing retroreflectors: bodies with a specular surface that reverses the direction of any incident beam of light.

Aveiro, Portugal Alexander Plakhov

Contents

Chapter 1
Notation and Synopsis of Main Results

In this chapter, we introduce the main mathematical notation that will be used throughout the book and state the main results of the book. The proofs of these results are given in Chaps. 2–9.

1.1 Definition of Resistance

Consider Euclidean space $\mathbb{R}^d$, $d \geq 2$.

Definition 1.1. A bounded subset of $\mathbb{R}^d$ with a piecewise smooth boundary is called a *body* and is denoted by B. As usual, a *convex body* is a convex set with nonempty interior. Throughout what follows, convex bodies are assumed to be *bounded* and are denoted by C.

Remark 1.1. According to this definition, but contrary to physical intuition, a body is not necessarily connected. This is because we do not require this in most of the results presented in the book. When we nevertheless need the condition, we speak of a "connected body."

Remark 1.2 In Sects. 2.5 and 9.1.1, we consider *unbounded* sets with a piecewise smooth boundary; in this case we use the term *unbounded body*.

Note that a convex body does not necessarily have a piecewise smooth boundary, so a convex body is not necessarily a "body."

For a regular point $\xi \in \partial C$ we denote the unit outward normal to ∂C at ξ by $n(\xi)$ and supply $\partial C \times S^{d-1}$ with the measure $\mu = \mu_{\partial C}$ by the formula $d\mu(\xi, v) = b_d |n(\xi) \cdot v| \, d\xi \, dv$, where the dot means the inner product and $d\xi$ and dv are the $(d-1)$-dimensional Lebesgue measures on ∂C and S^{d-1}, respectively. The quantity $b_d = \Gamma(\frac{d+1}{2})\pi^{(1-d)/2}$ is a normalizing coefficient chosen so that $\mu\left(\partial C \times S^{d-1}\right) = 2|\partial C|$. It is the reciprocal of the volume of the unit $(d-1)$-dimensional ball; in particular, $b_2 = 1/2$ and $b_3 = 1/\pi$. We consider the measurable spaces

$$(\partial C \times S^{d-1})_\pm := \{(\xi, v) \in \partial C \times S^{d-1} : \pm n(\xi) \cdot v \geq 0\}$$

A. Plakhov, *Exterior Billiards: Systems with Impacts Outside Bounded Domains*, DOI 10.1007/978-1-4614-4481-7_1,

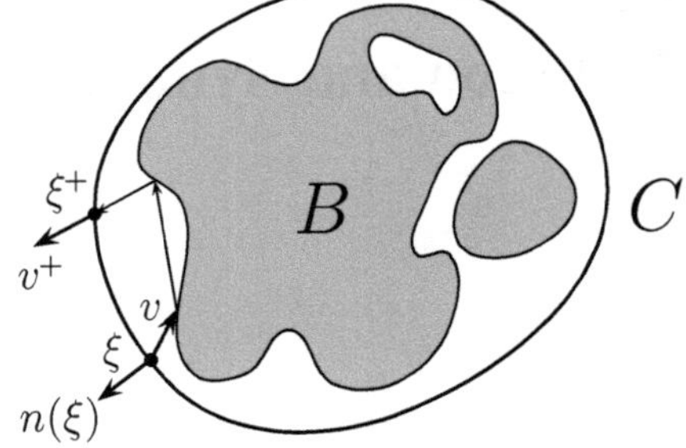

Fig. 1.1 Billiard in $\mathbb{R}^d \setminus B$

with induced measure μ. Informally speaking, $(\partial C \times S^{d-1})_-$ and $(\partial C \times S^{d-1})_+$ are sets of particles coming into C and going out of C, respectively, and μ measures the (normalized) number of incoming or outgoing particles. We have $\mu\left((\partial C \times S^{d-1})_\pm\right) = |\partial C|$, so that the number of particles incoming across ∂C, as well as the number of outgoing particles, is equal to the surface area of C.

In the sequel, we will also use the notation

$$(\partial C \times A)_\pm := \{(\xi, v) \in \partial C \times A : \pm n(\xi) \cdot v \geq 0\},$$

where A is a subset of $\mathbb{R}^d$.

The involutive map $\mathcal{I} = \mathcal{I}_C : (\xi, v) \mapsto (\xi, -v)$ is defined on $\partial C \times S^{d-1}$, and it maps $(\partial C \times S^{d-1})_-$ one to one onto $(\partial C \times S^{d-1})_+$ (and vice versa).

Consider a body $B \subset C$ and the billiard in $\mathbb{R}^d \setminus B$. We define a map $T_{B,C}$: $(\xi, v) \mapsto (\xi^+_{B,C}(\xi, v), v^+_{B,C}(\xi, v))$ between the subspaces $(\partial C \times S^{d-1})_-$ and $(\partial C \times S^{d-1})_+$ as follows.

Let $(\xi, v) \in (\partial C \times S^{d-1})_-$. A billiard particle starts its motion from point ξ with velocity v, moves in $C \setminus B$ for some time, possibly reflecting from the boundary of B (this may not happen), and finally crosses ∂C again, at a point $\xi^+_{B,C}(\xi, v)$ and with velocity $v^+_{B,C}(\xi, v)$, and leaves C (Fig. 1.1). In particular, if ξ happens to be a regular point of ∂B, then the period of time when the particle stays in C reduces to a point; in this case, we set $\xi^+_{B,C}(\xi, v) = \xi$ and $v^+_{B,C}(\xi, v) = v - 2(n(\xi) \cdot v)\, n(\xi)$.

The map $T_{B,C}$ thus defined establishes a one-to-one correspondence between full-measure subsets of the spaces $(\partial C \times S^{d-1})_-$ and $(\partial C \times S^{d-1})_+$. In addition, it preserves μ and satisfies the equality $T^{-1}_{B,C} = \mathcal{I}\, T_{B,C}\, \mathcal{I}$. In fact, this map determines the billiard scattering in $\mathbb{R}^d \setminus B$.

Notice that $v^+_{B,C}$ can be extended to a function v^+_B on a full-measure subset of $\mathbb{R}^d \times S^{d-1}$ that specifies the velocity of the reflected particle whose position and velocity at an arbitrary moment t before being reflected are equal to $\xi + vt$ and v, respectively. The function v^+_B is translation invariant: $v^+_B(\xi + v\tau, v) = v^+_B(\xi, v)$ for real τ.

We shall consider functionals of the form

$$R_\chi[T_{B,C}] = \int_{(\partial C \times S^{d-1})_-} c(v, v^+_B(\xi, v)) \cdot |n(\xi) \cdot v|\, d\xi\, d\chi(v),$$

where χ is a Borel probability measure on S^{d-1} and

$$c : S^{d-1} \times S^{d-1} \to \mathbb{R}^q, \quad q \geq 1$$

is a (generally vector-valued) continuous function satisfying the condition

$$c(v, v) = 0. \tag{1.1}$$

Thus, the functional $R[T_{B,C}]$ also takes values in $\mathbb{R}^q$.

Proposition 1.1. *If $B \subset C_1$ and $B \subset C_2$, then $R_\chi[T_{B,C_1}] = R_\chi[T_{B,C_2}]$.*

Proof. Let $(\partial C \times S^{d-1})_-^B$ be the set of values $(\xi, v) \in (\partial C \times S^{d-1})_-$ such that the corresponding billiard particle reflects from ∂B at least once, and let $T_{B,C}^{\text{ess}}$ be the restriction of the map $T_{B,C}$ to $(\partial C \times S^{d-1})_-^B$. The restriction of this map to the complementary subset preserves the second component v, that is, $v_B^+(\xi, v) = v$.

For each $(\xi, v) \in (\partial C_1 \times S^{d-1})_-^B$ the line $\xi + vt$, $t \in \mathbb{R}$ has a nonempty intersection (namely, one or two points) with ∂C_2. Let ξ' be a point in this intersection such that $(\xi', v) \in (\partial C_2 \times S^{d-1})_-$, and let $\mathcal{T}_{C_1,C_2,B}(\xi, v) := (\xi', v)$. The map

$$\mathcal{T}_{C_1,C_2,B} : (\partial C_1 \times S^{d-1})_-^B \to (\partial C_2 \times S^{d-1})_-^B$$

thus defined is one-to-one and leaves invariant the second component v. Moreover, it satisfies the relation $\mathcal{T}_{C_1,C_2,B}^{-1} = \mathcal{T}_{C_2,C_1,B}$ and preserves the measure $|n(\xi) \cdot v|\, d\xi\, d\chi(v)$ for any such χ. Finally,

$$T_{B,C_1}^{\text{ess}} = \mathcal{I}\, \mathcal{T}_{C_2,C_1,B}\, \mathcal{I}\, T_{B,C_2}^{\text{ess}}\, \mathcal{T}_{C_1,C_2,B}. \tag{1.2}$$

Since $c(v, v) = 0$, it follows that

$$R_\chi[T_{B,C_1}] = \int_{(\partial C_1 \times S^{d-1})_-^B} c(v, v_B^+(\xi, v))\, |n(\xi) \cdot v|\, d\xi\, d\chi(v).$$

We make the change of variables $(\xi, v) \mapsto (\tilde{\xi}, v) = \mathcal{T}_{C_1,C_2,B}(\xi, v)$ in this integral. By (1.2),

$$T_{B,C_1}^{\text{ess}}(\xi, v) = \mathcal{I}\mathcal{T}_{C_2,C_1,B}\mathcal{I}T_{B,C_2}^{\text{ess}}(\tilde{\xi}, v),$$

and we have $v_B^+(\xi, v) = v_B^+(\tilde{\xi}, v)$ since the second component v is unchanged under $\mathcal{I}\mathcal{T}_{C_2,C_1,B}\mathcal{I}$. Furthermore,

$$|n(\xi) \cdot v|\, d\xi\, d\chi(v) = |n(\tilde{\xi}) \cdot v|\, d\tilde{\xi}\, d\chi(v).$$

Thus,

$$R_\chi[T_{B,C_1}] = \int_{(\partial C_2 \times S^{d-1})_-^B} c(v, v_B^+(\tilde{\xi}, v))\, |n(\tilde{\xi}) \cdot v|\, d\tilde{\xi}\, d\chi(v),$$

so that $R_\chi[T_{B,C_1}] = R_\chi[T_{B,C_2}]$. The proof of Proposition 1.1 is complete. □

This proposition shows that the value of $R_\chi[T_{B,C}]$ depends only on B and not on the choice of the ambient convex body C. Hence we may write $R_\chi(B)$ in place of $R_\chi[T_{B,C}]$,

$$R_\chi(B) := \int_{(\partial C \times S^{d-1})_-} c(v, v_B^+(\xi, v))\, |n(\xi) \cdot v|\, d\xi\, d\chi(v). \tag{1.3}$$

The functional R is interpreted as the force of resistance of the medium acting on the body, where the distribution of the particles over velocities (in a reference system connected with the body) is given by χ. The concrete value of the integrand c is defined by the concrete mechanical model serving as a prototype for the problem under consideration. Thus, the function $c(v, v^+) = v - v^+$ corresponds to the case where a flow of particles falls on a resting body, and in addition the distribution of velocities in the flow is given by χ. In this case, R_χ in (1.3) is the force of resistance of the body to the flow. The integrand $v - v_B^+(\xi, v)$ is proportional to the momentum transmitted to the body by an individual particle.

The function $c(v, v^+) = (v - v^+) \cdot v$ corresponds to the case of a parallel flow of particles impinging on the resting body, with the direction of the flow being a random variable on S^{d-1} with distribution χ. In this case, the value R_χ is the expectation (mean value) of the component of pressure force of the flow along the direction of the flow. The integrand $(v - v_B^+(\xi, v)) \cdot v$ is proportional to the projection of the momentum transmitted to the body by an individual particle on the direction of the flow.

In other settings of mechanical problems, one must take other functions c (both scalar and vector-valued). Some of these functions are considered in Chap. 7, which is dedicated to problems of resistance optimization for rapidly rotating rough bodies.

1.2 Newton's Aerodynamic Problem

Here we describe Newton's aerodynamic problem (or the problem of minimal resistance) and its generalizations and state some new results obtained in this area in the 1990s and 2000s.

We consider the three-dimensional case, $d = 3$. Let $c(v, v^+) = (v - v^+) \cdot v$, and let δ_{v_0} be the probability measure on S^2 concentrated at a point $v_0 \in S^2$. The functional $R_{\delta_{v_0}}$ determines the longitudinal component of the resistance of the medium to the translational motion of a body with velocity $-v_0$ (or, which is the same, the longitudinal component of the pressure force of a parallel flow of particles at the velocity v_0 impinging on the resting body).

Consider a right circular cylinder C_h of height h and unit radius and a unit vector v_0 parallel to the axis of the cylinder. The problem is to find the minimum value of $R_{\delta_{v_0}}(B)$ in the class of convex bodies B lying in C_h and tangent to all its elements.

We can represent $R_{\delta_{v0}}$ in a convenient analytic form. Take an orthonormal system of coordinates x_1, x_2, x_3 such that the cylinder has the form $C_h = \{(x_1, x_2, x_3) : x_1^2 + x_2^2 \le 1, -h \le x_3 \le 0\}$, and $v_0 = (0, 0, -1)$. The upper half of the surface of B is the graph of a function $-f_B$, where the opposite function $f_B : \mathrm{Ball}_1(0) \to [0, h]$ is convex and $\mathrm{Ball}_1(0) \subset \mathbb{R}^2$ is the unit ball $x_1^2 + x_2^2 \le 1$. In view of (1.3), the functional $R_{\delta_{v0}}$ takes the form

$$R_{\delta_{v0}}(B) = \int_{\mathrm{Ball}_1(0)\times\{0\}} (v_0 - v_B^+(\xi, v_0)) \cdot v_0 \, d\xi, \tag{1.4}$$

where $\xi = (x_1, x_2, 0)$ and $d\xi$ is a two-dimensional Lebesgue measure. Considering that each particle impinging on the body hits it precisely once, so that

$$v_B^+(\xi, v_0) = v_0 + 2(1 + |\nabla f_B(x_1, x_2)|^2)^{-1} \left(-\frac{\partial f_B}{\partial x_1}(x_1, x_2), \, -\frac{\partial f_B}{\partial x_2}(x_1, x_2), \, 1 \right),$$

we see that $R_{\delta_{v0}}(B) = 2\mathcal{R}(f_B)$, where

$$\mathcal{R}(f) = \iint_{\mathrm{Ball}_1(0)} \frac{dx_1\, dx_2}{1 + |\nabla f(x_1, x_2)|^2}. \tag{1.5}$$

Thus, the problem of minimum resistance takes the following form.

Problem 1. Find $\inf \mathcal{R}(f)$ in the class of convex functions $f : \mathrm{Ball}_1(0) \to [0, h]$.

Initially, the problem of minimum resistance was considered by Newton [40] for a narrower class of convex bodies B that do not merely lie in the cylinder C_h and touch its lateral surface but are also *symmetric* relative to the vertical axis Ox_3. In that case, the function f_B describing the upper half of the surface of B is radial: $f_B(x_1, x_2) = \varphi_B(r)$, where $r = \sqrt{x_1^2 + x_2^2}$, and the problem takes the following form.

Problem 2. Find

$$\inf \int_0^1 \frac{r\, dr}{1 + \varphi'^2(r)} \tag{1.6}$$

in the class of convex nondecreasing functions $\varphi : [0, 1] \to [0, h]$.

The solution to Problem 2 (which Newton presented in geometric form and without proof) has the following form in modern notation:

$$\varphi(r) = 0 \quad \text{for} \quad 0 \le r \le r_0,$$

and for $r_0 \le r \le 1$ the function φ is described parametrically by

$$\begin{cases} r = \frac{r_0}{4}\left(u^3 + 2u + \frac{1}{u}\right) \\ \varphi(r) = \frac{r_0}{4}\left(\frac{3}{4}u^4 + u^2 - \ln u - \frac{7}{4}\right) \end{cases}, \qquad 1 \le u \le u_0.$$

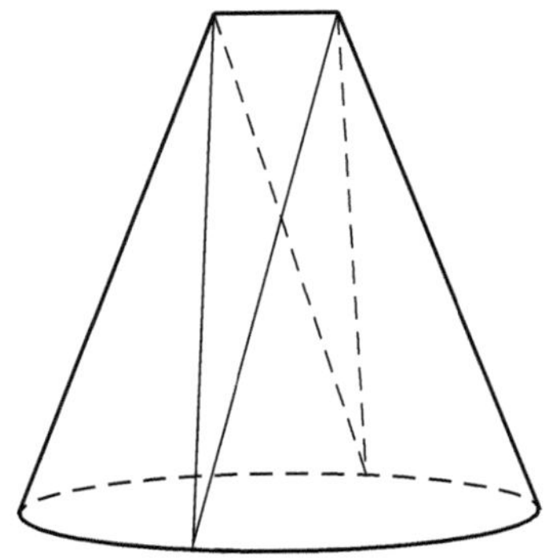

Fig. 1.2 A convex nonaxisymmetric body in the class $\mathfrak{D}(h)$, $h = 2$ having minimum resistance

Here $r_0 = r_0(h)$ and $u_0 = u_0(h)$ can be found from the system of equations

$$\frac{r_0}{4}\left(u_0^3 + 2u_0 + \frac{1}{u_0}\right) = 1, \quad \frac{r_0}{4}\left(\frac{3}{4}u_0^4 + u_0^2 - \ln u_0 - \frac{7}{4}\right) = h, \quad u_0 \geq 1.$$

A brief exposition and an elementary (accessible to high-school students) solution of Newton's problem can be found in Tikhomirov's paper [69] and in his book [70].

The solution to Newton's problem is a body bounded from above and from below by flat discs and recalls a truncated cone with a slightly inflated lateral surface. Figure 1.3a gives a good idea of its shape for $h = 2$. The lateral surface forms a 135° angle with the front surface (upper disc) along the disc boundary.

Problem 1 has been intensively studied since the early 1990s (see [8, 9, 11, 16–18, 18, 29–31]). It is known to be soluble, and the solution does not coincide with Newton's radial solution. It was found numerically in [29]; however, the properties of the solution have not been well understood until now. In addition, the solution of the problem $\inf_{f\in\mathfrak{D}(h)} \mathcal{R}(f)$ in a narrower class $\mathfrak{D}(h)$ was found analytically in [30]. The functions g in this class have the form $g_K = f_{B(K)}$, where $-f_{B(K)}$ describes the upper half of the surface of the set

$$B(K) = \mathrm{Conv}[(\mathrm{Ball}_1(0) \times \{-h\}) \cup (K \times \{0\})]$$

and $K \subset \mathrm{Ball}_1(0)$ is an arbitrary two-dimensional convex set. Here and in what follows, "Conv" denotes the convex hull. Thus, $B(K)$ is the convex hull of the union of the circular base $\mathrm{Ball}_1(0) \times \{-h\}$ and the convex set $K \times \{0\}$ contained in the horizontal plane Ox_1x_2. Notice that $\mathfrak{D}(h)$ contains the class of convex radially symmetric functions from $\mathrm{Ball}_1(0)$ to $[-h, 0]$.

We depict the solution of this problem for $h = 2$ in Fig. 1.2, where set K is a horizontal interval with midpoint at the origin.

Some results in the problem of least resistance have also been obtained for *nonconvex* bodies under the condition that each particle hits the body at most once (this assumption about the shape of the body is called the *single-impact assumption*); see [12, 16–18].

Further, in this book we consider problems related to the optimization of resistance in various classes of bodies, mostly nonconvex. In general, particles collide with a nonconvex body several times, so one cannot use simple analytic formulae like (1.5) or (1.6) to calculate the resistance. Instead, we must study billiards in the exterior of the body; in addition, in several cases, the optimization problems are reduced to special problems of optimal mass transfer.

Sections 1.3–1.9 present a synopsis of the main results of the book; to each chapter corresponds a separate section.

1.3 Problems of Least Resistance to Translational Motion of Nonconvex Bodies

If $\chi = \delta_{v_0}$, then we have a parallel flow of particles at velocity v_0 falling on a resting body. The value of the corresponding functional $R_{\delta_{v_0}}(B)$ in an appropriate reference frame has the form

$$R_{\delta_{v_0}}(B) = \int_{\mathbb{R}^2\times\{0\}} c(v_0, v_B^+(\xi, v_0))\, d\xi. \tag{1.7}$$

We choose a reference frame such that $v_0 = (0, 0, -1)$ and the body B lies in the half-space $x_3 \le 0$. In fact, the body lies in a sufficiently large cylinder $\mathrm{Ball}_r(0) \times [-H,\, 0]$, and in (1.7) we integrate over the top base of the cylinder $\mathrm{Ball}_r(0) \times \{0\}$, while outside the base we have $v_B^+(\xi, v_0) = v_0$, so the integrand vanishes.

The function c is continuous and nonnegative and satisfies $c(v, v) = 0$. In the special case of $c(v, v^+) = (v - v^+) \cdot v$, the integral (1.7) has a straightforward physical interpretation: this is the resistance produced by the medium to the translational motion of a body with velocity $-v_0$.

Note that although the function v_B^+ is measurable on a full-measure subset of $\mathbb{R}^3 \times S^2$, its restriction to the subspace $v = v_0$ of measure zero is not necessarily defined. So we assume in addition that the restriction of v_B^+ to the subspace $v = v_0$ is a function defined almost everywhere and measurable with respect to a Lebesgue measure in $\mathbb{R}^3 \times \{v_0\}$. In effect, this means that the scattering of particles falling in the direction of v_0 is regular. We assume that this condition holds for all bodies considered throughout this section.

In Chap. 2, we consider a generalized Newton problem of a body of least resistance—generalized because we are looking for the minimum in wider classes $\mathcal{P}(h)$ and $\mathcal{S}(h)$ of *nonconvex* bodies inscribed in a fixed cylinder.

Let $\mathcal{P}(h)$ be the class of connected (nonconvex in general) sets B lying in the cylinder $C_h = \mathrm{Ball}_1(0) \times [-h,\, 0]$ and such that the orthogonal projection of B on the plane Ox_1x_2 is the disc $\mathrm{Ball}_1(0)$. This is a broader class than those discussed previously in Sect. 1.2. The problem of minimum resistance in this class has an unexpected answer:

$$\inf_{B\in\mathcal{P}(h)} R_{\delta_{v_0}}(B) = 0. \tag{1.8}$$

That is, the resistance of bodies inscribed in a fixed cylinder can be made arbitrarily small.

We also consider the class $\mathcal{S}(h)$ of connected sets lying in the cylinder C_h and containing at least one section $\text{Ball}_1(0) \times \{c\}$, $-h \le c \le 0$ of it. This is a subclass of the previous class, $\mathcal{S}(h) \subset \mathcal{P}(h)$, but nevertheless is broader than the classes considered previously. The answer in this class is the same:

$$\inf_{B \in \mathcal{S}(h)} R_{\delta_{v_0}}(B) = 0. \tag{1.9}$$

Next we consider a cylinder with an arbitrary (not necessarily circular) base and show that the infimum of resistance of bodies inscribed in this cylinder is also equal to zero.

Further, we consider the class of connected bodies B such that $C_1 \subset B \subset C_2$, where C_1 and C_2 are fixed bounded connected bodies in $\mathbb{R}^3$ such that $C_1 \subset C_2$ and $\partial C_1 \cap \partial C_2 = \emptyset$. Again,

$$\inf_{C_1 \subset B \subset C_2} R_{\delta_{v_0}}(B) = 0. \tag{1.10}$$

Relation (1.10) can be interpreted as follows. Any convex body can be transformed within the ε-neighborhood of its boundary so that when the resulting body moves in the prescribed direction in a medium of resting particles, it encounters a resistance smaller than an arbitrarily prescribed quantity $\varepsilon > 0$.

Then we consider the minimization problem for analogs of these classes in the two-dimensional case, $d = 2$. In this case, the least resistance is always positive. We find it explicitly for $c(v, v^+) = (v - v^+) \cdot v$.

Finally, we consider the problem of least specific resistance for unbounded bodies. This problem was first stated by Comte and Lachand-Robert in [18] under the single-impact assumption. We do not impose this assumption; thus, in our setting a particle may collide several times with the body's surface. We find, in particular, that the infimum of the specific resistance of bodies containing a fixed half-space in a flow perpendicular to the boundary of the half-space equals one half of the resistance of the half-space itself.

1.4 Generalized Newton's Problem in Media with Positive Temperature

In Chap. 3, we address the problem of minimum resistance to the translational motion of bodies in a medium with *thermal motion* of particles. This problem, like the classical Newton's problem, is considered in the class of convex axisymmetric bodies with fixed length and width.

While the solution of this problem is conventional, the solutions are more diverse. Unlike in Newton's original problem, one must take into account the composition of the medium: the solution for a homogeneous (and, therefore, containing molecules

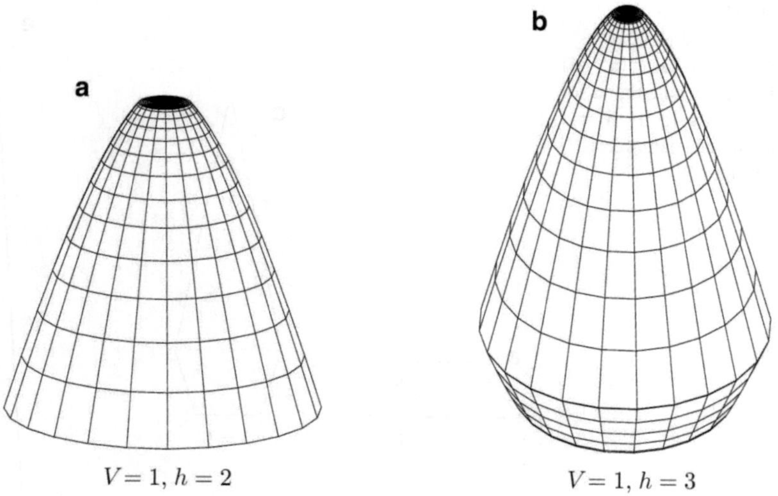

Fig. 1.3 Solutions of the three-dimensional problem in the class of convex bodies of revolution of height h whose maximal cross section is a unit circle. The motion proceeds in a rarefied homogeneous monatomic ideal gas; the velocity of the body is V. The mean square velocity of the gas molecules is 1

of the same mass) gas is not the same as for a gas consisting of several homogeneous components (and, thus, containing molecules of different masses).

In the three-dimensional case, there are two distinct kinds of solution. A solution of the first kind is similar to the solution of the classical Newton's problem, that is, its surface can be described in the same way as the surface of Newton's solution. Notice that, unlike in the Newton solution, the angle between the lateral surface and the front disc at the points of the disc boundary is not generally 135°.

A solution of the second kind is a union of two bodies similar to Newton's solution "glued together" along the rear parts of their surfaces. The length (along the direction of the motion) of the front body is always larger than that of the rear "reversed" body.

Letting h and the velocity distribution of the particles be fixed and changing the velocity V of the body, we find that a solution of the first kind is realized for $V \geq V_c$ and of the second kind for $V < V_c$, where $V_c = V_c(h)$ is a certain critical value depending on h. We present examples of solutions of the first and second kinds in Fig. 1.3a, b, respectively. In these figures and in what follows, the body is assumed to move upward.

In the two-dimensional case, $d = 2$, the classification of solutions is somewhat more complicated. There exist five kinds of solution (Fig. 1.4a–e):

(a) A trapezium,
(b) An isosceles triangle,
(c) The union of a triangle and a trapezium,
(d) The union of two isosceles triangles,

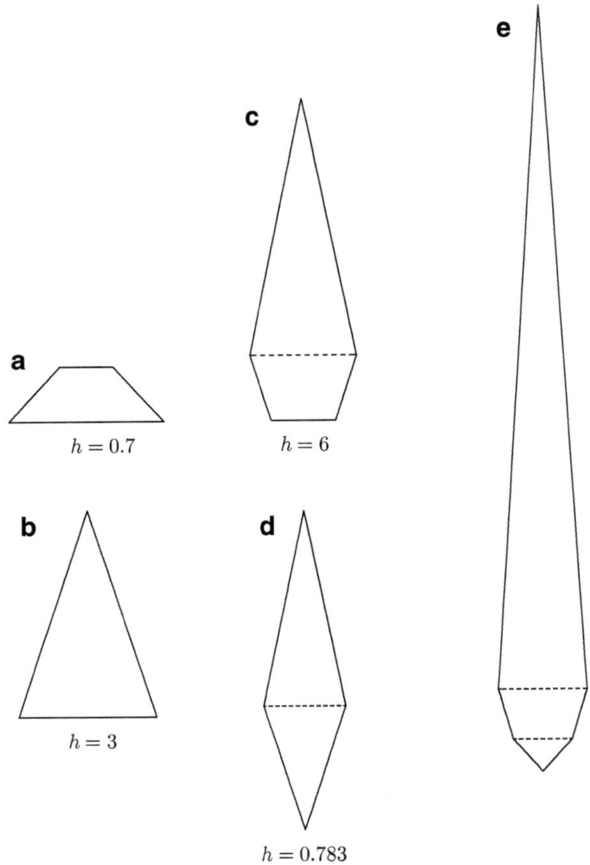

Fig. 1.4 The two-dimensional problem. The solutions in cases (**a**)–(**d**) are numerically calculated for a motion with velocity $V = 1$ in a gas; the gas parameters are as in Fig. 1.3

(e) The union of two triangles and a trapezium.

Solutions (a)–(d) are realized for arbitrary velocity distributions of the particles and for arbitrary V; solution (e) is realized only for some of them. The optimal shapes (a)–(d) appear in the simplest case of homogeneous monatomic gas, while shape (e) can appear in the case where the gas is a mixture of at least two homogeneous components. The numerical computation of solution (e) is a hard task and remains unsolved. We note that in the two-dimensional analog of Newton's problem (that is, with zero temperature), there are only two optimal shapes corresponding to cases (a) and (b).

In the limit cases, when the speed of the body is large or small in comparison with the mean speed of the particles, the shape of the body of least resistance is universal: it depends only on the length h but does not depend on the velocity distribution of the particles. In the first limit case ($V \to +\infty$), the optimal body coincides with the solution of the classical Newton's problem. In the second limit case ($V \to 0$), for $d = 3$, the optimal body is a solution of the second kind symmetric with respect to a plane perpendicular to the direction of motion, and the angle between this plane and the lateral surface at its upper and lower points is always 51.8°, whereas for $d = 2$ the optimal body is one of the following four figures:

(a) A trapezium if $0 < h < 1.272$,
(b) An isosceles triangle if $h = 1.272$,
(c) The union of an isosceles triangle and a trapezium if $1.272 < h < 2.544$,
(d) A rhombus if $h \geq 2.544$.

In cases (a)–(c), the inclination of the lateral sides of these figures to the base is 51.8°, and in case (d) it is larger.

In a homogeneous monatomic ideal gas, the velocities of the molecules are distributed in accordance with the Gaussian law. Assume that the mean square velocity of the molecules is 1; then the type of the solution is determined by two parameters: the velocity V of the body and its length h. We define numerically the regions in the parameter plane corresponding to different kinds of solutions; in addition, for some values of the parameters we determine the shape of the optimal body and calculate the corresponding resistance. We carry out this work separately in the two- and three-dimensional cases.

1.5 Scattering in Billiards

In Chaps. 4 and 5, we do some preparatory work before passing (in Chaps. 6 and 7) to the case where bodies perform both translational and rotational motion.

Chapter 4 is devoted to billiard scattering by nonconvex and rough obstacles. First we consider a billiard in the exterior of a two-dimensional connected body B. The *law of billiard scattering* on B is the probability measure η_B describing the distribution of the pair (φ, φ^+), where φ is the incidence angle and φ^+ is the angle of emission going away of a randomly chosen particle incident on the body (Fig. 4.1). The angles are counted counterclockwise from the normal to $\partial(\text{Conv } B)$ and belong to $[-\pi/2, \pi/2]$.

The law of scattering on a body admits a convenient representation as follows. We define a sequence of *hollows* on the boundary of a body (there are three hollows in Fig. 4.1); the *law of billiard scattering in a hollow* is a probability measure defining the joint distribution of (φ, φ^+) for a randomly chosen particle going into the hollow. Further, the scattering law on the convex part of the boundary of the body is determined by the rule "the angle of incidence is equal to the angle of reflection" and is a measure concentrated on the diagonal $\varphi^+ = -\varphi$. The scattering law η_B is a weighted sum of the scattering laws in all the hollows of the body and on the convex part of its boundary.

In an arbitrary dimension, we define a *rough convex* body. The law of scattering on such a body is the joint distribution of a triple of vectors (v, v^+, n)—the initial and final velocities and the outer normal at the point of collision—for a randomly chosen particle hitting the body. Thus, a scattering law on a rough body $\mathcal{B}$ is a (not necessarily probability) measure $\nu_{\mathcal{B}}$ on $S^{d-1}_{\{v\}} \times S^{d-1}_{\{v+\}} \times S^{d-1}_{\{n\}}$. It is also convenient to consider the scattering law at a point on the surface of a rough body; it is the conditional measure $\nu_{\mathcal{B}}\rfloor_{n=n_0}$ defined on $S^{d-1}_{\{v\}} \times S^{d-1}_{\{v+\}}$, where n_0 is the outer normal to the body surface at that point.

In informal terms, we can describe a rough body as follows. The surface of a convex body is pocked with microscopic hollows (e.g., grooves, cracks), so that macroscopically the resulting (rough) body with hollows looks precisely convex, but billiard scattering on it can be utterly different. The mathematical definition is as follows: a rough body is associated with a sequence of bodies with hollows of size approaching zero. In addition, a sequence of such bodies must satisfy the condition of convergence of the sequence of corresponding scattering laws. Furthermore, an equivalence relation between such sequences is defined, and the convention is that equivalent sequences of bodies represent the same rough body.

Otherwise, we can say that a rough body is obtained by *grooving* a fixed convex body. Clearly, a convex body can be grooved in infinitely many ways, differing (informally speaking) by the shape of their hollows.

In Theorems 4.1–4.5 we give a complete characterization of scattering laws. Each statement has roughly the same form: we assert that a measure is a scattering law if and only if it has fixed marginals and possesses a certain symmetry property.

As examples we give two such statements (Theorem 4.3 and Corollary 4.2).

First, a measure on $[-\pi/2, \pi/2]^2$ can be weakly approximated by scattering laws η_B if and only if it is invariant relative to the exchange of variables $(\varphi, \varphi^+) \mapsto (\varphi^+, \varphi)$ and both its natural projections on $[-\pi/2, \pi/2]$ (that is, its marginals) coincide with the measure λ given by $d\lambda(\varphi) = \frac{1}{2} \cos\varphi \, d\varphi$.

Second, a measure on $S^{d-1}_{\{v\}} \times S^{d-1}_{\{v+\}}$ is a scattering law at a point of a rough body if and only if it is invariant relative to the transformation $(v, v^+) \mapsto (-v^+, -v)$ and its natural projections on $S^{d-1}_{\{v\}}$ and $S^{d-1}_{\{v+\}}$ are probability measures λ_{-n} and λ_n with the densities $b_d (v \cdot n)_-$ and $b_d (v^+ \cdot n)_+$, respectively. The normalizing coefficient b_d is defined in Sect. 1.1, and n is the outward normal to the body surface at the given point. Notice that the measures λ_{-n} and λ_n define the distributions of the incident and reflected flows of particles over velocities.

1.6 Problems of Optimal Mass Transportation

We will see in Chap. 6 that some problems of resistance optimization for rough surfaces, using the aforementioned Theorems 4.1–4.5, can be reduced to a problem of finding a measure with fixed marginals (that is, the scattering law) on $[-\pi/2, \pi/2]^2$ or $(S^{d-1})^2$ minimizing a certain linear functional. This problem in general is as follows.

Consider measurable spaces (X, λ_1) and (Y, λ_2) such that $\lambda_1(X) = \lambda_2(Y)$ and a continuous function $c : X \times Y \to \mathbb{R}$ (usually called the *cost function*). Let $\Gamma(\lambda_1, \lambda_2)$ be the set of measures ν on $X \times Y$ whose marginals (projections on X and Y) are, respectively, λ_1 and λ_2. (This means that for any two measurable sets $A_1 \subset X$ and $A_2 \subset Y$ holds $\nu(A_1 \times Y) = \lambda_1(A_1)$ and $\nu(X \times A_2) = \lambda_2(A_2)$.) The problem of minimization

$$\inf_{\nu \in \Gamma(\lambda_1, \lambda_2)} \iint_{X \times Y} c(x, y)\, d\nu(x, y) \tag{1.11}$$

is called the *problem of optimal mass transportation*, or the *Monge–Kantorovich problem.*

This problem can be interpreted as follows. We have two mass distributions given by the measures λ_1 and λ_2 on X and Y, respectively, and a function $c(x, y)$ defining the cost of transfer of a unit mass from $x \in X$ to $y \in Y$. A plan of mass transfer from the initial position λ_1 to the final position λ_2 (or just a transport plan) is given by a measure ν with marginals λ_1 and λ_2, and the total cost of the transfer with this plan is equal to the integral in (1.11). One needs to find the optimal transfer plan, that is, the measure ν_* minimizing the transfer cost.

In general, it seems impossible to provide an exact solution for an optimal transportation problem. The known cases of exactly soluble problems are quite rare; they are rather exceptions to the general rule. The case of the one-dimensional transport, where X and $Y \subset \mathbb{R}$, $c(x, y) = f(x \pm y)$, and f is strictly convex or concave, is the simplest one; then the optimal plan is monotone, that is, is given by a measure supported on the graph of a monotone function (see, e.g., [37]). We note a very interesting case considered by McCann [37] where $c(x, y) = f(|x - y|)$ and f is a positive strictly concave function. Note also the case where the measures λ_1 and λ_2 coincide and are uniform on the segment $X = Y = [0, 1]$, with $c(x, y) = h(x + y)$ or $h(x - y)$, where h has three intervals of monotonicity [72].

Below we cite some explicitly soluble cases of the two-dimensional mass transportation problem, where λ_1 and λ_2 are Lebesgue measures on compact sets X and Y in $\mathbb{R}^2$, and the cost function is the Euclidean distance, $c(x, y) = |x - y|$. The following examples are taken from papers by Levin [33–35].

1. The set Y is obtained by shifting X by a vector $b \in \mathbb{R}^2$, that is, $Y = X + b$. This shift actually produces optimal transportation; in other words, the measure supported on $\{(x, y) : x \in X,\ y = x + b\} \subset \mathbb{R}^4$ is an optimal transport plan.
2. X is a rectangle of size 1×2, and Y is the rectangle obtained by rotating X by $90°$ about its center.
3. X is an equilateral triangle, and Y is the triangle obtained by rotating X by $60°$ about its center.
4. X is an equilateral triangle, and Y is the triangle obtained by reflecting X relative to one of its sides.
5. X is a square, and Y is the square obtained by rotating X by $45°$ about its center.

In all these cases, the optimal transfer is generated by piecewise isometrical transformations.

In Chap. 5 some special optimal transfer problems are explicitly solved. First we consider a problem of mass transport from $\mathbb{R}$ to $\mathbb{R}$ with a cost function of the form $c(x, y) = f(x + y)$, where f is an odd function strictly concave on $\mathbb{R}_+ = \{x \leq 0\}$

(and therefore strictly convex on $\mathbb{R}_- = \{x \le 0\}$), in the case where the initial mass distribution coincides with the final one, $\lambda_1 = \lambda_2$. We impose some additional technical conditions on λ_1.

We show that the optimal measure is uniquely defined by its support, which belongs to the union of two lines on a plane: the ray $x = y \ge 0$ and a curve symmetric relative to this ray (Fig. 5.1). The curve belongs to a finite- or countable-parameter family of curves that does not depend on f and is defined merely by λ_1, whereas the choice of the optimal curve from this family is defined by f.

In an important particular case, the family is one-parameter, and therefore the problem reduces to minimization of a function of a real variable.

Further, we consider a special problem of mass transport on the unit sphere in $\mathbb{R}^d$. The initial and final spaces are complementary hemispheres, $X = S^{d-1}_{-n} := \{x \in S^{d-1} : x \cdot n \le 0\}$ and $Y = S^{d-1}_n := \{x \in S^{d-1} : x \cdot n \ge 0\}$, and the measures λ_1 and λ_2 are defined by the following condition: the orthogonal projection of each measure on the plane $x \cdot n = 0$ coincides with the Lebesgue $(d-1)$-dimensional measure on the circle $\{x : x \cdot n = 0,\ |x| \le 1\}$. The cost function is the squared distance, $c(x, y) = \frac{1}{2}|x - y|^2$.

This problem can be naturally interpreted in terms of billiard scattering by rough surfaces. Fix a point on a rough surface, and let n be the outward normal to the surface at this point. The flows of incident and reflected particles are identified with the hemispheres X and Y, respectively. With this identification, to any incident or reflected particle we assign its velocity v or v^+ (we have $v \cdot n < 0$ and $v^+ \cdot n > 0$). The measures λ_1 and λ_2 describe the densities of the incident and reflected flows. Each admissible measure $\eta \in \Gamma(\lambda_1, \lambda_2)$ defines a billiard scattering at that point (more precisely, the symmetrized measure $\eta_{\text{symm}} = \frac{1}{2}(\eta + \pi_{\text{d}}^{\#}\eta)$ is a scattering law at a point; here π_{d} exchanges the arguments, $\pi_{\text{d}}(x, y) = (y, x)$, and $\pi_{\text{d}}^{\#}$ is the induced map of measures). The cost function $\frac{1}{2}|v - v^+|^2$ is the (normalized) momentum transmitted to the body by a particle with the corresponding velocities of incidence and reflection, and the total cost of the transfer is the specific resistance at that point.

Since this problem possesses an axial symmetry relative to n, one can show that the optimal transfer is performed along the meridians (we take the sphere poles to be n and $-n$) and is axially symmetric. As a result, one comes to a one-dimensional problem identical to that considered earlier in the chapter.

Two schemes of mass transfer along the meridians are depicted in Fig. 1.5a and b. The transfer shown in Fig. 1.5a is induced by the law $\varphi^+ = \varphi$ ("the angle of incidence = the angle of reflection") and corresponds to reflection from a smooth surface. It is instructive to consider an argument showing that it is not optimal. Consider two small arcs I_1 and I_2 adjoining the equator and reverse the monotonicity of the transfer from I_1 to I_2, that is, replace monotone increasing with monotone decreasing. Since these arcs are "almost" rectilinear and the cost function equals the squared distance, the transfer cost will decrease under this reversal.

In Fig. 1.5b, the optimal transfer scheme is depicted in the case $d = 2$. Both the upper and lower halves of the meridian are divided into pairs of arcs, left and right. The transfer between the left arcs is monotone increasing, $\varphi^+ = \varphi$, whereas the transfer between the right arcs is monotone decreasing. The left and right arcs

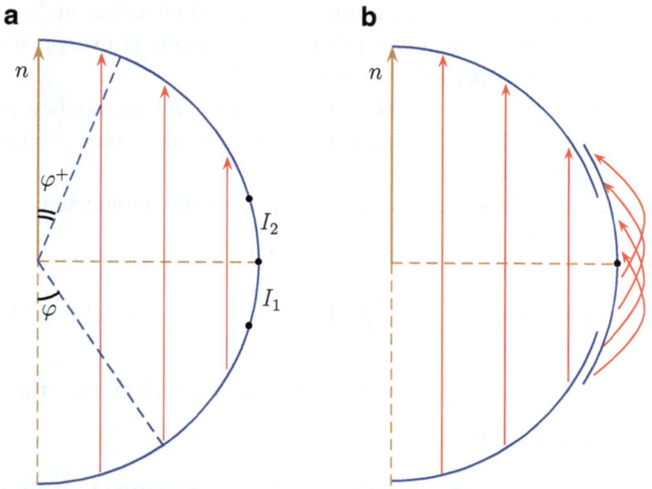

Fig. 1.5 Mass transfer along the meridian (from the lower part of the meridian to the upper one) is marked by *arrows*. The transfer induced by the shift along n is shown in panel (**a**), whereas the optimal transfer is shown in panel (**b**)

partly overlap; this means that the mass at each point of the "overlapping zone" splits into two parts, which are then transported to two different points. In terms of optimal transfer, this means that the transfer solves the Monge–Kantorovich, but not the Monge, problem.

In higher dimensions, $d \geq 3$, the scheme of optimal transfer is roughly the same; the most significant difference is that the overlapping of the two arcs disappears (or, more precisely, reduces to a point).

1.7 Optimizing the Mean Resistance

The value of the functional $R_u(B)$ in (1.3) with the cost function $c(v, v^+) = (v - v^+) \cdot v$ and the uniform probability measure u on S^{d-1} is interpreted as follows. A body B starts a translational motion in $\mathbb{R}^d$ in a medium of resting particles, with velocity v randomly chosen from a uniform distribution in S^{d-1}. The resistance $R_{\delta_v}(B)$ to this motion (more precisely, the projection of the resistance force on the direction of motion) is a random variable, and its mathematical expectation equals $R_u(B)$. The cost function can be written as $c(v, v^+) = \frac{1}{2}\,|v - v^+|^2$.

We can propose another interpretation of this functional: the body B moves translationally with fixed velocity while at the same time slowly rotating. The rate of rotation is small enough that we can neglect it in interactions of the body with individual particles. In a reference system attached to the body, the velocity vector draws a curve on the sphere S^{d-1}, thereby inducing a (singular) probability measure on the sphere: the measure of a subset of S^{d-1} is the (normalized) total time when the vector lies in this subset within a certain period of observation. We assume that as the period of observation extends, this measure weakly converges to u. Then the mean resistance over this period approaches $R_u(B)$.

The problems of minimizing and maximizing $R_u(B)$ are studied in Chap. 6 in different classes of bodies. The mathematical tools necessary for this study are elaborated in Chaps. 4 and 5.

In the two-dimensional case, the mean resistance of a connected body B can be represented in the form

$$R_u(B) = |\partial(\mathrm{Conv} B)| \iint_{\square} (1 + \cos(\varphi - \varphi^+))\, d\eta_B(\varphi, \varphi^+),$$

where $\square = [-\pi/2, \pi/2] \times [-\pi/2, \pi/2]$. We state the following problem.

Problem 3. Find $\inf R_u(B)$

(a) In the class of connected (generally speaking, **nonconvex**) bodies B of fixed area.
(b) In the class of **convex** bodies B of fixed area.

Using the results of Chap. 4 on the characterization of the measures η_B, we reduce Problem 3(a) to the following special optimal transportation problem:

$$\inf_{\eta \in \Gamma(\lambda, \lambda)} \iint_{\square} (1 + \cos(\varphi - \varphi^+))\, d\eta(\varphi, \varphi^+)$$

(where λ is the measure on $[-\pi/2, \pi/2]$ given by $d\lambda(\varphi) = \frac{1}{2} \cos\varphi\, d\varphi$), which we then solve using the results of Chap. 5. A minimizing sequence of bodies is constructed that can be identified with a rough disc of prescribed area. The solution in Problem 3(b) is a (standard) disc of the same area, and

$$\frac{\text{(the least resistance in the class of } \textbf{nonconvex} \text{ bodies)}}{\text{(the least resistance in the class of } \textbf{convex} \text{ bodies)}} = m_2 \approx 0.9878.$$

Allowing some freedom of speech, one can say that the body of least resistance in the class of convex bodies is a *disc*, and in the class of nonconvex bodies it is a *rough disc*, and the resistance of the latter body is approximately 1.22 % smaller than that of the former one.

Next we consider the following problem. Let C_1 and C_2 be bounded convex bodies such that $C_1 \subset C_2 \subset \mathbb{R}^2$ and $\partial C_1 \cap \partial C_2 = \emptyset$. We consider the class of convex bodies B such that $C_1 \subset B \subset C_2$.

Problem 4. Find (a) $\inf_{C_1 \subset B \subset C_2} R_u(B)$ and (b) $\sup_{C_1 \subset B \subset C_2} R_u(B)$.

The solution of Problem 4(a) essentially repeats that of Problem 3(a).

In cases (a) and (b), minimizing and maximizing sequences can be identified with rough bodies obtained by grooving C_1 and C_2, respectively, and we have

$$\frac{\inf_B R_u(B)}{R_u(C_1)} = m_2 \approx 0.9878 \quad \text{and} \quad \frac{\sup_B R_u(B)}{R_u(C_2)} = 1.5.$$

Next we show that the resistance of a body in a medium with thermal motion of particles is proportional to the resistance in a medium consisting of resting particles, with a coefficient that is larger than 1 and depends only on the nature of motion of the particles: the higher the temperature, the larger the coefficient, the larger the resistance. Hence, in a medium with positive temperature, Problems 3 and 4 have the same solution as before.

In the case of an arbitrary dimension $d \geq 2$, we solve problems of optimizing resistance in the class of rough bodies obtained by grooving a fixed convex body C.

Problem 5. Find

(a) $\frac{1}{R(C)} \sup\{R_u(\mathcal{B}) : \mathcal{B} \text{ is obtained by grooving } C\}$.
(b) $\frac{1}{R(C)} \inf\{R_u(\mathcal{B}) : \mathcal{B} \text{ is obtained by grooving } C\}$.

These ratios appear to depend only on the dimension d and not on the particular body C. Using the results on the characterization of measures generated by rough bodies we can reduce Problem 5 to the problem of optimal mass transportation on S^{d-1} considered in Chap. 5. One finds that

$$\frac{\sup_{\mathcal{B}} R_u(\mathcal{B})}{R(C)} = \frac{d+1}{2} \quad \text{and} \quad \frac{\inf_{\mathcal{B}} R_u(\mathcal{B})}{R(C)} = m_d,$$

where, in particular,

$$m_2 \approx 0.9878, \quad m_3 \approx 0.9694 \quad \text{and} \quad \lim_{d\to\infty} m_d = \frac{1}{2}\left(1 + \int_0^1 \sqrt{\ln z \ln(1-z)}\, dz\right) \approx 0.791.$$

We illustrate these results by the following example. Consider a spherical artificial satellite rotating around the Earth and being decelerated by the thin atmosphere. Assume that the surface of the satellite is made of materials ensuring that molecules of the atmosphere reflect from it elastically. The twofold problem consists in (a) reducing or (b) increasing the resistance to the motion by appropriate grooving the surface of the satellite. It follows from our results that the resistance can be reduced by at most 3.05 % or can be at most doubled.

1.8 Dynamics of a Spinning Rough Disc

In Chap. 7, we will study the resistance and dynamics of rotating bodies. Unlike Chap. 6, here we consider the case of *fast rotation*. This means that the product of the angular velocity and the diameter of the body has the same order of magnitude as its translational velocity.

We limit ourselves to the simplest case where a rough disc rotates around the center and at the same time moves through a rarefied medium on a plane. The center of mass of the disc coincides with its geometric center. In this case, a simple scheme of scattering is realized, where an incident particle interacts with the body at the point of collision and then goes away forever. Note that any other shape of a convex body (which is not necessarily rough) and any other location of the center of mass might lead to a more complicated scheme, where a particle is reflected from the body several times at two or more points of the boundary.

On the other hand, the dynamics of *nonconvex* bodies seems to be even more complicated. The point is that it is natural to consider the interaction of the particle with the body in a reference system attached to the body, but in such a reference system, the motion of a particle becomes curvilinear between consecutive collisions and, therefore, is difficult to study.

The main feature of the dynamics of a rapidly rotating disc is that the force of resistance is not parallel to the velocity of the body. This phenomenon, called the *Magnus effect*, is well known to physicists. The transversal component of the resistance force can be codirectional to the instant velocity of the body's front point, or it can have opposite directions relative to this point. In these cases, one speaks of the *proper* or *inverse* Magnus effect, respectively. It is well known to physicists that in relatively dense gases the proper effect takes place, while in rarefied media usually the inverse effect is realized.

The Magnus effect in rarefied media is usually derived from *nonelastic* interaction of gas particles with a body [7,27,74,75]. In contrast, in our model, this effect is due to multiple reflections of particles from the body. The magnitude and direction of the resistance force, as well as the kind of effect (proper or inverse), depend on the shape of roughness in a complicated way. In our model, both effects are realized, but the inverse effect dominates in a sense. For any fixed value of relative angular velocity we represent the force acting on the disc and the moment of this force as functionals depending on the "shape of roughness" (Theorem 7.1).

The set of admissible forces is a convex set formed by the resistance forces (R_1, R_2) corresponding to all possible roughness shapes. The problem of finding the set of admissible forces reduces to a *vector-valued* problem of optimal mass transfer and is then numerically solved for some values of angular velocity (Figs. 7.5 and 7.6). Each of these sets is divided by the vertical line $R_1 = 0$ into two unequal parts; the greater part corresponds to the inverse Magnus effect, the smaller part to the proper effect.

In some simple cases, the disc trajectory is found explicitly (Figs. 7.5 and 7.6). In particular, as shown by numerical simulation, a single disc with roughness formed by equilateral triangles can demonstrate three different kinds of behavior, depending only on the initial data. If the initial angular velocity is sufficiently small, then the disc trajectory is a curve approaching a straight line. If it is sufficiently large, the trajectory is a converging spiral, and if it takes an intermediate value, the trajectory coincides with or approaches a circumference.

The following problem remains unsolved. Find all curves that can be drawn by the center of mass of a spinning rough disc (or, more generally, an arbitrary body) that moves in a rarefied medium.

1.9 Billiards Possessing Extremal Aerodynamic Properties

In Chaps. 8 and 9 we will study bodies that have the best and the worst aerodynamic properties. In Chap. 8 we will concentrate on invisibility and related problems of perfectly streamlined bodies. Interest in invisibility (e.g., creation of "invisibility coats") has increased dramatically due to recent developments in metamaterials with unusual optical properties. In contrast, we examine the effect of invisibility achieved using only mirror surfaces.

First we define three classes of bodies: bodies having zero resistance when moving in a fixed direction, bodies leaving no trace when moving in one direction, and bodies invisible in one direction. We say, in particular, that a body is invisible in one direction if a particle falling on it along a certain straight line in this direction, after making several reflections, will eventually move along the same line. We show that these three classes are nonempty, do not coincide, and are embedded into one another.

The very fact of the existence of bodies having zero aerodynamic resistance when moving in a medium is surprising. We provide explicit constructions of such bodies (Figs. 8.1–8.7). Figure 8.7, for instance, shows how to obtain an invisible body by making a hole inside a cylinder along its axis.

A body of zero resistance is supposed to move uniformly in a medium with zero temperature and constant density. If, say, a spaceship having zero resistance turns its engines on and performs a maneuver, the medium will produce a force resisting the maneuver. Further, when flying into a zone with larger density, the ship experiences a decelerating force, and upon exiting this zone it experiences a compensatory accelerating force.

Next we design a body invisible in two mutually orthogonal directions (Figs. 8.12 and 8.13) and a body invisible from one point (Fig. 8.17). It is impossible, however, to design a body invisible *in all directions* (Theorem 8.4). There remain many unsolved questions, for example, how many directions or points of invisibility can be realized?

In Chap. 9, we will study bodies with the worst aerodynamic properties: retroreflectors.

A retroreflector is an optical device that reverses the direction of any incident beam of light. A perfect retroreflector using *refraction of light rays* is well known in optics: it is the Eaton lens, a transparent ball with refractive index growing from 1 on the ball boundary to infinity at its center. It is unknown, however, if there exist perfect retroreflectors that use only *reflection* of light rays, that is, *billiard* retroreflectors. Instead, we construct in two dimensions several *asymptotically perfect* billiard retroreflectors, that is, families of bodies whose reflective properties approach the property of retroreflection.

The reflective properties of connected two-dimensional bodes are derived from the properties of hollows on their boundary; therefore, we concentrate on constructing hollows. Three families of asymptotically retroreflecting hollows are constructed: mushroom (Fig. 4.7), tube (Fig. 9.6), and notched angle (Fig. 9.9). The first of these admits generalization to higher dimensions. The fourth hollow, considered in Chap. 9, is called a helmet (Fig. 9.12); it possesses very good properties of retroreflection, yet it is not perfect.

The four resulting bodies—two-dimensional retroreflectors with the corresponding hollows on their boundary—are depicted in Fig. 9.14.

Each of the proposed shapes has its own drawbacks. The number of reflections in a tube and a notched angle is very large and goes to infinity when the reflective properties of the corresponding shape approach retroreflection. On the other hand, most particles make only one reflection in a mushroom; however, there always exists a nonzero difference between the directions of incidence and reflection. In addition, as noted by Protasov,[1] in practice it is impossible to produce a good-quality retroreflector with mushroom-shaped hollows since the size of the smallest hollows should be much smaller than that of atoms (the corresponding estimates are given in Sect. 9.7.3). The helmet seems to be the best in practical applications, especially for the purpose of recognizing the body contour.

[1]Personal communication.

Chapter 2
Problem of Minimum Resistance to Translational Motion of Bodies

Newton's aerodynamic problem consists in minimizing the resistance to the *translational motion* of a three-dimensional body moving in a homogeneous medium of resting particles. The particles do not interact with each other and reflect off elastically in collisions with the body. This problem has been considered for various classes of admissible bodies. In Newton's initial setting [40], the class of admissible bodies consisted of convex axisymmetric bodies of fixed length and width, that is, bodies inscribed in a fixed right circular cylinder. The problem was later considered for various classes of (convex and axisymmetric) bodies, for example, for bodies whose front generator has a fixed length (and whose width is also fixed) [4, 32], for bodies of fixed volume [5], and so on. A major step forward was made in the 1990s, when unexpected and striking results were obtained for some classes of *nonaxisymmetric* bodies and, later, for *nonconvex* bodies [8, 11, 12, 16–18, 29–31]). However, the authors kept the initial assumption that the body must have a fixed length and width, that is, can be inscribed in a fixed right circular cylinder.

A further constraint imposed on all classes of bodies was as follows. A particle cannot hit a body more than once. Here we do not impose this constraint. In Sect. 2.1, we consider two classes of (generally speaking, nonconvex and nonsymmetric) bodies inscribed in a circular cylinder. These classes differ in accordance with how one understands the expression "inscribed in a cylinder." We show that in each class the infimum of the resistance is zero, that is, there exist "almost perfectly streamlined" bodies. In Sect. 2.2, this result is generalized to right cylinders with an arbitrary (not only circular) section. In Sect. 2.3, we demonstrate that any convex body can be modified within a small neighborhood of its boundary so that the resulting body displays a resistance less than an arbitrarily small $\varepsilon > 0$. In fact, any body can be made "almost perfectly streamlined" by making microscopic longitudinal "grooves" in its surface. In Sect. 2.4, we consider an analog of Newton's problem in the two-dimensional case. Here the minimum resistance is always positive but smaller than that of convex bodies. In Sect. 2.5, we consider the problem of minimum *specific resistance* for *unbounded bodies* in a parallel flow of particles.

Most of the results presented in this chapter were first published in [43, 44, 47, 52, 55].

A. Plakhov, *Exterior Billiards: Systems with Impacts Outside Bounded Domains*, DOI 10.1007/978-1-4614-4481-7_2,

2.1 Bodies Inscribed in a Circular Cylinder

We consider the problem of minimizing the functional $R_{\delta_{v_0}}$ in (1.7) with $v_0 = (0, 0, -1)$ and continuous function c satisfying $c(v, v) = 0$, over various classes of bodies. In the important case $c(v, v^+) = (v - v^+) \cdot v$, integral (1.7) denotes the resistance to the translational motion of a body with velocity $-v_0$ in a medium of resting particles.

Throughout this chapter, we assume without further mention that the measurable function $v_B^+(\cdot, v_0)$ is defined almost everywhere on $\mathbb{R}^3$ (in view of the translational invariance, this means that it is defined almost everywhere on $\mathbb{R}^2 \times \{0\}$ and is measurable there). This ensures the existence of the integral $R_{\delta_{v_0}}(B)$ in (1.7). In essence, the condition means the regularity of the scattering of particles falling in the direction of v_0. In Figs. 2.1 and 2.2, we present two examples of bodies for which the condition of regularity fails.

A set B not satisfying the regularity condition is obtained by revolution of a plane set S through 360° about the vertical axis AB (Fig. 2.1). The set S is obtained from the rectangle $ABCD$ by removing a subset part of whose boundary is an arc of a parabola with a vertical axis and with a focus at a singular point F of the boundary of this subset. The velocity v_0 of the flow of particles is directed downward: $v_0 = (0, 0, -1)$. The particles reflecting from the parabolic part of the boundary go to the singular point F, and after hitting this point their further motion is not defined. Thus, the function $v_B^+(\cdot, v_0)$ is not defined on a positive-measure subset corresponding to these particles.

Another mechanism to break regularity is due to the fact that a positive-measure set of particles can remain within a bounded set forever, that is, be trapped, and in that case the function $v_B^+(\cdot, v_0)$ is not defined on the set corresponding to these particles. An example of a trap made from arcs of ellipse and parabola is given in Sect. 2.2 of [68]; a similar construction is reconstructed here in Fig. 2.2.

Below we consider the minimization problem in classes of (generally) nonconvex bodies inscribed in a given cylinder. The notion of a body inscribed in a cylinder can

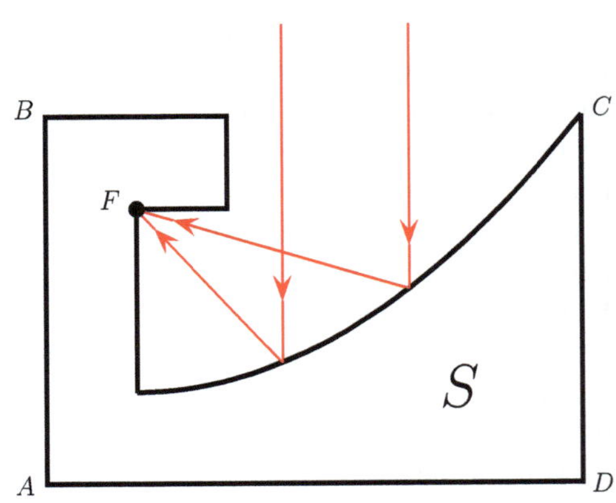

Fig. 2.1 Example of a body with irregular scattering. A positive-measure set of particles hit a singular point of the body's boundary

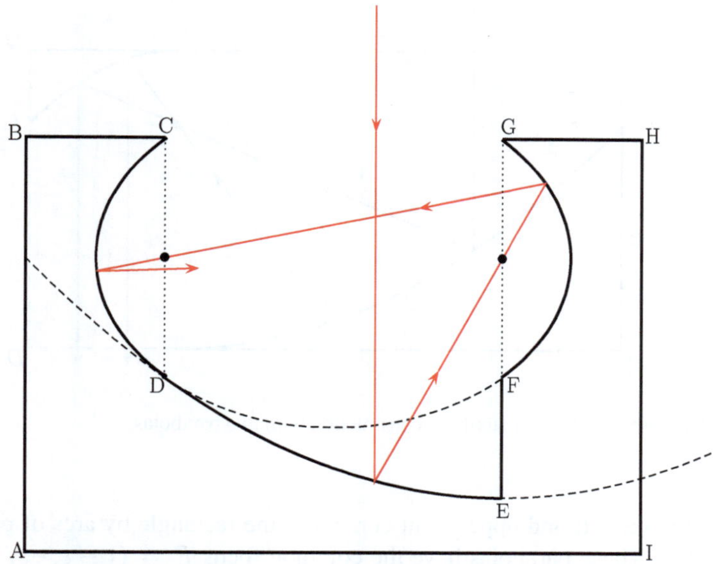

Fig. 2.2 The body ABCDEFGHI is a trap. Curves CD and GF are arcs of an ellipse, and DE is an arc of the parabola with a vertical axis and focus at one of the foci of the ellipse. The set of particles falling downward on arc DE will remain forever in the hollow formed by curve $CDEFG$

be defined in different ways. We introduce two distinct classes of bodies inscribed in a right circular cylinder of radius 1 and height h: the class $\mathcal{P}(h)$ of bodies whose projection on the horizontal plane is a disc and the class $\mathcal{S}(h)$ of bodies containing at least one horizontal section of the cylinder. Note that the class $\mathcal{P}(h)$ is wider than $\mathcal{S}(h)$: $\mathcal{S}(h) \subset \mathcal{P}(h)$. The infima over both classes are equal to 0. We first prove this for $\mathcal{P}(h)$ and then give a sketch of the proof for $\mathcal{S}(h)$.

2.1.1 The Class of Bodies with Fixed Horizontal Projection

Definition 2.1. We denote by $\mathcal{P}(h)$ the class of connected bodies B lying in the cylinder $C_h = \text{Ball}_1(0) \times [-h, 0]$ and such that the orthogonal projection of B on the plane Ox_1x_2 is the unit disc $\text{Ball}_1(0)$.

Proposition 2.1. $\inf_{B\in\mathcal{P}(h)} R_{\delta_{v_0}}(B) = 0$.

Proof. We shall construct a one-parameter family $B_\varepsilon \in \mathcal{P}(h)$, $\varepsilon > 0$, of bodies such that $R_{\delta_{v_0}}(B_\varepsilon) \to 0$ as $\varepsilon \to 0$. First we carry out the construction for $h \geq 1/2$ and then for $h < 1/2$.

(i) For $h \geq 1/2$ we fix ε and consider in the plane Ox_1x_3 the two curvilinear triangles in the rectangle $AOCD = [0, 1] \times [-1/2, 0]$ obtained by cutting

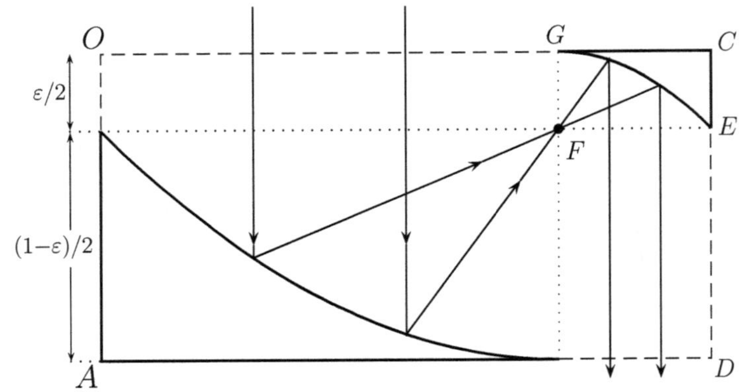

Fig. 2.3 The case $h \geq 1/2$: an auxiliary construction with two parabolas

off the lower left and upper right corners of the rectangle by arcs of parabolas (Fig. 2.3). These parabolas have the common focus $F = (1-\varepsilon, -\varepsilon/2)$ and a common vertical axis. The arc of the first parabola has endpoints $(1-\varepsilon, -1/2)$ and $(0, -\varepsilon/2)$, and it cuts off the larger triangle. The arc of the second parabola has endpoints $G = (1-\varepsilon, 0)$ and $E = (1, -\varepsilon/2)$, and it cuts off the smaller triangle CEG with size of order ε.

Let $B_\varepsilon = B_\varepsilon(h) = B'_\varepsilon \cup B''_\varepsilon$, where the set B'_ε is obtained by revolution of both triangles about the vertical axis Ox_3 (containing side AO). We add the set B''_ε to make the resulting set B_ε connected. The scattering of particles is determined by the set B'_ε, whereas B''_ε perturbs the scattering slightly and can be selected in various ways. For instance, we can take B''_ε to be a cylindrical sector with a small opening: the intersection of the cylinder C_h with the set $|x_1| \leq \varepsilon|x_2|$.

It is easy to calculate $v^+_{B_\varepsilon}(\xi, v_0)$, where $\xi = (x_1, x_2, 0)$, $x_1^2 + x_2^2 \leq 1$. If $|x_1| < \varepsilon|x_2|$ or $1-\varepsilon < \sqrt{x_1^2 + x_2^2} < 1$, then a particle is reflected vertically upward, so that $v^+_{B_\varepsilon}(\xi, v_0) = -v_0$. On the other hand, if $|x_1| > \varepsilon|x_2|$ and $\sqrt{x_1^2 + x_2^2} < 1 - \varepsilon$, then a particle is reflected from the larger parabolic arc, passes through the common focus, is reflected from the smaller parabolic arc, and then moves vertically downward. Thus, on leaving cylinder C the particle has velocity $v^+_{B_\varepsilon}(\xi, v_0) = v_0$. The quantity $R_{\delta_{v_0}}(B_\varepsilon)$ is the product of $c(v_0, -v_0)$ and the area of the upper cross section of the set B_ε, which is the union of the annulus $1-\varepsilon \leq \sqrt{x_1^2 + x_2^2} \leq 1$ and the two sectors $|x_1| < \varepsilon|x_2|$. Hence $R_{\delta_{v_0}}(B_\varepsilon) = O(\varepsilon)$.

(ii) For $h < 1/2$ the construction is slightly more complicated. We take an integer n such that $1/n < 2h$ [for instance, $n = n(h) = \lfloor 1/(2h) \rfloor + 1$], and let $b = 1/n - 3\varepsilon/2$. Below we define a set in the plane Ox_1x_3 that is the union of n pairs of curvilinear triangles and one right isosceles triangle with side $n\varepsilon$ (Fig. 2.4). First we consider the sequence of n rectangles with horizontal side

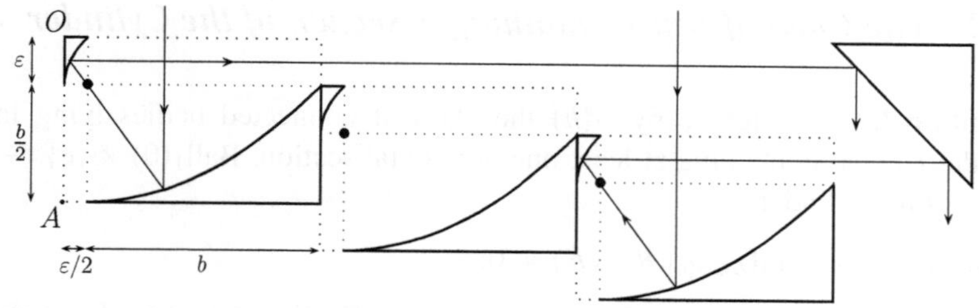

Fig. 2.4 The case $h < 1/2$: an auxiliary construction with many parabolas

$b + \varepsilon/2$ and vertical side $b/2 + \varepsilon$. The curvilinear triangles in each pair are located at the lower left and upper right angles of the corresponding rectangle. First we define the leftmost rectangle and the corresponding pair of curvilinear triangles; all the other rectangles and pairs of triangles are obtained from these by several successive translations by the vector $(b + \varepsilon/2, -\varepsilon)$. Thus, each successive rectangle is adjacent to the previous one along a vertical side.

The leftmost rectangle is $[0, b + \varepsilon/2] \times [-b/2 - \varepsilon, 0]$. We cut two triangles off its upper left and lower right corners by arcs of parabolas. These are parabolas with common focus at $(\varepsilon/2, -\varepsilon)$. The left-hand parabola has a horizontal axis and passes through the points $(\varepsilon/2, 0)$ and $(0, -\varepsilon)$ bounding the corresponding arc (which has length of order ε). The right-hand parabola has a vertical axis and passes through the points $(\varepsilon/2, -b/2 - \varepsilon)$ and $(b + \varepsilon/2, -\varepsilon)$ bounding the corresponding (larger) arc. Finally, the right isosceles triangle has vertices at $(1, 0)$, $(1 - n\varepsilon, 0)$, and $(1, -n\varepsilon)$.

As in case (i), we set $B_\varepsilon = B_\varepsilon(h) = B'_\varepsilon \cup B''_\varepsilon$. The set B'_ε is obtained by rotating all these triangles about the vertical axis Ox_3 containing the left side OA of the left-hand rectangle. The set B''_ε is as in case (i). Again, it is easy to find the function $v^+_{B_\varepsilon}(\xi, v_0)$ for $\xi = (x_1, x_2, 0)$, $x_1^2 + x_2^2 \le 1$. Denote by $\mathcal{N}(h, \varepsilon)$ the set of ξ for which $|x_1| \le \varepsilon |x_2|$ or $\sqrt{x_1^2 + x_2^2} - i(b + \varepsilon/2) \in [0, \varepsilon/2]$ for some i, $0 \le i \le n - 1$. We have $\mathrm{Area}(\mathcal{N}(h, \varepsilon)) \to 0$ as $\varepsilon \to 0$. If ξ lies in the interior of $\mathcal{N}(h, \varepsilon)$, then the corresponding particle reflects off vertically upward, so that $v^+_{B_\varepsilon}(\xi, v_0) = -v_0$. On the other hand, if $\xi \notin \mathcal{N}(h, \varepsilon)$, then the particle is reflected from a large parabolic arc and, passing through the common focus, is reflected from the corresponding smaller parabolic arc, moves horizontally, and then is reflected from the hypotenuse of the right triangle and finally moves vertically downward. Thus, after all these reflections the velocity of the particle is $v^+_{B_\varepsilon}(\xi, v_0) = v_0$. Hence

$$R_{\delta_{v_0}}(B_\varepsilon) = c(v_0, -v_0) \cdot \mathrm{Area}(\mathcal{N}(h, \varepsilon)) \to 0 \quad \text{as } \varepsilon \to 0.$$

The proof of Proposition 2.1 is complete. □

2.1.2 The Class of Sets Containing a Section of the Cylinder

Definition 2.2. We denote by $\mathcal{S}(h)$ the class of connected bodies lying in the cylinder C_h and containing at least one horizontal section, $\text{Ball}_1(0) \times \{c\}$, $-h \leq c \leq 0$, of the cylinder.

Proposition 2.2. $\inf_{B \in \mathcal{S}(h)} R_{\delta_{v0}}(B) = 0.$

Proof. As was done previously, we construct a family of bodies B_ε such that $R_{\delta_{v0}}(B_\varepsilon) \to 0$ as $\varepsilon \to 0$. We obtain this family by a slight modification of the construction used in the proof of Proposition 2.1, namely, we add a set containing the bottom base $\text{Ball}_1(0) \times \{-h\}$ of the cylinder. In addition, for $h > 1/2$ we rotate slightly the axis of the upper (smaller) parabolic arc about the fixed focus, and in the case $h \leq 1/2$, we make the slope of the hypotenuse of the right triangle slightly less than 45°. All these modifications are required to make a particle reflected from the upper parabolic arc or from the hypotenuse of the right triangle move along a slightly inclined (not vertical) line and to go past the lower base of the cylinder without further collisions.

(i) For $h > 1/2$ we consider two sets in place of the two curvilinear triangles in the previous construction (Fig. 2.3). First, the arc of the lower parabola is extended to the right until it intersects the right lateral side of the rectangle $\boxminus = [0, 1] \times [-h, 0]$, and we consider the set M_1 cut off of the rectangle $\boxminus$ by this arc and lying under the arc. (Note that $\boxminus$ contains the rectangle $OADC$ and generates the cylinder C_h when revolved about the axis OA.)

Next, the line joining this intersection point with the focus F is the axis of the upper parabola. Thus, the upper parabola has the same axis $F = (1 - \varepsilon, -\varepsilon/2)$ and passes through the same point $E = (1, -\varepsilon/2)$ as in the previous construction [part (i) of the proof of Proposition 2.1], but now the axis of the parabola will form an angle of order ε with the vertical direction. This parabola intersects the boundary of the smaller rectangle $CEFG = [1-\varepsilon, 1] \times [-\varepsilon/2, 0]$ at two points: one is point E and the other lies on side FG. The second set M_2 is the part of rectangle $CEFG$ cut off by the parabola and lying over it.

The set B'_ε is the result of the revolution of these two sets $M_1 \cup M_2$ about the axis Ox_3, and the set B''_ε is as above. The set $B_\varepsilon = B'_\varepsilon \cup B''_\varepsilon$ contains the section $\text{Ball}_1(0) \times \{-1/2\}$ of the cylinder. A particle reflected from the arc of the lower parabola passes through the common focus F, is reflected from the arc of the upper parabola, and then moves freely at an angle of magnitude $O(\varepsilon)$ with the vertical direction. The contribution of such particles to the functional is $o(1)$, whereas the contribution of the particles reflecting back vertically upward is $O(\varepsilon)$, as before. Thus, $R_{\delta_{v0}}(B_\varepsilon) = o(1)$.

(ii) For $h \leq 1/2$ we add the rectangle $[0, 1] \times [-h, -h + \varepsilon]$ to the system of triangles ($2n$ curvilinear triangles and one right triangle). The right triangle from the system should now be modified as follows. It should have the same two vertices $(1, 0)$ and $(1, -n\varepsilon)$ as before, whereas the third vertex now has coordinates $(1 - n\varepsilon', 0)$. Here we choose ε' such that each particle moving horizontally to the right, after reflection from the hypotenuse of the triangle, passes to the right of $(1, -h + \varepsilon)$ and so does not hit the indicated (added) rectangle. For this we can take $\varepsilon' = \varepsilon/\sqrt{1 - 2n\varepsilon/(h - \varepsilon)} = \varepsilon + O(\varepsilon^2)$. Finally, we set $b = 1/n - \varepsilon' - \varepsilon/2$. Taking account of this modification, we construct n pairs of curvilinear triangles as before.

A particle falling on one of the larger arcs of parabolas in this construction will be successfully reflected by a larger and a smaller arc of parabolas, and then it is reflected from the hypotenuse of the right triangle and moves freely at an angle $\arctan \frac{\varepsilon'^2 - \varepsilon^2}{2\varepsilon\varepsilon'} = O(\varepsilon)$ with the vertical direction. Thus, the difference between the initial and final velocities of the particle is $O(\varepsilon^2)$.

The set B'_ε is obtained by the revolution of n pairs of triangles, the right triangle, and the additional rectangle about the axis AO. The set B''_ε is as above. The set $B_\varepsilon = B'_\varepsilon \cup B''_\varepsilon$ contains the section $\text{Ball}_1(0) \times \{-h\}$ of the cylinder. As before (see the proof of Proposition 2.1), we define the set $\mathcal{N}(h, \varepsilon)$ and show that $\text{Area}(\mathcal{N}(h, \varepsilon)) \to 0$ as $\varepsilon \to 0$. Furthermore, for $\xi \in \mathcal{N}(h, \varepsilon)$ the particle reflects back vertically upward, so that $v^+_{B_\varepsilon}(\xi, v_0) = -v_0$, while if $\xi \notin \mathcal{N}(h, \varepsilon)$, then $v^+_{B_\varepsilon}(\xi, v_0) = v_0 + o(1)$. Thus, $R_{\delta_{v_0}}(B_\varepsilon) = c(v_0, -v_0) \cdot \text{Area}(\mathcal{N}(h, \varepsilon)) + o(1) = o(1)$ as $\varepsilon \to 0$. The proof of Proposition 2.2 is complete. □

Remark 2.1. Note that in case (i) each particle hits the body B_ε at most twice, and in case (ii) at most three times. An open question is: for which h are two hits enough, that is, what is the minimal h_0 such that for $h > h_0$ there exists a sequence of sets with resistance tending to zero such that particles in the flow have only one or two collisions with them? This question can be posed for the classes of sets $\mathcal{P}(h)$ and $\mathcal{S}(h)$. It follows from the proofs of Propositions 2.1 and 2.2 that $0 \le h_0 \le 1/2$ in both cases.

Remark 2.2. The body in Fig. 2.4 is actually a (disconnected) *two-dimensional* body of arbitrarily small resistance. It is contained in the rectangle $[0,\ 1] \times [-h,\ 0] \subset \mathbb{R}^2_{x_1 x_3}$, and its projection on the x_1-axis is $[0,\ 1]$.

Substituting the segment $[0,\ 1]$ with a generic set $I \subset \mathbb{R}_{x_1}$, we come to the following more general statement, which will be used in the next section.

Proposition 2.3. *Let* $I = \cup_i I_i \subset \mathbb{R}_{x_1}$ *be the union of a finite number of disjoint compact segments. Set* $x_0 = (0, -1)$. *Consider a family of segments* J_δ, $\delta > 0$ *and assume that* $|J_\delta| = \delta$, $I \cap J_\delta = \emptyset$ *and* $\text{dist}(\text{Conv}(I), J_\delta) \to 0$ *as* $\delta \to 0$. *Then the following holds.*

(a) For any ε *there exists a two-dimensional body* $B \subset I \times [-h,\ 0] \subset \mathbb{R}^2_{\{x_1 x_2\}}$ *such that its projection on the* x_1*-axis coincides with* I *and its resistance is smaller than* ε, *that is,* $R_{\delta_{x_0}}(B) < \varepsilon$.

(b) For δ *sufficiently small there exists a body* $B_\delta \subset I \times [-h,\ 0]$ *such that the projection of* B_δ *on the* x_1*-axis coincides with* I, $\lim_{\delta \to 0} R_{\delta_{x_0}}(B_\delta) = 0$, *and no particle of the vertical flow intersects* $(\mathbb{R} \setminus J_\delta) \times [-h,\ -h + \delta]$.

The proof of this proposition is obtained by a slight modification of the construction of Fig. 2.4 and is omitted here.

2.2 Bodies Inscribed in an Arbitrary Cylinder

Here we state a generalization of the results of the previous section. Consider a connected body $\Omega \subset \mathbb{R}^2$.

Definition 2.3. (a) We denote by $\mathcal{P}(\Omega, h)$ the class of connected sets B contained in the cylinder $\Omega \times [-h,\ 0]$ and such that the orthogonal projection of B on the plane Ox_1x_2 is Ω.

(b) We denote by $\mathcal{S}(\Omega, h)$ the class of connected sets B contained in the cylinder $\Omega \times [-h,\ 0]$ and containing a section of the cylinder $\Omega \times \{c\},\ -h \le c \le 0$.

Theorem 2.1. *(a)* $\inf_{B \in \mathcal{P}(\Omega,h)} |R_{\delta_{v0}}(B)| = 0$.
(b) $\inf_{B \in \mathcal{S}(\Omega,h)} |R_{\delta_{v0}}(B)| = 0$.

Proof. Clearly, $\mathcal{S}(\Omega, h) \subset \mathcal{P}(\Omega, h)$; therefore (b) implies (a). Nevertheless, we will first provide a proof for (a) that makes clearer the reasoning in the proof of (b).

Generally speaking, the boundary $\partial\Omega$ is the union of a finite number of closed non-self-intersecting (and not intersecting each other) piecewise smooth curves of finite length. We shall assume that $\partial\Omega$ is a single curve; the general case is obtained by a slight modification of the argument. Let l be the length of this curve, $|\partial\Omega| = l$. Consider the square lattice formed by the lines $x_1 = n_1\delta,\ x_2 = n_2\delta,\ n_1,\ n_2 \in \mathbb{Z}$ in the plane Ox_1x_2 containing Ω. The squares of this grid have size $\delta \times \delta$. The squares that have a nonempty intersection with $\partial\Omega$ will be called *boundary squares*. Let us show that for δ sufficiently small the number of boundary squares does not exceed $5l/\delta$.

Indeed, divide the curve $\partial\Omega$ into $n = \lfloor l/(2\delta)+1 \rfloor$ pieces of equal length (clearly, this length does not exceed 2δ) and mark the midpoint on each piece; this point divides the piece into two parts of length δ. Take the square containing this point and eight squares surrounding it. We get a large square composed of these nine squares, which will be called a *tile*. The distance from the midpoint to each point of the piece does not exceed δ, and the distance from the midpoint to the boundary of the tile is at least δ; this implies that the piece lies in the tile. Hence the curve is contained in the union of at most n tiles and, thus, in the union of at most $9n$ squares. Taking into account that $n \le l/(2\delta) + 1$ and choosing δ sufficiently small, we get that the number of boundary squares does not exceed $5l/\delta$. Thus the area of the union of boundary squares does not exceed $5l\delta$.

(a) Let Ω_δ be the union of the squares contained in Ω, and let Ω_δ^n be the intersection of Ω with the strip $n\delta \le x_1 \le (n+1)\delta$. Thus, Ω_δ^n is composed of squares of size $\delta \times \delta$ lying in a strip of width δ. Note that the sets Ω_δ^n are nonempty only for a finite number of values n and have the form $\Omega_\delta^n = [n\delta,\ (n+1)\delta] \times \times I_\delta^n$, where I_δ^n is the union of a finite number of segments whose length is a multiple of δ. One has $\Omega_\delta = \cup_n \Omega_\delta^n$. Let Ω'_δ be the intersection of Ω with the union of boundary squares; then one has $\Omega = \Omega_\delta \cup \Omega'_\delta$.

The required set of small resistance is $B_\delta = \tilde{B} \cup \hat{B}$, where $\tilde{B} = \tilde{B}_\delta$ and $\hat{B} = \hat{B}_\delta$ are constructed as follows. First we define the intersection of $\tilde{B}$ with $\Omega'_\delta \times \mathbb{R}$ and, second, the intersection of $\tilde{B}$ with each set $\Omega_\delta^n \times \mathbb{R}$, $n \in \mathbb{Z}$. By this, $\tilde{B}$ will be uniquely defined. Further, $\hat{B}$ is the intersection of the union of "walls" $W_n = W_n^\delta = \{x = (x_1, x_2, x_3) : |x_1 - n\delta| < \delta^2\}$ with the cylinder $\Omega \times [-h, 0]$, $\hat{B} = (\cup_{n\in\mathbb{Z}} W_n) \cap (\Omega \times [-h, 0])$. The "sandwich" B_δ is the union of these sets; adding the union of "layers" $\hat{B}$ makes it connected and changes the resistance only by $O(\delta)$.

Set $\tilde{B} \cap (\Omega'_\delta \times \mathbb{R}) = \Omega'_\delta \times [-h, -h+\delta]$; this is a cylinder of height δ. This is a part of the body; it makes a small contribution (of order δ) to the resistance.

Further, using the statement of Proposition 2.3(a), for each value n such that Ω_δ^n is nonempty we construct a body $D_\delta^n \subset I_\delta^n \times [-h, 0]$. The resistance of this body is smaller than δ, and its projection on the x_1-axis is I_δ^n. Then we set

$$\tilde{B} \cap (\Omega_\delta^n \times \mathbb{R}) = [n\delta, (n+1)\delta] \times D_\delta^n.$$

Thus we have $R_{\delta_{x0}}(\tilde{B} \cap (\Omega_\delta^n \times \mathbb{R})) < \delta^2$. As a result one obtains

$$R_{\delta_{x0}}(B_\delta) = R_{\delta_{x0}}(\tilde{B} \cap (\Omega'_\delta \times \mathbb{R})) + \sum_n R_{\delta_{x0}}(\tilde{B} \cap (\Omega_\delta^n \times \mathbb{R})) + R_{\delta_{x0}}(\hat{B}) = O(\delta), \quad \delta \to 0.$$

Finally, it is clear from the construction that $B \subset \mathcal{P}(\Omega, h)$.

(b) A row of squares is called *special* if it contains more than $\sqrt{5l/\delta}$ boundary squares. Clearly, the number of special rows does not exceed $\sqrt{5l/\delta}$; otherwise we would have a number of boundary squares larger than $5l/\delta$.

The intersection of Ω with the union of all boundary squares and all special rows is denoted by Ω''_δ. Clearly, the area of Ω''_δ is $O(\sqrt{\delta})$.

For each n corresponding to a nonspecial row we select a square of the lattice $[n\delta, (n+1)\delta] \times J_\delta^n$ that does not intersect Ω and has the distance at most $\sqrt{5l\delta}$ from $\mathrm{Conv}(\Omega_\delta^n)$. Note that this selection can be made due to the property of a nonspecial row: there are at most $\sqrt{5l/\delta}$ boundary squares between the square $[n\delta, (n+1)\delta] \times J_\delta^n$ and the rectangle $\mathrm{Conv}(\Omega_\delta^n)$. Then, using Proposition 2.3(b), one constructs a set $\tilde{D}_\delta^n$ lying in $I_\delta^n \times [-h, 0]$, with the resistance $o(1)$ as $\delta \to 0$, and such that no particle reflected from $\tilde{D}_\delta^n$ intersects $(\mathbb{R} \setminus J_\delta^n) \times [-h, -h+\delta]$.

Let

$$B_\delta = \tilde{B} \cup \hat{B} \cup B',$$

where $\hat{B} = \hat{B}_\delta$ is as in (a),

$$B' = B'_\delta = \Omega \times [-h, -h+\delta]$$

and

$$\tilde{B} = \tilde{B}_\delta = \cup'_n([n\delta, (n+1)\delta] \times \tilde{D}^n_\delta),$$

where the union is taken over nonspecial n.

By adding $\hat{B}$, as in (a), we ensure that the resulting body B_δ is connected, and adding B' ensures that the body belongs to the class $\mathcal{S}(\Omega, h)$. Finally, the set of small resistance $\tilde{B}$ is constructed in such a way that particles reflected from it do not hit B'. Thus, we have $R_{\delta_{x0}}(B_\delta) = o(1),\ \delta \to 0$. □

The topic of this section will be continued in Chap. 8 (Theorem 8.2).

2.3 Bodies Modified in a Neighborhood of Their Boundary

Here we show that a convex body can be "modified" in a small neighborhood of its boundary so that the resulting body displays an arbitrarily small resistance to the parallel flow of particles falling on it.

Let us first provide an "astronautical" interpretation of the problem. Suppose we are traveling in a spaceship $C_2 \subset \mathbb{R}^3$, which is a bounded convex set. The inner space of the spaceship coincides with another convex set $C_1 \subset C_2$ (Fig. 2.5). The spaceship body is then $C_2 \setminus C_1$; it is natural to require that $\partial C_1 \cap \partial C_2 = \emptyset$ (meaning that the thickness of the spacecraft body is everywhere positive).

We are going to process the metallic body of the spaceship aiming to minimize the velocity slowdown when going through space clouds. The processing may result in dimples, hollows, grooves, etc. on the spaceship surface. In general, we assume that any body B satisfying the inclusions $C_1 \subset B \subset C_2$ can be obtained by such a processing. We pose the following question.

Question: Given the convex bodies C_1 and C_2, the spaceship velocity v, and the cloud density, what is the minimum resistance of the resulting body B?

Note in passing that the resistance of the original body C_2 can be very easily decreased just by making dimples. Indeed, let the direction of v be vertical and consider a region $\mathcal{U} \subset \partial C_2$ in the upper part of the surface ∂C_2 whose inclination

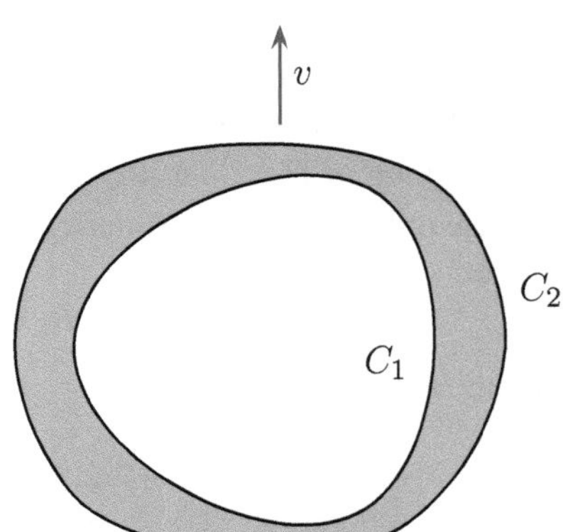

Fig. 2.5 Spaceship. The direction of motion is indicated by v

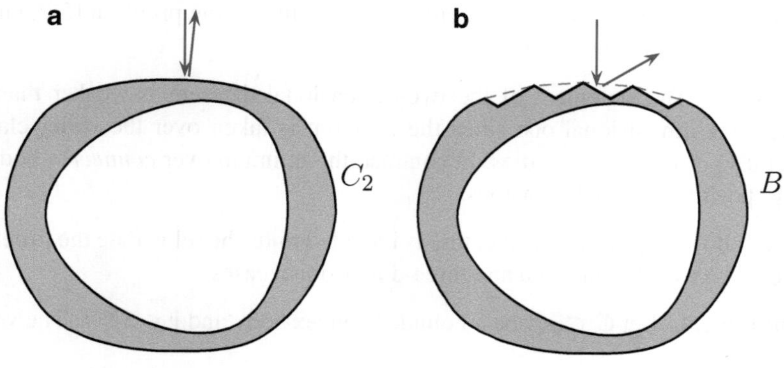

Fig. 2.6 Original spaceship C_2 (**a**) and spaceship B with dimples (**b**)

relative to the horizontal plane is less than 30°. (We assume that $\mathcal{U}$ is not empty.) Then make several conical dimples in this region, with the inclination of the cone surface being exactly 30° (Fig. 2.6b).

The resistance of the resulting body B is smaller than that of the original body C_2. Indeed, in a reference system attached to the body we observe a parallel flow of particles falling vertically downward. A particle hitting C_2 in the region $\mathcal{U}$ is reflected at an angle smaller than 60° relative to the vertical (Fig. 2.6a). In contrast, a particle hitting B in a conical dimple will be reflected exactly at a 60° angle (Fig. 2.6b). Therefore, the momentum transmitted by the particle to the body is smaller in the latter case than in the former one, and summing up all the transmitted momenta, we get that the resistance of B is smaller than that of C_2.

In a similar way, the resistance of Newton's optimal body can be decreased by making dimples on its front (flat) surface. This observation was first made by Buttazzo and Kawohl in [12]. The techniques of making dimples and grooves were further developed by Comte and Lachand-Robert in [16–18] in their study of generalizations of Newton's problem in classes of nonconvex bodies.

The answer to our question is surprising: the resistance of the resulting body can be made arbitrarily small. That is, by processing the surface of our spaceship, one can make it almost perfectly streamlined!

Theorem 2.2. *In the three-dimensional case we have*

$$\inf\{|R_v(B)| : C_1 \subset B \subset C_2,\ B \mathit{connected}\} = 0.$$

The proof of this theorem is based on the following auxiliary two-dimensional result. Let C_1 and C_2 be bounded convex bodies, $C_1 \subset C_2 \subset \mathbb{R}^2$, $\partial C_1 \cap \partial C_2 = \emptyset$, and $v \in S^1$.

Theorem 2.3. *In the two-dimensional case we have*

$$\inf\{|R_v(B)| : C_1 \subset B \subset C_2\} = 0.$$

These results were announced, with a brief outline of the proof, in [52], and the authors' detailed exposition is given in [55].

Remark 2.3. The statement of the two-dimensional theorem is weaker than that of the three-dimensional one since the infimum is taken over the wider class of (generally) *disconnected* bodies. In contrast, the infimum over *connected* bodies is always positive in two dimensions.

The following plausible conjecture is intended to further elucidate the difference between the two-dimensional and three-dimensional cases.

Definition 2.4. Let $C \subset \mathbb{R}^d$ be a bounded convex body and $v \in S^{d-1}$. The value

$$\mathcal{R}_v(C) := \sup_{C_1} \inf\{|R_v(B)| : C_1 \subset B \subset C,\ B \text{connected}\},$$

where the supremum is taken over all convex bodies C_1 such that $C_1 \subset C$ and $\partial C_1 \cap \partial C_2 = \emptyset$, is called the *minimal resistance of bodies obtained by roughening* C.

Conjecture 2.1. (a) For any $v \in S^2$ and any convex $C \subset \mathbb{R}^3$,

$$\mathcal{R}_v(C) = 0$$

holds.

(b) For any $v \in S^1$ and any convex $C \subset \mathbb{R}^2$,

$$1/4 < \frac{\mathcal{R}_v(C)}{|R_v(C)|} \leq 1/2 \tag{2.1}$$

holds.

Statement (a) of this conjecture is a direct consequence of Theorem 2.2, but statement (b) has not been proved yet. Note, however, that in the particular case where C is symmetric with respect to an axis parallel to v, the double inequality (2.1) can be easily derived from the proof of Theorem 2 in [52].

2.3.1 Preliminary Constructions

Here we explain the basic two-dimensional construction that will be used in the proofs of Theorems 2.3 and 2.2. The main idea of these proofs consists in making a large number of "diversion channels" penetrating the body near its boundary. Each channel is the union of three sets: a *front funnel*, a *tube*, and a *rear funnel*. The front funnel is turned to the flow, and the rear one, to the opposite direction. Each particle of the flow goes into the front funnel of a channel, then moves through the channel along the body boundary, and finally leaves through the rear funnel, and its final velocity only slightly differs from the velocity of incidence.

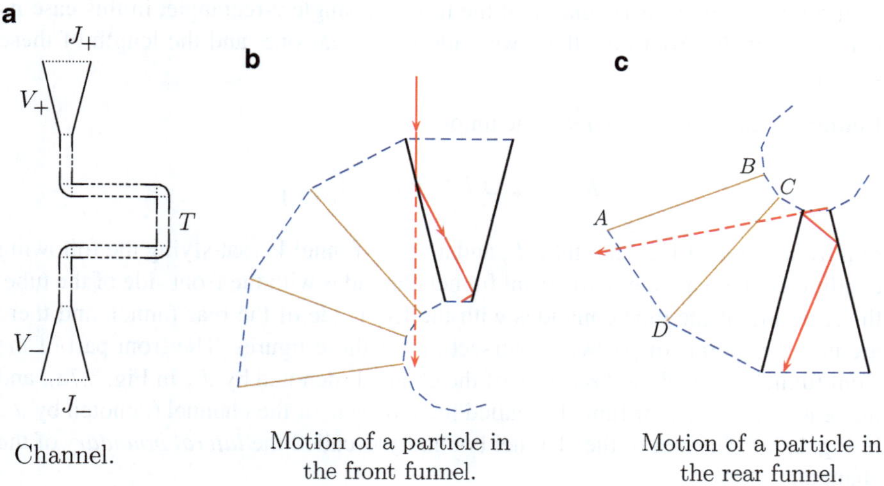

Fig. 2.7 Dynamics of a particle in a channel

Introduce the coordinates $x_1,\ x_2$ on a plane such that $v = (0, -1)$, the x_1-axis being considered horizontal, and the x_2-axis, vertical. Fix the parameter $0 < \varepsilon < 1$. The front and rear ε-funnels $V_\pm$ are the trapezoids $|x_1| \le \varepsilon |x_2|,\ \varepsilon^2 \le \pm x_2 \le \varepsilon$, respectively. The point $(0, \pm\varepsilon^2)$ is called the vertex of the corresponding funnel. The front and rear sides of the front funnel are, respectively, its larger and smaller bases, that is, the segments $\{|x_1| \le \varepsilon^2,\ x_2 = \varepsilon\}$ and $\{|x_1| \le \varepsilon^3,\ x_2 = \varepsilon^2\}$. In contrast, the front and rear sides of the rear funnel are its smaller and larger bases, that is, the segments $\{|x_1| \le \varepsilon^3,\ x_2 = -\varepsilon^2\}$ and $\{|x_1| \le \varepsilon^2,\ x_2 = -\varepsilon\}$. A parallel translation of the front (rear) funnel is also called a front (rear) funnel (Fig. 2.7a).

An ε-tube is a finite sequence of figures: rectangles and circle sectors. These figures are called *elements* of the tube. The rectangles are vertically or horizontally oriented; they are called v- and *h-rectangles*, respectively. In a v-rectangle, one of the horizontal (upper and lower) sides is considered the *front* side, and the other horizontal side is the *rear* one. Their length equals $2\varepsilon^3$. In an h-rectangle, the length of the vertical (left and right) sides equals $2\varepsilon^3$; one of these sides is the front one, and the other side is the rear one. Each circle sector has an angular size of 90°; it is a quarter of a circle of radius $2\varepsilon^3$. One of the radii bounding the sector is vertical, and the other one is horizontal; one of these radii is called the front one, and the other the rear one. In the sequence of figures forming the tube, rectangles and circle sectors alternate; the first and last figures are v-rectangles, the upper side of the first rectangle is the front one, and the lower side of the last rectangle is the rear one (Fig. 2.7a). Further, in the subsequence composed of rectangles, the v- and h-rectangles alternate. Finally, in the sequence of figures (rectangles and circle sectors) forming the tube, the rear side of the preceding figure coincides with the front side of the subsequent figure, and there are no other points of pairwise intersection of the figures.

It may happen, in particular, that the tube is a single v-rectangle; in this case its upper side is the front one, the lower side is the rear one, and the length of these sides is $2\varepsilon^3$.

Definition 2.5. *An ε-channel* is the union

$$K = V_+ \cup T \cup V_- \subset \mathbb{R}^2_{\{x_1,x_2\}}$$

of a front ε-funnel V_+, an ε-tube T, and a rear ε-funnel V_- satisfying the following conditions: the rear side of the front funnel coincides with the front side of the tube, the rear side of the tube coincides with the front side of the rear funnel, and there are no other points of pairwise intersection for these figures. The front part of the front funnel is called the *front side* of the channel (denoted by J_+ in Fig. 2.7a), and the rear side of the rear funnel is called the *rear side* of the channel (denoted by J_- in Fig. 2.7a). The rest of the channel boundary is called the *lateral boundary* of the channel.

Lemma 2.1. *Consider a billiard in an ε-channel. If a particle starts its motion with velocity $v = (0, -1)$ at a point of the front side of the channel, then after making a finite number of reflections from the lateral boundary it finally comes into a point of the rear side of the channel with velocity $v + O(\varepsilon)$, $\varepsilon \to 0$. Here $O(\varepsilon)$ is uniform with respect to all ε-channels and all initial positions.*

Proof. First we prove that the particle, after a finite number of reflections from the lateral boundary, crosses the rear side of the channel (and not its front side). The proof of this statement is inductive. That is, for each figure forming the channel (trapezoid, rectangle, circle sector) we will prove the following: if the particle enters the figure through its front side, then after a while it will leave the figure through its rear side.

The motion in a rectangle is unidirectional, from the front to the rear side; this is obvious. Further, notice that when moving in a circle, the angular coordinate of the particle changes monotonically. This implies that if the particle intersects the front radius of a sector, then after several (maybe none) reflections from the arc it will intersect the rear radius.

It remains to consider the motion in the funnels. The particle starts moving vertically down from the front side of the front funnel (that is, from the larger side of the trapezoid). Apply the method of unfolding of the trajectory; (Fig. 2.7b). In a convenient reference system, the trapezoid takes the form $|x_1| \le \varepsilon x_2$, $\varepsilon^2 \le x_2 \le \varepsilon$. The unfolded trajectory is a vertical line at a distance less than ε^2 from the origin; therefore, it intersects the circle $\mathrm{Ball}_{\varepsilon^2}(0)$. On the other hand, the sequence of images of the smaller side of the trapezoid under the unfolding forms a broken line winding around the origin and touching the same circle. (Notice that this broken line is contained in the larger circle $\mathrm{Ball}_{\varepsilon^2\sqrt{1+\varepsilon^2}}(0)$; we will use it later.) Hence the unfolded trajectory intersects the broken line; this means that the original trajectory, after several reflections from the lateral sides of the trapezoid, will intersect its smaller side.

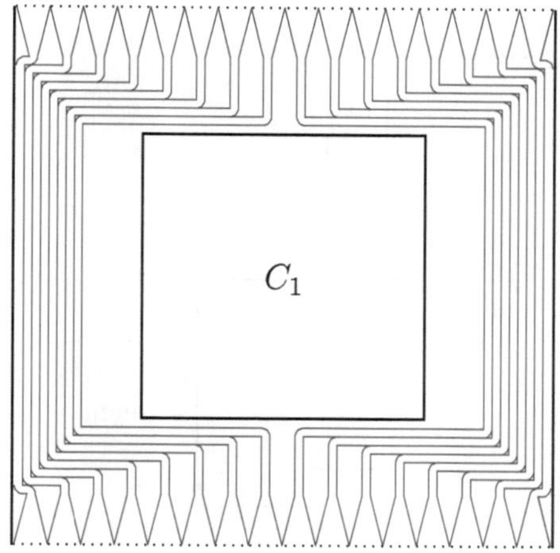

Fig. 2.8 Two concentric squares with a built-in channel system

Finally, when considering the motion in the rear funnel we again use the unfolding method. This time we unfold the final part of the trajectory starting from the point of intersection with the front side of the funnel (that is, the smaller base of the trapezoid; see Fig. 2.7c). The unfolded trajectory intersects one of the images, under the unfolding, of the larger base of the trapezoid; this image is AD in the figure. This means that the particle, after several reflections from the lateral sides of the trapezoid, finally reaches the rear side of the channel. Using Fig. 2.7c one gets an estimate for the particle velocity at the point of intersection with the rear side of the funnel. The angle the velocity vector forms with the vertical is obviously smaller than the largest angle formed by the symmetry axis of $ABCD$ with the tangent lines from A to the circle $\mathrm{Ball}_{\varepsilon^2\sqrt{1+\varepsilon^2}}(0)$. The latter quantity equals $\arctan\varepsilon + \arcsin\varepsilon$. Thus, the difference between the initial and final velocities, v and v^+, of a particle in an ε-channel can be estimated from above as follows:

$$|v - v^+| \le 4\sin^2((\arctan\varepsilon + \arcsin\varepsilon)/2) = O(\varepsilon).$$

□

Figure 2.8 shows what the channel system might look like in the case where C_1 and C_2 are concentric squares. A body of small resistance is obtained by removing the channels from the larger square C_2.

The construction of the channel system in the general case is more complicated. We will start with a method of constructing a special ε-channel that will be used later in this section. Consider two rectangles Π^+ and Π^- with horizontal sides of length $2\varepsilon^2$ and vertical sides of length $\varepsilon - \varepsilon^2$, let A^+ be the midpoint of the lower side of Π^+, and let A^- be the midpoint of the upper side of Π^-. A broken line joining points A^+ and A^- and satisfying the conditions stated later in this paragraph will be called an *ε-axis*, and these points will be called the front and rear endpoints of the axis. The broken line consists of a finite number of vertical and horizontal segments. The initial and final segments are vertical ones of lengths greater than or equal to ε^3,

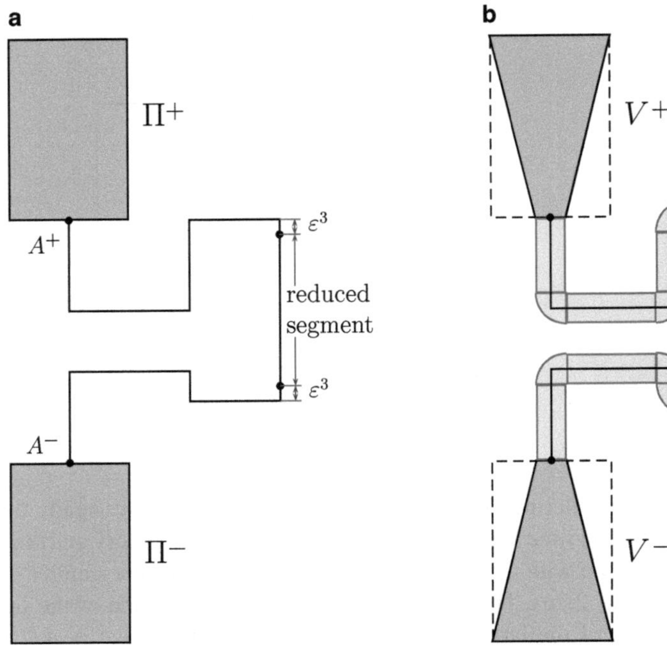

Fig. 2.9 An ε-contour (**a**) and the channel generated by this contour (**b**)

and the lengths of the other segments are greater than or equal to $2\varepsilon^3$. The broken line does not have points of self-intersection and or points of intersection with Π^+ and Π^- other than endpoints A^+ and A^-. The endpoints of the segments, except for A^+ and A^-, are called *vertices*. Thus, both the initial and final segments have only one vertex, and the other segments have two vertices. A shortened segment of the broken line, with (one or two) segments of length ε^3 adjacent to its vertices removed, is called a *reduced segment*. A reduced segment may be a true segment, or it may be a single point. The union of the rectangles Π^+ and Π^- and the ε-axis is called an ε-contour (Fig. 2.9a).

Now suppose that we have an ε-contour. To each reduced segment of the ε-axis we assign a rectangle of width $2\varepsilon^3$ such that the segment is a midline of the rectangle and divides it into two rectangles of width ε^3 (Fig. 2.9b). In the degenerated case, where the reduced segment is a point, the assigned rectangle is a segment of length $2\varepsilon^3$. To each vertex we assign a circle sector of radius $2\varepsilon^3$ such that the two radii bounding the sector coincide with sides of the rectangles assigned to the adjacent reduced segments. Finally, to the rectangles Π^+ and Π^-we assign the inscribed trapezoids V^+ and V^- such that the midpoints of two sides of these rectangles, A^+ and A^-, are also midpoints of smaller bases of length $2\varepsilon^3$ of the trapezoids, and the opposite sides of the rectangles coincide with the larger bases of the trapezoids. If the obtained figures (rectangles, sectors, and trapezoids) do not mutually intersect, then their union is an ε-channel. It will be called a *channel generated by the given ε-contour.*

2.3.2 *Proof of Theorem 2.3*

Consider two plane convex bodies C_1 and C_2. Without loss of generality we assume that $v = (0, -1)$. The proof of Theorem 2.3 amounts to constructing a family of bodies B_ε, $C_1 \subset B_\varepsilon \subset C_2$ with resistance going to zero, $\lim_{\varepsilon \to 0} R_v(B_\varepsilon) = 0$. In what follows, we will write R instead of R_v, omitting the subscript v.

It suffices to provide a family B_ε in the special case where

$$\mathrm{dist}(\partial C_1, \partial C_2) > 4\sqrt{2}. \tag{2.2}$$

Indeed, in the general case, take $k > 0$ large enough so that $\mathrm{dist}(\partial(kC_1), \partial(kC_2)) > 4\sqrt{2}$ and find a family $\tilde{B}_\varepsilon$, $kC_1 \subset \tilde{B}_\varepsilon \subset kC_2$ such that $\lim_{\varepsilon \to 0} R(\tilde{B}_\varepsilon) = 0$. Then the family $B_\varepsilon = k^{-1}\tilde{B}_\varepsilon$ satisfies the required relations $C_1 \subset B_\varepsilon \subset C_2$ and $R(B_\varepsilon) = k^{-1}R(\tilde{B}_\varepsilon) \to 0$ as $\varepsilon \to 0$. Thus, the general case is reduced to the special case (2.2).

Consider the partition of $\mathbb{R}^2$ into (closed) squares of size 2×2 with vertices in $2\mathbb{Z} \times 2\mathbb{Z}$ and denote by D the union of squares contained in the interior of C_2. One easily sees that $C_1 \subset D$. The squares of the partition that are contained in D and have nonempty intersection with ∂D will be called *boundary squares*. The boundary squares do not intersect C_1.

Indeed, each boundary square (let it be Q) has a nonempty intersection with another square of a partition (say, Q') that does not belong to D. By definition of D, Q' contains a point from ∂C_2. The diameter of the union $Q \cup Q'$ does not exceed $4\sqrt{2}$, and therefore by (2.2) $Q \cup Q'$ does not contain points of ∂C_1. This implies that $Q \cap C_1 = \emptyset$.

In Fig. 2.10, C_1 and C_2 are bounded by black closed curves, and D is bounded by a thick polygonal line. The boundary squares are situated between the thick and thin black solid polygonal lines.

For future convenience we will use the small parameter ε of the form $\varepsilon = 1/(2n+1)$, where n is a positive integer, and impose the restriction $\varepsilon < \mathrm{dist}(D, \partial C_2)$. Denote by $l = l(D) = \max\{|x_1 - y_1| : (x_1, x_2), (y_1, y_2) \in D\}$ the width of D and impose one more restriction $\varepsilon < 1/l$.

Denote by $\partial_+ D$ and $\partial_- D$ the upper and lower parts of the boundary ∂D, that is, the intersection of ∂D with the union of upper (lower) sides of the squares forming D. Introduce the metric $\bar{d}$ in $\mathbb{R}^2$ by $\bar{d}(x, y) = \max\{|x_1 - y_1|, |x_2 - y_2|\}$, where $x = (x_1, x_2)$ and $y = (y_1, y_2)$. In this metric a ball of radius r is a square of size $2r \times 2r$ with vertical and horizontal sides.

Take $d_i = (2i - 1)\varepsilon^3$, $N = l/(2\varepsilon^2)$, and denote by L_i the set of points $x \in D$ such that $\bar{d}(x, \partial D) = d_i$, $i = 1, \ldots, N$. The curve L_i will be called the *ith-level line*. Due to the choice of ε one always has $d_i < 1$, so each curve L_i is contained in the union of boundary squares and, therefore, does not intersect C_1. The curves L_i are closed, do not have self-intersections, and are composed of vertical and horizontal segments. Let us divide each level line L_i into two curves by two points with maximal and minimal x_1-coordinates; then the x_1-coordinate will change monotonically along each of these curves. Finally, the x_1-coordinate

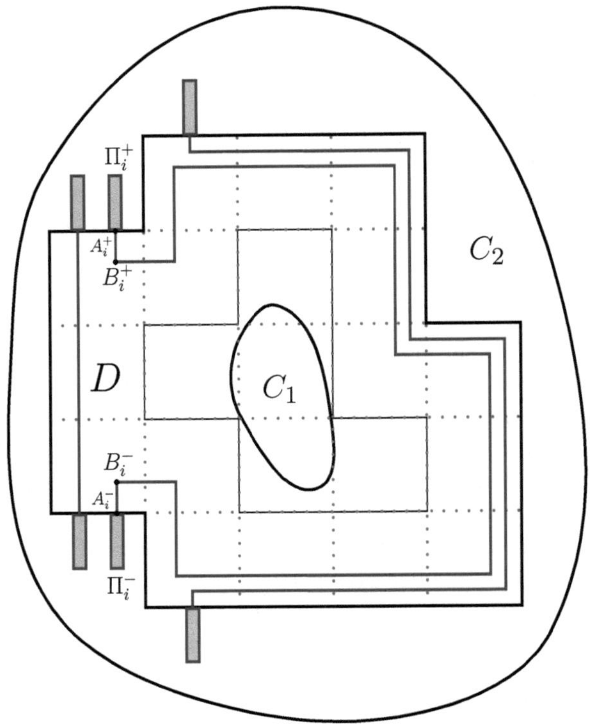

Fig. 2.10 The construction of a built-in channel system: the ε-contours for three values of i are shown. The *rectangles* Π_i^+, Π_i^- and the points A_i^+, A_i^-, B_i^+, B_i^- are indicated for a single value of i

of each vertical segment forming L_i differs by $(2i-1)\varepsilon^3$ from a multiple of 2; hence the difference of x_1-coordinates of any two vertical segments belonging to any two level lines is a multiple of $2\varepsilon^3$. The same is valid for the x_2-coordinate. These observations imply that the length of each segment in each level line is greater than or equal to $2\varepsilon^3$.

Divide the upper boundary, $\partial_+ D$, into segments of length $2\varepsilon^2$. The number of these segments is N, and the upper side of each square forming $\partial_+ D$ contains exactly ε^{-2} segments (recall that this number is an integer). Denote the segments from right to left (that is, from the larger to the smaller x_1-coordinate) by $I_1^+, \dots, I_N^+$, and construct the rectangles $\Pi_1^+, \dots, \Pi_N^+$ of height $\varepsilon - \varepsilon^2$ resting on these segments; that is, I_i^+ is the lower side of Π_i^+. Similarly, divide the lower boundary, $\partial_- D$, into segments of the same length, enumerate them from right to left, $I_1^-, \dots, I_N^-$, and take the rectangles $\Pi_1^-, \dots, \Pi_N^-$ of the same height $\varepsilon - \varepsilon^2$ resting on these resting on these segments; that is, each segment I_i^- is the upper side of Π_i^-. The rectangles Π_i^+ and Π_i^- corresponding to three different values of i are shown in Fig. 2.10.

All the rectangles $\Pi_i^\pm$ are contained in $C_2 \setminus D$. Denote by A_i^+ the midpoint of the segment I_i^+ and by A_i^- the midpoint of I_i^-. The x_1-coordinate of both A_i^+ and A_i^- equals

$$x_1(A_i^+) = \varepsilon^2(1 + 2N - 2i). \tag{2.3}$$

The set

$$\{x : x_1 = x_1(A_i^+),\ \bar{d}(x, \partial D) \le d_i\}$$

is either (i) the vertical segment $A_i^+ A_i^-$ or (ii) a union of two segments $A_i^+ B_i^+ \cup A_i^- B_i^-$ (Fig. 2.10). In case (ii), one has $\bar{d}(B_i^+, \partial D) = \bar{d}(B_i^-, \partial D) = d_i$.

Define the broken line Γ_i as follows. In case (i), take $\Gamma_i = A_i^+ A_i^-$. In case (ii), Γ_i is the union of the segments $A_i^+ B_i^+$ and $A_i^- B_i^-$ and the part of the curve L_i contained in the half-plane $x_1 > x_1(A_i^+)$. In this case, the length of each of the segments $A_i^+ B_i^+$ and $A_i^- B_i^-$ is greater than or equal to ε^3.

By formula (2.3) and by the choice of ε, the value $x_1(A_i^+) - \varepsilon^3$ is a multiple of $2\varepsilon^3$. In addition, it has already been established that each vertical segment of the broken line L_i also has this property: denoting by x_1 the first coordinate of the segment, we have that $x_1 - \varepsilon^3$ is a multiple of $2\varepsilon^3$. Thus, the distance from each of the vertical segments $A_i^+ B_i^+$, $A_i^- B_i^-$ to the nearest vertical segment of L_i is a multiple of $2\varepsilon^3$. This implies that the lengths of the first and last horizontal segments of Γ_i are greater than or equal to $2\varepsilon^3$. The other intermediate segments of Γ_i are at the same time segments of L_i and, therefore, also have lengths greater than or equal to $2\varepsilon^3$.

Thus, the broken line Γ_i is composed of vertical and horizontal segments, the lengths of the initial and final segments are greater than or equal to ε^3, and the lengths of the other segments are greater than or equal to $2\varepsilon^3$. In particular, this broken line may coincide with a single vertical segment. It starts at point A_i^+ and finishes at point A_i^- and has no points of self-intersection. Therefore, Γ_i is an ε-contour joining the rectangles Π_i^+ and Π_i^-. The $\bar{d}$-distance between different curves Γ_i and Γ_j is at least $2\varepsilon^3$:

$$\bar{d}(\Gamma_i, \Gamma_j) \ge 2\varepsilon^3 \quad \text{for } i \ne j.$$

Fix i and consider the rectangles and sectors generated by the reduced segments and vertices of the broken line Γ_i. Notice that any triple of consecutive elements—a v-rectangle, a sector, and an h-rectangle—contains elements that do not intersect pairwise. On the other hand, if a pair of elements does not belong to such a triple, then the minimal union of the squares of the partition containing one element does not intersect the minimal union of the squares containing the other element. Therefore, the elements do not mutually intersect; hence these elements, jointly with the trapezoids $V_i^+ \subset \Pi_i^+$ and $V_i^- \subset \Pi_i^-$ generated by the rectangles Π_i^+ and Π_i^-, form a channel; let it be denoted by $\tilde{K}_i$. We assume that the channel is open, that is, does not intersect its boundary.

Now let us show that the channels defined previously do not mutually intersect. Let K_i' be the union of the rectangles (or the rectangle) generated by the segments $A_i^+ B_i^+$, $A_i^- B_i^-$ and the adjacent sectors. Let K_i'' be the union of the other sets generated by the broken line Γ_i and forming the channel. One obviously has

$$\tilde{K}_i = K_i' \cup K_i'' \cup V_i^+ \cup V_i^- \subset K_i' \cup K_i'' \cup \Pi_i^+ \cup \Pi_i^-. \tag{2.4}$$

We have $K_i' \subset \{x : \bar{d}(x, \partial D) \le 2i\varepsilon^3\}$ and $K_i'' \subset \{x : (2i-2)\varepsilon^3 < \bar{d}(x, \partial D) < 2i\varepsilon^3\}$; hence

$$K_i'' \cap K_j'' = \emptyset \text{ for } i \ne j \quad \text{and} \quad K_i' \cap K_j'' = \emptyset \text{ for } i < j. \tag{2.5}$$

Further, $K_i' \subset \{x : x_1(A_i^+) - \varepsilon^3 < x_1 < x_1(A_i^+) + \varepsilon^3\}$ and $K_i'' \subset \{x : x_1 > x_1(A_i^+) + \varepsilon^3\}$, and hence

$$K_i' \cap K_j' = \emptyset \quad \text{and} \quad K_i'' \cap K_j' = \emptyset \text{ for } i < j. \tag{2.6}$$

Finally, neither the set $\Pi_i^+ \cup \Pi_i^-$ nor the set $\Pi_j^+ \cup \Pi_j^-$, $i \ne j$ intersects D and their projections on the horizontal axis do not intersect; therefore,

$$(\Pi_i^+ \cup \Pi_i^-) \cap (\Pi_j^+ \cup \Pi_j^-) = \emptyset \quad \text{and} \quad (\Pi_i^+ \cup \Pi_i^-) \cap (K_j' \cup K_j'') = \emptyset \text{ for } i \ne j. \tag{2.7}$$

Relations (2.4)–(2.7) imply that the channels $\tilde{K}_i$, $i = 1, \dots, N$ do not mutually intersect. Moreover, the union of the front sides of these channels is a horizontal segment of length l shielding the vertical flow of particles incident on D.

Denote

$$\tilde{B}_\varepsilon = D \cup (\cup_i \Pi_i^+) \cup (\cup_i \Pi_i^-) \setminus (\cup_i \tilde{K}_i). \tag{2.8}$$

Thus, the set $\tilde{B}_\varepsilon$ is obtained by adding all the rectangles $\Pi_i^\pm$ to the set D and then subtracting all the channels $\tilde{K}_i$. A particle incident on D with initial velocity $v = (0, -1)$ intersects the front side of a channel, passes through the channel in the positive direction, then intersects its rear side and continues to move with a velocity $v^+ = v + O(\varepsilon)$. Therefore, the resistance of $\tilde{B}_\varepsilon$ equals

$$R(\tilde{B}_\varepsilon) = O(\varepsilon), \quad \varepsilon \to 0. \tag{2.9}$$

Notice that, roughly speaking, the set $\tilde{B}_\varepsilon$ is not a body since it is locally one-dimensional. Indeed, the intersection of $\tilde{B}_\varepsilon$ with a neighborhood of a point on the common boundary of neighbor v- or h-rectangles is a rectilinear interval. To improve the construction, replace the ε-channels $\tilde{K}_i$ in formula (2.8) with ε'-channels K_i contained in $\tilde{K}_i$, with $\varepsilon' = \varepsilon + O(\varepsilon^2) < \varepsilon$. (We additionally require that the lateral boundaries of $\tilde{K}_i$ and K_i be disjoint.) The resulting set $B_\varepsilon = D \cup (\cup \Pi_i^+) \cup (\cup \Pi_i^-) \setminus (\cup K_i)$ is a true body, and it also satisfies the relation $R(B_\varepsilon) = O(\varepsilon)$, $\varepsilon \to 0$ and the inclusion $C_1 \subset B_\varepsilon \subset C_2$. The proof of Theorem 2.3 is complete.

2.3.3 *Proof of Theorem 2.2*

Fix bodies C_1 and C_2 and assume without loss of generality that $v = (0, 0, -1)$. As in the previous section, we construct here a family of connected bodies B_ε, $C_1 \subset B_\varepsilon \subset C_2$ with vanishing resistance, $\lim_{\varepsilon \to 0} R(B_\varepsilon) = 0$.

A typical body of the family is sandwich-shaped: it is the union of several thin sheets of two kinds: "sheets of small resistance" and "solid sheets." These two kinds of sheets alternate in the sandwich. The plane of the sheets is parallel to v. The sheets of small resistance are constructed with the use of Theorem 2.3 proved in the previous section. The solid sheets are much thinner than the sheets of small resistance and "glue them together," so that the resulting body is connected.

Let us proceed to a description of the construction. For a convex body $C \subset \mathbb{R}^3$ denote

$$C^t = \{(x_2, x_3) : (t, x_2, x_3) \in C\}.$$

In other words, C^t is the projection of the cross section $C \cap \{x_1 = t\}$ on the plane $\mathbb{R}^2_{\{x_2,x_3\}}$. Further, for a set $A \subset \mathbb{R}$ define the sets

$$\underline{C}^A = \cap_{t\in A} C^t \quad \text{and} \quad \overline{C}^A = \text{Conv}(\cup_{t\in A} C^t).$$

One easily sees that for any $t \in A$

$$\overline{C}^A \subset C^t \subset \underline{C}^A$$

holds.

Define the set

$$I = \{t : C_1^t \neq \emptyset\};$$

that is, I is the projection of C_1 on $\mathbb{R}_{\{x_1\}}$. Without loss of generality assume that the sets C_1 and C_2 are open. Then I is a bounded open interval, $I = (a, b)$.

Further, one has $C_1^t \subset C_2^t$. Moreover, for any $t \in I$ there exists an open interval $\mathcal{U}_t$ containing t such that

$$\overline{C}_1^{\mathcal{U}_t} \subset \underline{C}_2^{\mathcal{U}_t} \quad \text{and} \quad \partial(\overline{C}_1^{\mathcal{U}_t}) \cap \partial(\underline{C}_2^{\mathcal{U}_t}) = \emptyset.$$

Fix $\varepsilon > 0$ and choose a finite subset of intervals $\mathcal{U}_{t_i}$ covering the segment $[a+\varepsilon,\ b-\varepsilon]$. Next choose disjoint intervals $I_i^\varepsilon \subset \mathcal{U}_{t_i}$ such that

$$\cup_i I_i^\varepsilon = [a+\varepsilon,\ b-\varepsilon].$$

Due to the choice of these intervals, for each i one has

$$\overline{C}_1^{I_i} \subset \underline{C}_2^{I_i} \quad \text{and} \quad \partial(\overline{C}_1^{I_i}) \cap \partial(\underline{C}_2^{I_i}) = \emptyset.$$

Define

$$C(\varepsilon) = C_2 \cap \{x_1 \in (a,\ a+\varepsilon) \cup (b-\varepsilon,\ b)\},$$

and denote by $C_{x_1,x_2}(\varepsilon)$ the projection of $C(\varepsilon)$ on the plane $\mathbb{R}^2_{\{x_1,x_2\}}$. More precisely, one has

$$C_{x_1,x_2}(\varepsilon) = \{(t,\tau) : t \in (a,\ a+\varepsilon) \cup (b-\varepsilon,\ b) \text{ and } C_2^t \cap \{x_2 = \tau\} \neq \emptyset\}.$$

Let a_ε be the area of $C_{x_1,x_2}(\varepsilon)$; one has $\lim_{\varepsilon\to 0} a_\varepsilon = 0$. The resistance of $C(\varepsilon)$ can be estimated as

$$|R(C(\varepsilon))| \le 2a_\varepsilon. \tag{2.10}$$

Using Theorem 2.3 we select plane bodies B_i^ε such that

$$\overline{C}_1^{I_i} \subset B_i^\varepsilon \subset \underline{C}_2^{I_i} \quad \text{and} \quad |R(B_i^\varepsilon)| < \varepsilon. \tag{2.11}$$

Then the resistance of the three-dimensional set $I_i \times B_i^\varepsilon$ can be estimated as

$$|R(I_i \times B_i^\varepsilon)| = |I_i| \cdot |R(B_i^\varepsilon)| < \varepsilon |I_i|, \tag{2.12}$$

where $|I|$ means the length of interval I. Define the body

$$\tilde{B}_\varepsilon = C(\varepsilon) \cup \left(\cup_i I_i \times B_i^\varepsilon\right).$$

Using the first relation in (2.11) and the definition of $C(\varepsilon)$ one concludes that

$$C_1 \subset \tilde{B}_\varepsilon \subset C_2.$$

Consider a particle incident on $\tilde{B}_\varepsilon$. If its x_1-coordinate belongs to $(a, a+\varepsilon) \cup (b-\varepsilon, b)$, then it makes a single reflection at a point of $\partial C(\varepsilon)$. If the x_1-coordinate belongs to an interval I_i, then the particle makes several reflections at points of $I_i \times B_i^\varepsilon$ and never hits any other subset constituting the body $\tilde{B}_\varepsilon$. Therefore, the resistance of $\tilde{B}_\varepsilon$ is the sum of resistances of its subsets:

$$R(\tilde{B}_\varepsilon) = R(C(\varepsilon)) + \sum_i R(I_i \times B_i^\varepsilon).$$

Using (2.10) and (2.12) one obtains the estimate

$$|R(\tilde{B}_\varepsilon)| \le 2a_\varepsilon + \varepsilon(b-a),$$

that is, $\lim_{\varepsilon\to 0} R(\tilde{B}_\varepsilon) = 0$.

However, the body $\tilde{B}_\varepsilon$ is not connected. Let us therefore modify it in the following way. Take an open set $J_\varepsilon \subset \mathbb{R}$ and require that it be the disjoint union of open intervals of total length less than ε and contains the endpoints of all intervals I_i, that is,

$$\cup_i \partial I_i \subset J_\varepsilon \quad \text{and} \quad |J_\varepsilon| < \varepsilon.$$

Define

$$D(\varepsilon) = C_2 \cap \{x_1 \in J_\varepsilon\}.$$

Then the body

$$B_\varepsilon = \tilde{B}_\varepsilon \cup D(\varepsilon)$$

is connected and satisfies the relations

$$C_1 \subset B_\varepsilon \subset C_2 \quad \text{and} \quad \lim_{\varepsilon\to 0} R(B_\varepsilon) = 0.$$

The proof of Theorem 2.2 is complete.

2.4 Two-Dimensional Problem

The results obtained in the three-dimensional case can be easily generalized to higher dimensions $d > 3$. The infimum of the functional $R_{\delta_{v0}}$ over the corresponding classes of nonconvex bodies is also equal to 0. By contrast, in the two-dimensional case, the infimum (whether taken over the class of convex or nonconvex bodies) is positive.

2.4.1 Minimum Resistance of Convex Bodies

We discuss briefly the known results for the two-dimensional analog of Newton's problem with $c(v, v^+) = (v - v^+) \cdot v$. Here, as above, we consider a flow of particles falling vertically downward with velocity $v_0 = (0, -1)$. Let $C = C_h = [-1, 1] \times [-h, 0]$ be a rectangle in a plane with variables x_1, x_2, and let B be a convex body inscribed in C. The upper part of the boundary of B is the graph of a concave function $-f_B$, and the resistance can be represented by the functional $R_{\delta_{v0}}(B) = 2\int_{-1}^{1}(1 + f_B'^2(x))^{-1}\, dx$. Thus, the problem of minimizing the resistance assumes the following form:

$$\text{Find} \quad \inf_{f \in \text{Convex}_2(h)} \mathcal{R}_2(f), \quad \text{where } \mathcal{R}_2(f) = \int_{-1}^{1} \frac{dx}{1 + f'^2(x)}. \tag{2.13}$$

Here $\text{Convex}_2(h)$ is the class of convex functions $f : [-1, 1] \to [0, h]$.

The solution to this problem is well known (see, e.g., [12]): the body of minimum resistance is an isosceles triangle if $h \geq 1$ and a trapezoid if $h < 1$ (Fig. 2.11). The smallest value of the resistance is

$$2 \inf_{f \in \text{Convex}_2(h)} \mathcal{R}_2(f) = \begin{cases} 4 - 2h, & \text{if } h < 1, \\ 4/(1 + h^2), & \text{if } h \geq 1. \end{cases} \tag{2.14}$$

Finally, in the two-dimensional case (in contrast to the three-dimensional case), the solution in the class of arbitrary convex bodies inscribed in C is the same as in the narrower class of convex bodies symmetric relative to a vertical axis.

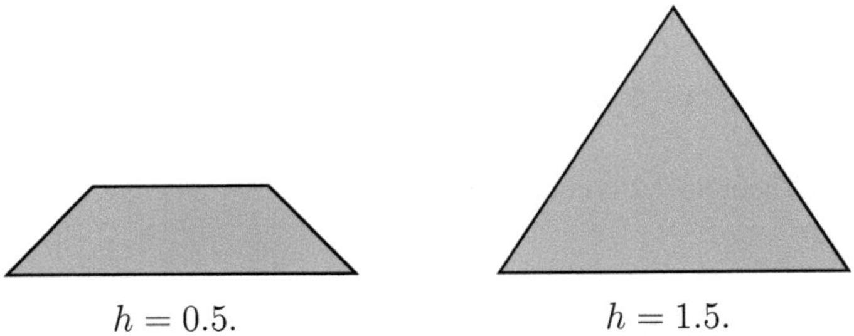

Fig. 2.11 Solution to two-dimensional problem for convex bodies

2.4.2 Minimum Resistance of Nonconvex Bodies

Let $\mathcal{P}_2(h)$ denote the class of connected bodies lying in the rectangle C_h and intersecting both its left and right sides $\{-1\} \times [-h, 0]$ and $\{1\} \times [-h, 0]$. For the resistance of such bodies we have the formula

$$R_{\delta_{v_0}}(B) = \int_{[-1,1]\times\{0\}} \langle v_0 - v_B^+(\xi, v_0),\, v_0 \rangle \, d\xi. \tag{2.15}$$

It is clear that if $B \in \mathcal{P}_2(h)$, then also $\mathrm{Conv} B \in \mathcal{P}_2(h)$.

Theorem 2.4. $\inf_{B\in\mathcal{P}_2(h)} R_{\delta_{v_0}}(B) = 2(1 - h/\sqrt{1+h^2})$.

Remark 2.4. The minimum value is attained on the subclass of $\mathcal{P}_2(h)$ of bodies that particles in the flow hit at most twice. This is clear from the proof below.

Remark 2.5. It follows from this theorem that the least resistance over the class of *nonconvex* bodies is positive but still much smaller than over the class of *convex* bodies. Comparing the statement of the theorem with (2.14) we conclude that the ratio

$$\frac{\text{the least resistance of } \textbf{nonconvex} \text{ bodies}}{\text{the least resistance of } \textbf{convex} \text{ bodies}}$$

decreases monotonically from $1/2$ to $1/4$ as h increases from 0 to $+\infty$.

Proof. Recall that $\mathrm{Convex}_2(h)$ is the set of convex functions $f : [-1, 1] \to [0, h]$. Consider the functional $\hat{R}$ on this set defined by the formula

$$\hat{R}(f) = \int_{-1}^{1} \left(1 - \frac{|f'(x)|}{\sqrt{1 + f'^2(x)}}\right) dx. \tag{2.16}$$

We have the following lemmas, which we prove subsequently.

Lemma 2.2. *The functional $\hat{R}$ takes its minimum value on the function $f_h(x) = h|x|$. This value is $2(1 - h/\sqrt{1+h^2})$. In other words,*

$$\inf_{f \in Convex_2(h)} \hat{R}(f) = \hat{R}(f_h) = 2\left(1 - \frac{h}{\sqrt{1+h^2}}\right).$$

Lemma 2.3. *For each $B \in \mathcal{P}_2(h)$,*

$$R_{\delta_{v0}}(B) \geq \hat{R}(f_{Conv B}).$$

Let Δ_h be the triangle $-h \leq x_2 \leq -f_h(x_1)$. This is the isosceles triangle in the rectangle C_h with vertices at $(1, -h)$, $(-1, -h)$, and $(0, 0)$.

Recall that the variable $\xi = (x, 0)$ ranges over $[-1, 1] \times \{0\}$.

Lemma 2.4. *There exists a family of bodies $B_\varepsilon \in \mathcal{P}_2(h)$ such that*

$$Conv\, B_\varepsilon = \Delta_h,$$

and the corresponding family of functions $v^+_{B_\varepsilon}(\xi, v_0)$ converges in measure to the function $(\operatorname{sgn} x, -h)/\sqrt{1+h^2}$ *as $\varepsilon \to 0^+$.*

These three lemmas immediately yield the required result. Indeed, by Lemmas 2.2 and 2.3,

$$\inf_{B \in \mathcal{P}_2(h)} R_{\delta_{v0}}(B) \geq 2\left(1 - \frac{h}{\sqrt{1+h^2}}\right),$$

whereas by Lemma 2.4 and (2.15),

$$\lim_{\varepsilon \to 0+} R_{\delta_{v0}}(B_\varepsilon) = 2\left(1 - \frac{h}{\sqrt{1+h^2}}\right),$$

so that the reverse inequality holds:

$$\inf_{B \in \mathcal{P}_2(h)} R_{\delta_{v0}}(B) \leq 2\left(1 - \frac{h}{\sqrt{1+h^2}}\right).$$

□

Proof of Lemma 2.2. Let $p(u) = 1 - u/\sqrt{1+u^2}$. In what follows we use the observations that for $u > 0$ the functions $p(u)$ and $u \cdot p(1/u)$ are convex and $p(u)$ is decreasing.

The result of the lemma is an immediate consequence of the relations

$$\int_{-1}^{1} p(|f'(x)|)\, dx \geq 2p(h) \quad \text{for each} \quad f \in \text{Convex}_2(h), \tag{2.17}$$

$$\int_{-1}^{1} p(|f_h'(x)|)\, dx = 2p(h). \tag{2.18}$$

Relation (2.18) follows from the identity $|f_h'(x)| \equiv h$. To prove (2.17), we consider two cases.

(i) The function f is (not strictly) monotone. Assuming without loss of generality that f is nondecreasing, and taking into account the convexity of p, we obtain from Jensen's integral inequality that

$$\frac{1}{2}\int_{-1}^{1} p(f'(x))\,dx \geq p\left(\frac{1}{2}\int_{-1}^{1} f'(x)\,dx\right) = p\left(\frac{f(1)-f(-1)}{2}\right) \geq p(h/2) > p(h).$$

This proves (2.17) in this case.

(ii) The function f is not monotone, so there exists x_0, $-1 < x_0 < 1$ such that f is monotonically nonincreasing for $x \leq x_0$ and nondecreasing for $x \geq x_0$. We have

$$\int_{-1}^{1} p(|f'(x)|)\,dx = \int_{-1}^{x_0} p(-f'(x))\,dx + \int_{x_0}^{1} p(f'(x))\,dx.$$

Applying Jensen's inequality to both terms on the right hand side of this equation, we obtain

$$\frac{1}{1+x_0}\int_{-1}^{x_0} p(-f'(x))\,dx \geq p\left(\frac{1}{1+x_0}\int_{-1}^{x_0}(-f'(x))\,dx\right),$$

$$\frac{1}{1-x_0}\int_{x_0}^{1} p(f'(x))\,dx \geq p\left(\frac{1}{1-x_0}\int_{x_0}^{1} f'(x)\,dx\right).$$

Since $\int_{-1}^{x_0}(-f'(x))\,dx \leq h$ and $\int_{x_0}^{1} f'(x)\,dx \leq h$, it follows that

$$\int_{-1}^{1} p(|f'(x)|)\,dx \geq (1+x_0)\,p\left(\frac{h}{1+x_0}\right) + (1-x_0)\,p\left(\frac{h}{1-x_0}\right). \quad (2.19)$$

The function $u \cdot p(1/u)$ is convex; therefore, the right-hand side of (2.19) is a convex function of x_0, which is also even. Hence at $x_0 = 0$ it takes a minimum value, which is $2p(h)$. The proof of (2.17) is complete. □

Proof of Lemma 2.3. We introduce the shorter notation $f_{\mathrm{Conv}B} =: f$ and $v_B^+(\xi, v_0) =: v^+(x) = (v_1^+(x), v_2^+(x))$, where $\xi = (x, 0)$. Then formula (2.15) assumes the form

$$R_{\delta_{v_0}}(B) = \int_{-1}^{1}(1 + v_2^+(x))\,dx. \quad (2.20)$$

We define the following modified law of reflection from the convex body $\mathrm{Conv}B$: the velocity of the particle after reflection is the unit vector with a nonpositive second component that is tangent to $\partial(\mathrm{Conv}B)$ at the point of reflection (Fig. 2.12). If the tangent is horizontal, then we agree to choose, for instance, the vector pointing to the right, that is, $(1, 0)$. In other words, the velocity after reflection makes the

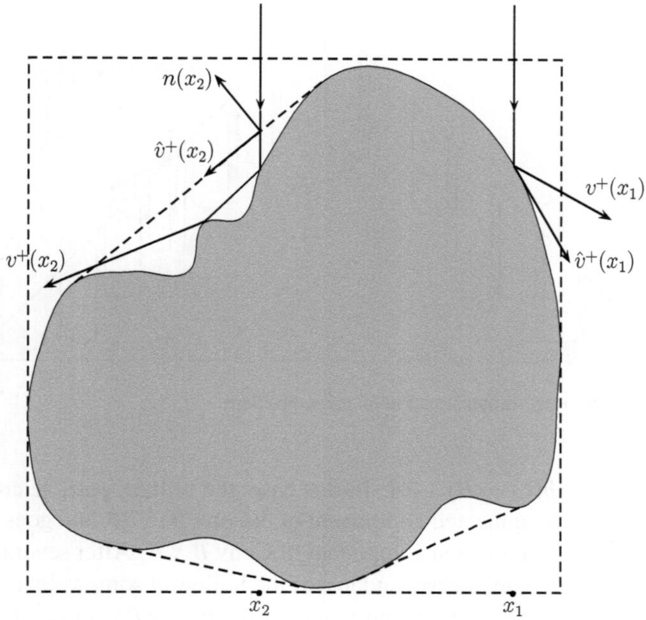

Fig. 2.12 Ordinary (billiard) and modified laws of reflection

smallest possible angle with the initial velocity $v_0 = (0, -1)$. Thus, if the reflection point is $(x, -f(x))$, then in accordance with the modified law the velocity of the reflected particle is

$$\hat{v}^+(x) = (\hat{v}_1^+(x), \hat{v}_2^+(x)) = \frac{(\operatorname{sgn} f'(x), \, -|f'(x)|)}{\sqrt{1 + f'^2(x)}}.$$

Here $\operatorname{sgn} z = 1$ if $z \geq 0$ and -1 otherwise. Using this notation, we can represent the functional $\hat{R}(f)$ as

$$\hat{R}(f) = \int_{-1}^{1} (1 + \hat{v}_2^+(x)) \, dx. \tag{2.21}$$

This is the resistance of the body Conv$B$ if the modified law of reflection holds.

We assert that $v_2^+(x) \geq \hat{v}_2^+(x)$. Then from (2.20) and (2.21) we immediately obtain the required result. We consider two cases.

(i) $(x, -f(x)) \in \partial(\mathrm{Conv} B) \cap \partial B$. In this case, the billiard particle makes a unique elastic collision with ∂B, after which the second component of the velocity becomes

$$v_2^+(x) = \frac{1 - f'^2(x)}{1 + f'^2(x)} > -\frac{|f'(x)|}{\sqrt{1 + f'^2(x)}} = \hat{v}_2^+(x)$$

(see Fig. 2.12 for $x = x_1$).

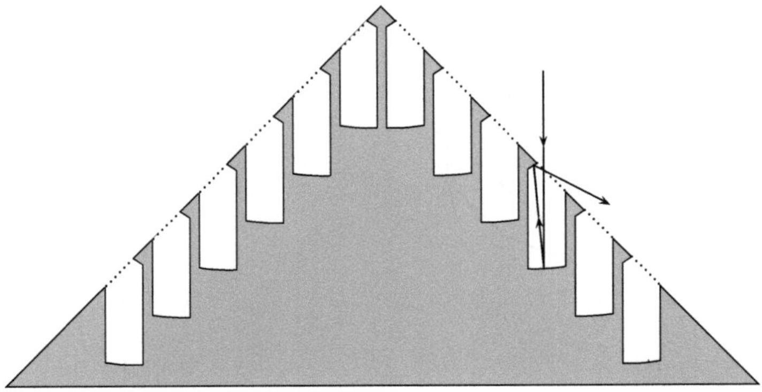

Fig. 2.13 Solution to two-dimensional nonconvex problem

(ii) $(x, -f(x)) \in \partial(\text{Conv}B) \setminus \partial B$. In this case, the billiard particle crosses a line interval that is a connected component of $\partial(\text{Conv } B) \setminus \partial B$ and goes into a "hollow," which is a connected component of Conv $B \setminus B$. After several reflections it intersects the same interval and leaves the hollow at some velocity $v^+(x)$ that makes an acute angle with the normal $n(x)$ to $\partial(\text{Conv } B)$ at $(x, -f(x))$, that is, $\langle v^+(x), n(x)\rangle > 0$, where $n(x) = (f'(x), 1)/\sqrt{1 + f'^2(x)}$ (see Fig. 2.12 for $x = x_2$). This easily yields the required inequality. □

Proof of Lemma 2.4. We set $B_\varepsilon = \Delta_h \setminus \left(\cup_{-n_0 \le n(\neq 0) \le n_0} \Omega_\varepsilon^n\right)$, where the value $n_0 = n_0(\varepsilon)$ will be defined below, the sets Ω_ε^n, $n = 1, \dots, n_0$ are translations of a set Ω_ε also defined below, and each set Ω_ε^{-n} is symmetric to Ω_ε^n relative to the vertical axis Ox_2 (Fig. 2.13). The projections of $\Omega_\varepsilon^{-n_0}, \dots, \Omega_\varepsilon^{-1}, \Omega_\varepsilon^1, \dots, \Omega_\varepsilon^{n_0}$ on the horizontal axis Ox_1 are disjoint line segments of length ε^2; each successive segment lies at a distance of ε^3 from the preceding one. The part of the boundary of Ω_ε^n indicated by a dotted line lies on the corresponding lateral side of the triangle Δ_h. The quantity n_0 is the largest n such that $\Omega_\varepsilon^n \subset \Delta_h \cap \{x_2 > -h\}$, so that the sets Ω_ε^n lie in the triangle but are disjoint from its base.

The set Ω_ε is the part of the vertical strip $0 \le x_1 \le \varepsilon^2$ bounded by the segments GC and EG and the arc AD of a parabola (Fig. 2.14). (Note that the positions of points $G, C, \dots$ depend on ε, so strictly speaking we should write G_ε, C_ε, and so on.) We set $O = (0, 0)$, $C = (\varepsilon^2, -h\varepsilon^2)$, $D = (\varepsilon^2, -\varepsilon)$, $F = (\varepsilon^3/2, -h\varepsilon^3)$. Thus, segment OC lies on a lateral side of the triangle Δ_h. Arc AD is a piece of a parabola with upward branches, focus at F, and a vertical axis; EG is a segment containing F such that a "billiard particle" coming from D and reflected from EG at F goes to C (we indicate the corresponding trajectory by a dotted line). Side GC is indicated by a dashed line.

Consider the billiard in Ω_ε represented in Fig. 2.14. The billiard particle moves vertically downward starting from some point in the segment GC. It is reflected from the arc AD, then from EG at point F, and again goes to some point in GC, with velocity before reflection making an angle of magnitude $O(\varepsilon)$ with FC. Since

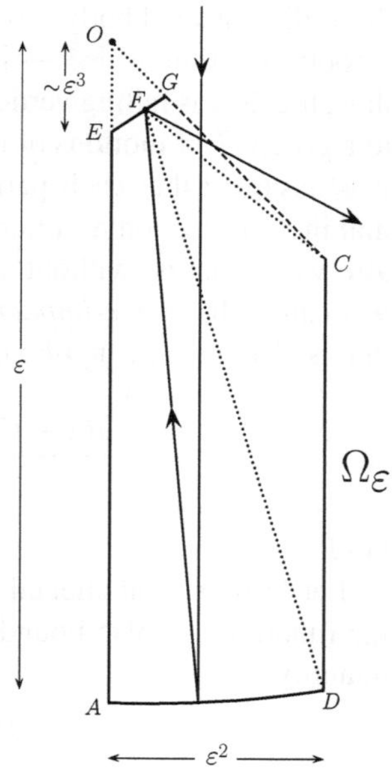

Fig. 2.14 Solution of two-dimensional nonconvex problem; the hollow Ω_ε

$\measuredangle FCO = O(\varepsilon)$ and the vector $\overrightarrow{OC}$ is proportional to $(1, -h)$, we conclude that this velocity is $(1, -h)/\sqrt{1+h^2} + O(\varepsilon)$ as $\varepsilon \to 0^+$, where $O(\varepsilon)$ is a uniform estimate for all the initial positions on the segment GC.

We now return to the billiard in $\mathbb{R}^2 \setminus B_\varepsilon$. For a set of values of ξ of total length $O(\varepsilon)$ the corresponding billiard particle is reflected from the part of a side of Δ_h indicated by a solid line [that is, from $\partial(\mathrm{Conv} B_\varepsilon) \cap \partial B_\varepsilon$]. After the reflection its velocity is

$$v^+_{B_\varepsilon}(\xi, v_0) = \frac{(2h\,\mathrm{sgn}x,\ 1 - h^2)}{1 + h^2}.$$

For other values $\xi \in [-1, 1] \times \{0\}$ the billiard particle intersects the dashed line, reflects twice from the boundary of the corresponding set Ω^n_ε, and, finally, intersects the same dashed line in the opposite direction and moves freely afterward with velocity

$$v^+_{B_\varepsilon}(\xi, v_0) = \frac{(\mathrm{sgn}x,\ -h)}{\sqrt{1 + h^2}} + O(\varepsilon) \ \text{ as } \ \varepsilon \to 0,$$

where the estimate $O(\varepsilon)$ is uniform in ξ. The proof of Lemma 2.4 is complete. □

2.5 Minimum Specific Resistance of Unbounded Bodies

Recently Comte and Lachand-Robert [18] considered an interesting modification of Newton's problem. A parallel flow of particles falls with velocity $(0, 0, -1)$ on a three-dimensional body $z \le u(x)$ bounded from above by the graph of a piecewise smooth function $u : \mathbb{R}^2 \to \mathbb{R}$. We assume that the axis Oz is directed upward and, thus, the flow is falling vertically downward. The function u is invariant with respect to a group G of motions of the plane, that is, $u \circ g = u$ for any $g \in G$. In addition, it is assumed that each particle impinging on the body vertically downward and making a reflection at a nonsingular point of the body boundary will move freely afterward, that is, without any further reflection from the body. This condition on u is called the *single-impact assumption* (s.i.a.). It can also be written analytically; that is, for any nonsingular point $x \in \mathbb{R}^2$ and any $t > 0$

$$\frac{u(x - t\nabla u(x)) - u(x)}{t} \le \frac{1}{2}\left(1 - |\nabla u(x)|^2\right) \tag{2.22}$$

holds.

Let Ω be a fundamental region of G. We assume that Ω is bounded; therefore, the function u is also bounded. By $|\Omega|$ we denote the area of Ω. By definition, the quantity

$$F(u) = \frac{1}{|\Omega|}\int_\Omega \frac{dx}{1 + |\nabla u(x)|^2} \tag{2.23}$$

is called the *specific resistance* of the body. Clearly, it does not depend on the choice of the fundamental region. This definition has a simple physical interpretation: $F(u)$ is the force of pressure of a flow of large cross section divided by the area of this section.

The greatest value of specific resistance is $F(u) = 1$, and the maximizers are constants $u \equiv \text{const}$; in this case the body is a half-space. On the other hand, for any body satisfying the s.i.a. we have $|\nabla u| \le 1$. Indeed, assuming that $|\nabla u(x_0)| > 1$ at a point x_0 and taking into account (2.22), we get that u tends to $-\infty$ along the ray $x - t\nabla u(x)$, $t > 0$, which is impossible since it is bounded. It follows from (2.23) that $F(u) \ge 0.5$, so we have

$$0.5 \le F(u) \le 1.$$

This formula is in agreement with physical intuition. Indeed, consider a particle with mass m that falls with velocity $(0, 0, -1)$ on a body and is reflected by it. The vertical component of velocity after the reflection is nonnegative; therefore, the vertical component of the momentum transmitted to the body by the particle is at least m. On the other hand, the maximal value of this component is $2m$ and is achieved when the gradient of u at the point of reflection is zero. Thus, the vertical component of the transmitted momentum is at least one half its maximum value; therefore, the specific resistance of a body is at least one half its maximum value.

The problem consists in finding the minimal value of specific resistance under the single-impact condition. It has not been solved until now. Note that the least known value is approximately $F(u_*) \approx 0.580778$; the corresponding function u_* was constructed in [53]. Therefore, one has

$$0.5 \le \inf_u F(u) \le 0.581.$$

Now consider a slightly modified problem. That is, we shall consider bodies whose boundary is not necessarily the graph of a function; in addition, we shall allow multiple reflections of the flow particles from the body. Note that in this broader class of bodies the explicit formula (2.23) for the specific resistance is no longer valid. Nevertheless, the problem in this form becomes easier and admits an explicit solution.

Let us pass to rigorous mathematical definitions. Following [47], we consider the problem in arbitrary dimension $d \geq 2$. Fix a unit vector $n = (n_1, \ldots n_d)$ such that $n_d > 0$ in Euclidean space $\mathbb{R}^d$ with orthogonal coordinates $x = (x_1, \ldots, x_d)$, and consider a set B with a piecewise smooth boundary and containing the half-space $x \cdot n < 0$. Further, we consider a flow of particles with velocity $v = (0, \ldots, 0, -1)$ impinging on the body. Initially (prior to reflections from the body) the coordinate of the particle is $x(t) = x + vt$; then it is reflected finitely many times from B at regular points of the boundary ∂B, and it finally moves freely. Let $v_B^+(x)$ denote its final velocity. The momentum transmitted by the particle to the body is proportional to $v - v_B^+(x)$, where the proportionality ratio is equal to the mass of the particle. Thus, we have the mapping $x \mapsto v_B^+(x)$ defined on a subset of $\mathbb{R}^d$ and translationally invariant in direction v. We also impose the regularity condition on the scattering of particles by the body.

Let $\mathcal{B}_n$ be the set of bodies $B \subset \mathbb{R}^d$ such that

(a) B contains the half-space $\{x \cdot n < 0\}$ and
(b) The scattering of particle flow by B is regular.
For a set $A \subset \{x_d = 0\}$ with a finite $(d-1)$-dimensional Lebesgue measure denote

$$R(B, A) = \frac{1}{|A|} \int_A (v - v_B^+(x))\, dx.$$

The quantity $R(B, A)$ admits a natural interpretation: it is the resistance of B to the flow with cross section A divided by the area of this section or, in other words, the flow pressure averaged over A.

Notice that $v_B^+(x) \cdot n \geq 0$, hence $(v - v_B^+(x)) \cdot n \leq -n_d$. On the other hand, v and $v_B^+(x)$ are unit vectors; therefore, $|v - v_B^+(x)| \leq 2$. This implies that

$$n_d \leq |R(B, A)| \leq 2 \tag{2.24}$$

for any $B \in \mathcal{B}_n$ and any Borel set A of a finite Lebesgue measure.

The upper estimate in (2.24) is exact; let, for instance, $\mathbb{R}^{d-1} = \cup_i A_i$, where A_i are pairwise disjoint bounded Borel sets with boundaries of measure zero, and let $\tilde{n} = (n_1, \ldots n_{d-1})$ and

$$B^* = \cup_i (A_i \times (-\infty,\, c_i)),$$

where $c_i \geq -\frac{1}{n_d} \inf\{\tilde{x} \cdot \tilde{n} : \tilde{x} \in A_i\}$. Then $B^* \in \mathcal{B}_n$ and

$$|R(B, A)| = 2$$

for any Borel set $A \subset \{x_d = 0\}$ of finite measure.

Let us show that the lower estimate in (2.24) is also exact.

Theorem 2.5. *There exists a sequence of sets* $B_k \in \mathcal{B}_n$ *such that*

$$\lim_{k\to\infty} |R(B_k, A)| = n_d$$

for any Borel set A *with a finite Lebesgue measure. Moreover, each particle of the flow collides with* B_k *at most twice, and the sequence* B_k *approximates the half-space* $x \cdot n < 0$ *or, in more precise terms,*

$$\{x : x \cdot n < 0\} \subset B_k \subset \{x : x \cdot n < \alpha_k\}$$

where $\lim_{k\to\infty} \alpha_k = 0$.

The following corollary is a direct consequence of the theorem.

Corollary 2.1. *For any set* A *of finite Lebesgue measure*

$$\inf_{B\in\mathcal{B}_n} |R(B, A)| = n_d$$

holds.

Proof. The proof of Theorem 2.5 is based on a direct construction of bodies B_k, and the case of higher dimensions reduces to the two-dimensional case $d = 2$. Indeed, suppose that for any $n^{(2)} = (n_1, n_2) \in S^1$ there exists a sequence of bodies $B_k^{(2)}$ providing the solution in two dimensions. Then the sequence $B_k^{(d)} = \mathbb{R}^{d-2} \times B_k^{(2)}$ is a solution for the vector $n^{(d)} = (0, \dots, 0, n_1, n_2) \in S^{d-1}$ in the d-dimensional case, $d \ge 3$. Now, to obtain a solution for arbitrary unit vector $n = (n_1, \dots, n_d)$, one first constructs a sequence B_k' that provides a solution for $n' = (0, \dots, 0, \sqrt{\sum_1^{d-1} n_j^2}, n_d)$, and then applies to each B_k' an isometry sending n' to n and v to v. Let B_k be the image of B_k' under this isometry; then the sequence B_k is a solution for n.

It remains to consider the two-dimensional case. First we briefly describe the idea of construction. Consider the plane with coordinates x, y, and assume that the axis Oy is directed upward. Take the half-plane $x \cdot n < \alpha_k$ and make a series of identical dimples on its boundary; as a result, we obtain the body B_k. A part of the boundary of the dimple is formed by two pairs of arcs of confocal parabolas, where the length of one arc in each pair is much smaller that that of the other one. A parallel beam of incident particles is reflected by the larger arc, passes through the common focus, is reflected by the second arc, and transforms again into a parallel beam (of smaller width), with the direction almost parallel to the boundary of the half-plane.

Let us now state a rigorous proof. Choose $\varphi \in (-\pi/2, \pi/2)$ such that $n = (\sin\varphi, \cos\varphi)$. Fix $a > 0$ and $\varepsilon > 0$ and set

$$x_0 = \frac{1 - \sin\varphi}{2} a. \tag{2.25}$$

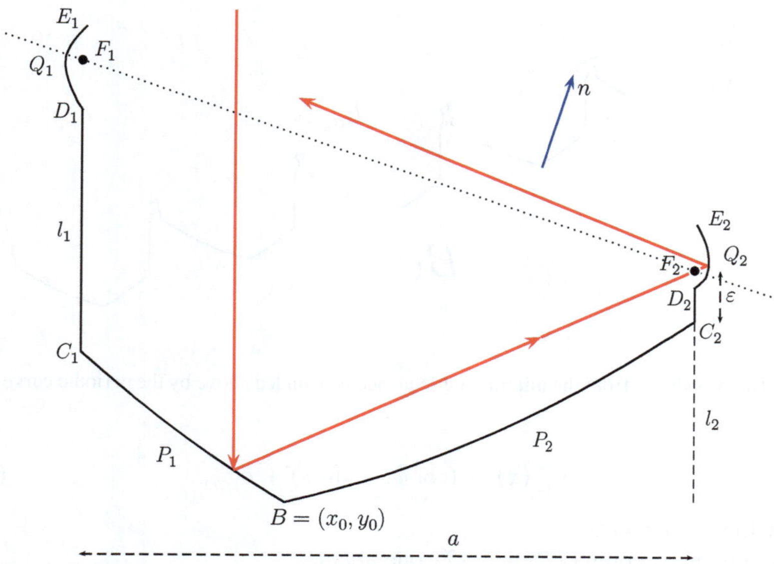

Fig. 2.15 A hollow

Consider the point $B = (x_0, y_0)$, with the value y_0 to be specified below. Draw two parabolas through B with upward branches, namely, parabola P_1 with focus at $F_2 = (a, -a\tan\varphi)$ and parabola P_2 with focus $F_1 = (0, 0)$. The axis of P_1 is denoted by l_2, and the axis of P_2 by l_1. The axes are vertical and pass through F_2 and F_1, respectively. Choose y_0 such that point C_1 of the intersection of P_1 with l_1 would lie below F_1, and point C_2 of intersection of P_2 with l_2 would lie below F_2, and additionally $\min_{i=1,2} |C_i F_i| = \varepsilon$. Take two points F_i', $i = 1, 2$ on the line l_i above F_i such that $|F_i F_i'| = \varepsilon$, and take two parabolas Q_1 and Q_2 such that

(i) Q_1 has its focus at F_1 and the axis $F_1 F_2'$, and Q_2 has its focus at F_2 and the axis $F_2 F_1'$;
(ii) $\max\{|D_i F_i|,\ |E_i F_i|,\ i = 1, 2\} = \varepsilon$, where D_i and E_i are, respectively, the lower and upper points of intersection of Q_i with l_i (Fig. 2.15).

Now consider the curve $\Gamma_{a,\varepsilon} = E_1 D_1 C_1 B C_2 D_2 E_2$ composed of the arcs of parabolas BC_i, $D_i E_i$ and the line segments $C_i D_i$, $i = 1, 2$. A particle falling vertically downward in accordance with $x(t) = (x, -t)$, $0 < x < x_0$, first reflects from the arc BC_1, passes through F_2, reflects from the arc $D_2 E_2$, and finally moves freely with velocity

$$v_{a,\varepsilon}^{+}(x) = (-\cos\varphi, \sin\varphi) + \beta_{\varepsilon/a}^{1}, \tag{2.26}$$

where $\lim_{\omega\to 0} \beta_\omega^1 = 0$. If $x_0 < x < a$, then the particle first reflects from the arc BC_2, then passes through F_1, reflects from $D_1 E_1$, and finally moves freely with velocity

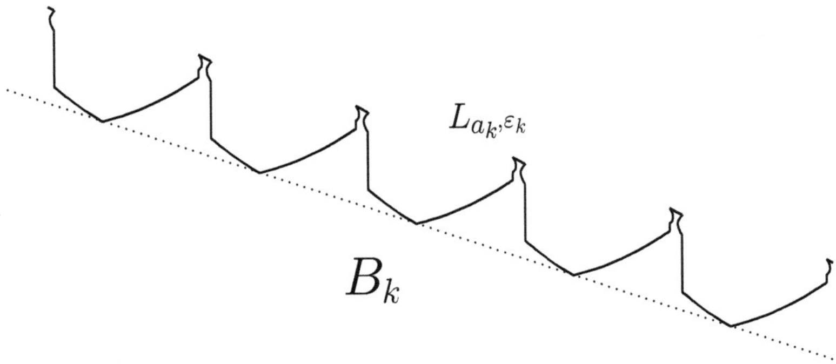

Fig. 2.16 A body B_k from the minimizing sequence is bounded above by the periodic curve L_{a_k,ε_k}

$$v^+_{a,\varepsilon}(x) = (\cos\varphi, \, -\sin\varphi) + \beta^2_{\varepsilon/a}, \tag{2.27}$$

where $\lim_{\omega\to 0}\beta^2_\omega = 0$.

Taking into account (2.25)–(2.27), one obtains

$$\int_0^a (v - v^+_{\varepsilon/a}(x))\, dx = -a\cos\varphi\ \varepsilon c n + o\,(1), \qquad \varepsilon/a \to 0.$$

Let $\Gamma^{(m)}_{a,\varepsilon}$ be the curve obtained by shifting $\Gamma_{a,\varepsilon}$ by the vector $-(x_0, y_0) + m\,(a + \varepsilon, \, -(a+\varepsilon)\,\tan\varphi)$, and consider the continuous line $L_{a,\varepsilon}$ composed of the curves $\Gamma^{(m)}_{a,\varepsilon}$, $m \in \mathbb{Z}$ and of line segments joining endpoints of neighboring curves. The line $L_{a,\varepsilon}$ is situated between the parallel straight lines $x \cdot n = 0$ and $x \cdot n = \alpha$, where $\alpha = x_0 \sin\varphi + y_0 \cos\varphi$. The parameters x_0 and y_0, and therefore the parameter α, depend on a and ε and approach 0 when a and ε go to 0.

Choose sequences a_k and ε_k such that $\lim_{k\to\infty} a_k = 0$, $\lim_{k\to\infty}(\varepsilon_k/a_k) = 0$, and take the sets B_k bounded above by L_{a_k,ε_k} (Fig. 2.16). Clearly, $B_k \in \mathcal{B}_n$ and $\{x \cdot n < 0\} \subset B_k \subset \{x \cdot n < \alpha_k\}$ for the sequence of corresponding values α_k approaching zero as $k \to \infty$. For any line segment $A \subset \Pi = \{y = 0\}$, and therefore, for any Borel set $A \subset \Pi$

$$R(B_k, A) = -\cos\varphi\, n + o\,(1), \qquad k \to \infty$$

holds. The proof of Theorem 2.5 is complete. □

Chapter 3
Newton's Problem in Media with Positive Temperature

In the original setting of Newton's aerodynamic problem, it is supposed that there is no thermal motion of medium particles, that is, in an appropriate coordinate system (and prior to collisions with the body), all particles are resting. However, it is much more realistic to assume that thermal motion is present. In this chapter, we address the generalization of Newton's problem for bodies moving in media with positive temperature. We shall see that the method of solution is quite conventional as compared with the original problem. On the other hand, a larger variety of optimal form is revealed here. In the three-dimensional case, an optimal body can (a) have a shape resembling the optimal Newtonian shape and (b) be the union of two Newton-like bodies "glued together" along their rear parts. Cases (a) and (b) are realized when the velocity of the body in the medium exceeds a critical value and when it is smaller than this value, respectively. In the two-dimensional case, there exist five different classes of solutions, while in the 2D analog of the original Newton problem, there are only two classes: an isosceles triangle and a trapezium.

In the low-temperature limit, the optimal shape becomes the Newton solution, while in the high-temperature limit (hot medium) the three-dimensional optimal shape becomes symmetric relative a plane orthogonal to the symmetry axis of the body.

The results of this chapter were first published in [48, 60].

3.1 Calculation of Resistance and Statement of Minimization Problem

3.1.1 *Description of the Medium*

We assume that the distribution of velocities in the medium is homogeneous and isotropic, that is, for arbitrary infinitesimal regions $\mathcal{X}$, $\mathcal{V} \subset \mathbb{R}^d$ that have

A. Plakhov, *Exterior Billiards: Systems with Impacts Outside Bounded Domains*, DOI 10.1007/978-1-4614-4481-7_3,

d-dimensional volumes $|\mathcal{X}| = dx$, $|\mathcal{V}| = dv$, the total mass of particles in $\mathcal{X}$ with velocities $v \in \mathcal{V}$ is $\sigma(|v|)\, dv\, dx$. Here σ is a function defined on $\mathbb{R}_+$.

A body moves in this medium with velocity $V > 0$. It is convenient to choose a reference system attached to the moving body and such that the dth coordinate vector, e_d, is codirectional with the velocity vector; the distribution of velocities of the particles in this coordinate system is given by $\sigma(|v + Ve_d|)$.

Below we give two important examples of the distribution σ.

Example 3.1. Consider a rarefied homogeneous monatomic ideal gas in $\mathbb{R}^3$ with absolute temperature $T > 0$. The density of distribution of the molecules' mass over velocities equals $\sigma_h(|v|)$, where

$$\sigma_h(r) = \nu \left(\frac{m}{2\pi kT}\right)^{3/2} e^{-\frac{mr^2}{2kT}}$$

(the Maxwell distribution); here k is Boltzmann's constant and ν is the density of the gas. A body moves in the gas with constant velocity of magnitude V in the direction of the third coordinate vector e_3. In a frame of reference attached to the body, the density of distribution over velocities equals $\sigma_h(|v + Ve_3|)$.

Example 3.2. Assume now that a rarefied ideal gas of temperature T is a mixture of n homogeneous components, where the ith component has density ν_i and consists of monatomic molecules of mass m_i. Then the density of distribution of the molecules' mass over velocities equals $\sigma_{nh}(|v|)$, where

$$\sigma_{nh}(r) = \sum_{i=1}^{n} \nu_i \left(\frac{m_i}{2\pi kT}\right)^{3/2} e^{-\frac{m_i r^2}{2kT}}.$$

A body moves as in the previous example. In a frame of reference attached to the body the density of distribution over velocities equals $\sigma_{nh}(|v + Ve_3|)$.

Definition 3.1. We denote by $\mathcal{A}_d$ the set of functions $\sigma \in C^1(\mathbb{R}_+)$ such that the function $\sigma'(r)/r$ is negative, bounded below, monotone increasing, and satisfies

$$\int_0^\infty r^2 \sigma(r)\, dr^d < \infty. \tag{3.1}$$

In the sequel, when solving optimization problems, we assume that $\sigma \in \mathcal{A}_d$.

Remark 3.1. From a physical viewpoint the integral in the left-hand side of (3.1) is proportional to the gas temperature.

Remark 3.2. Note that if $\sigma_1,\ \sigma_2 \in \mathcal{A}_d$, then $\sigma_1 + \sigma_2 \in \mathcal{A}_d$, while if $\tilde{\sigma}(r) = \alpha\sigma(\beta r)$, $\alpha,\ \beta > 0$, $\sigma \in \mathcal{A}_d$, then also $\tilde{\sigma} \in \mathcal{A}_d$.

Remark 3.3. It is easy to verify that $\sigma_h \in \mathcal{A}_3$, and taking Remark 3.2 into account one concludes that also $\sigma_{nh} \in \mathcal{A}_3$.

3.1.2 Calculation of Resistance

Here we derive formulas for the resistance. The momentum transmitted by an incident particle with initial velocity v and final velocity v^+ to the body B is proportional to the quantity $v - v^+$, and the resistance of the body, $R_{\text{therm}}(B)$, is obtained by summing up over all incident particles. The summation amounts to taking the integral

$$R_{\text{therm}}(B) = \int_{(\partial C \times \mathbb{R}^d)_-} (v - v_B^+(\xi, v))\, \sigma(|v + Ve_d|)\, |n(\xi) \cdot v|\, d\xi\, dv; \qquad (3.2)$$

here, as always, C is a convex body containing B.

The resistance $R_{\text{therm}}(B)$ is a vector. Note (but we will not use this subsequently) that it can be written in a "canonical" form $R_\chi(B)$ (1.3), where $c(v, v^+) = v - v^+$ and χ is the measure on the sphere S^{d-1} induced by the distribution $\sigma(|v + Ve_d|)$. In other words, denoting by $f : x \mapsto \frac{x}{|x|}$ the projection of $\mathbb{R}^d \setminus \{0\}$ to S^{d-1} and denoting by ζ the measure on $\mathbb{R}^d \setminus \{0\}$ with density $\sigma(|v + Ve_d|)$, we have $\chi = f_\# \zeta$. The density of χ at $v \in S^{d-1}$ is calculated by the formula $\rho_\chi(v) = \int_0^\infty \sigma(|rv + Ve_d|)\, r^{d-1} dr$.

Formula (3.2) is implicit (it contains the function v_B^+, which is generally very difficult to calculate) and therefore is not good to use. However it can be simplified in the case where B is a convex body. Then, taking $C = B$ and considering that for $(\xi, v) \in (\partial B \times \mathbb{R}^d)_-$ holds $v_B^+(\xi, v) = v - 2(v \cdot n(\xi))\, n(\xi)$ (the corresponding particle makes a single reflection at $\xi \in \partial B$), we find

$$R_{\text{therm}}(B) = \int_{(\partial B \times \mathbb{R}^d)_-} 2(v \cdot n(\xi))^2\, n(\xi)\, \sigma(|v + Ve_d|)\, d\xi\, dv.$$

Finally, denote by $\pi(n)$ the pressure of the flow in the direction $-n$; then we have[1]

$$\pi(n) = -2n \int_{\mathbb{R}^d} (v \cdot n)_-^2\, \sigma(|v + Ve_d|)\, dv \qquad (3.3)$$

and

$$R_{\text{therm}}(B) = \int_{\partial B} \pi(n(\xi))\, d\xi. \qquad (3.4)$$

Assuming that B is a convex body of revolution with the symmetry axis Oe_d, we can further simplify this formula. Assume additionally that the body is inscribed in a cylinder of height h and radius 1. In this case, by means of a translation along the dth coordinate axis, the body can be reduced to the following form:

$$B = \{(\xi', \xi_d) : |\xi'| \le 1,\ f_-(|\xi'|) \le \xi_d \le -f_+(|\xi'|)\},$$

[1] Here we adopt the notation $z_-^2 := (z_-)^2$, where $z_- = \max\{-z, 0\}$ is the negative part of the real number z.

where $\xi' = (\xi_1, \dots, \xi_{d-1})$ and f_+ and f_- are convex nonpositive nondecreasing functions defined on $[0,\ 1]$. The length h of the body along the symmetry axis is

$$h = -f_+(0) - f_-(0).$$

Let $\xi_+ = (\xi', -f_+(|\xi'|))$ be a regular point of the upper part of the boundary ∂B; the outward normal $n(\xi_+)$ at this point equals

$$n(\xi_+) = \frac{1}{\sqrt{f_+'(|\xi'|)^2 + 1}} \left(f_+'(|\xi'|) \frac{\xi'}{|\xi'|},\ 1 \right). \tag{3.5}$$

Using (3.3) and taking into account the axial symmetry of the function $\sigma(|v + Ve_d|)$ with respect to the dth coordinate axis, we can write the pressure of the flow at ξ_+ in the form

$$\pi(n(\xi_+)) = -p_+\left(f_+'(|\xi'|)\right) \cdot n(\xi_+), \tag{3.6}$$

where

$$p_+(u) := \left| \pi\left(\frac{1}{\sqrt{u^2+1}} (u,\ 0, \dots, 0,\ 1) \right) \right|. \tag{3.7}$$

In a similar fashion, the pressure of the flow at a regular point $\xi_- = (\xi', f_-(|\xi'|))$ in the lower part of ∂B is equal to

$$\pi(n(\xi_-)) = p_-\left(f_-'(|\xi'|)\right) \cdot n(\xi_-), \tag{3.8}$$

where

$$n(\xi_-) = \frac{1}{\sqrt{f_-'(|\xi'|)^2 + 1}} \left(f_-'(|\xi'|) \frac{\xi'}{|\xi'|},\ -1 \right) \tag{3.9}$$

and

$$p_-(u) := -\left| \pi\left(\frac{1}{\sqrt{u^2+1}} (u,\ 0, \dots, 0,\ -1) \right) \right|. \tag{3.10}$$

By (3.7), (3.10), and (3.3) one obtains

$$p_\pm(u) = \pm 2 \int_{\mathbb{R}^d} \frac{(v_1 u \pm v_d)_-^2}{1+u^2}\, \sigma(|v + Ve_d|)\, dv, \tag{3.11}$$

where the signs "+" and "−," in place of "±," are taken simultaneously. The integral in the right-hand side of (3.4) is the sum of the two integrals corresponding to the upper and lower parts of ∂B. Making the change of variable in each of these integrals and using formulae (3.5), (3.6), (3.8), and (3.9) one obtains

$$R_{\text{therm}}(B) = \int_{|x'|\le 1} p_+(f_+'(|x'|)) \cdot \left(-f_+'(|x'|) \frac{x'}{|x'|},\ -1 \right) dx' +$$
$$+ \int_{|x'|\le 1} p_-(f_-'(|x'|)) \cdot \left(f_-'(|x'|) \frac{x'}{|x'|},\ -1 \right) dx'.$$

Due to the axial symmetry of B, the resistance is parallel to the axis Oe_d, and therefore

$$R_{\text{therm}}(B) = -\left(\int_{|x'|\le 1} p_+(f'_+(|x'|))\, dx' + \int_{|x'|\le 1} p_-(f'_-(|x'|))\, dx' \right) e_d. \tag{3.12}$$

Denote

$$\mathcal{R}_\pm(f) = \int_0^1 p_\pm(f'(t))\, dt^{d-1} \tag{3.13}$$

and recall that b_d^{-1} is the volume of the unit ball in $\mathbb{R}^{d-1}$; then (3.12) can be written as

$$R_{\text{therm}}(B) = -b_d^{-1} \left(\mathcal{R}_+(f_+) + \mathcal{R}_-(f_-) \right) \cdot e_d.$$

3.1.3 Statement of Minimization Problem

Let $\mathcal{M}(h)$ be the class of convex nonpositive, nondecreasing continuous functions f on [0, 1] such that $f(0) = -h$. Note that each function $f \in \mathcal{M}(h)$ has a derivative everywhere with the possible exception of a countable subset, and f' is monotone; therefore, the integral in (3.13) is well defined for functions in $\mathcal{M}(h)$ Thus, the problem of the minimum resistance takes the following form.

Problem. Minimize $\mathcal{R}_+(f_+) + \mathcal{R}_-(f_-)$ under the hypothesis that f_+ and f_- are convex, nonpositive, nondecreasing continuous functions such that $-f_+(0) - f_-(0) = h$.

We solve this problem in two steps. First, for fixed $h_- \ge 0$ and $h_+ \ge 0$ we find

$$\inf_{f \in \mathcal{M}(h_-)} \mathcal{R}_-(f) \quad \text{and} \quad \inf_{f \in \mathcal{M}(h_+)} \mathcal{R}_+(f). \tag{3.14}$$

Second, given the solutions $f_{h_-}^-$ and $f_{h_+}^+$ of problems (3.14) we find

$$\mathrm{R}(h) := \inf_{h_+ + h_- = h} \left(\mathcal{R}_+(f_{h_+}^+) + \mathcal{R}_-(f_{h_-}^-) \right).$$

3.2 Auxiliary Minimization Problems

3.2.1 Two Lemmas on the Functions $p_\pm$

Consider an infinitesimal area element and let φ be the angle between the normal to the area and the direction of the flow. Denote $u = \tan\varphi$; then $p_+(u)$ and $p_-(u)$ in

(3.11) are, respectively, the pressure of the flow from the front and back sides. The total pressure on the area is $p_+(u) + p_-(u)$. Clearly, knowing the functions $p_+(u)$ and $p_-(u)$ is important for solving shape optimization problems.

It is not always easy to describe them, however; sometimes the derivation of one or another property requires laborious work. We will see that the function $p_+(u)$ behaves like $\frac{1}{1+u^2}+$ const: namely, its derivative $p'_+(u)$ is first monotone decreasing and then monotone increasing, when u runs along the semiaxis $[0, +\infty]$, and becomes zero when u equals 0 or $+\infty$. The behavior of $p_-(u)$ can be more complicated if the flow is a mixture of several component.

Lemmas 3.1 and 3.2 state the properties of these functions. Their proofs [especially of claim (b) of Lemma 3.2] are rather bulky and are presented in Sect. 3.5.

Lemma 3.1 states several properties of the functions p_+ and p_-, which we shall require in subsequent sections. In particular, (a) means that the pressure becomes zero when the surface area is parallel to the flow direction; (b) means that the pressure is not very sensitive to small rotations of the area perpendicular to the flow; (d) implies that the front pressure decreases monotonically when the area turns from a perpendicular to a parallel position relative to the flow.

Lemma 3.1. *Let $\sigma \in \mathcal{A}_d$. Then*

(a) There exist the limits $p_\pm(+\infty) = \lim_{u\to+\infty} p_\pm(u)$;
moreover, $\qquad p_+(+\infty) + p_-(+\infty) = 0.$
(b) $p_\pm \in C^1(\mathbb{R}_+)$ and $p'_\pm(0) = \lim_{u\to+\infty} p'_\pm(u) = 0$.
(c) For $u > 0$ $p'_+(u) < p'_-(u)$.
(d) $p'_+(u) < 0$ for $u > 0$ and for each $u \geq 0$, $p_-(u) > p_-(+\infty)$.

Lemma 3.2 specifies the form of the functions p_+ and p_- for $d = 2$. The function p_+ has a simple behavior in a certain sense, namely, it is concave on a certain interval $[0, \bar{u}]$ and convex on the complementary interval $[\bar{u}, +\infty)$; therefore, it resembles $\frac{1}{1+u^2}+$ const (Fig. 3.1a). The behavior of p_- can be complicated: in some cases, related to a two-component gas, p_- has more than two intervals of convexity or concavity (Fig. 3.1b). We shall use this specification in the next section in our construction of the body of least resistance in two dimensions.

Let $\sigma^{\alpha,\beta}(r) = \sigma(r) + \alpha\,\sigma(\beta r)$, where $\alpha \geq 0,\ \beta > 0$. By Remark 3.2, if $\sigma \in \mathcal{A}_2$, then $\sigma^{\alpha,\beta} \in \mathcal{A}_2$. Let $p_\pm^{\alpha,\beta}$ be the function corresponding to $\sigma^{\alpha,\beta}$ in accordance with formula (3.11), and let $\bar{p}_\pm^{\alpha,\beta}$ be the largest convex function on $\mathbb{R}_+$ majorized by $p_\pm^{\alpha,\beta}$.

Lemma 3.2. *Let $d = 2$.*

(a) If $\sigma \in \mathcal{A}_2$, then for some $\bar{u} > 0$ the derivative p'_+ is strictly decreasing on $[0, \bar{u}]$ and strictly increasing on $[\bar{u}, +\infty)$ (Fig. 3.1a).
(b) Let $\sigma \in \mathcal{A}_2$, and assume that for each $n > 0$ the function $r^n\sigma(r)$ decreases for sufficiently large r. Then there exist $\alpha \geq 0$ and $\beta > 0$ such that the set $\mathcal{O}^{\alpha,\beta} = \{u : p_-^{\alpha,\beta}(u) > \bar{p}_-^{\alpha,\beta}(u)\}$ has at least two connected components (Fig. 3.1b; the set $\mathcal{O}^{\alpha,\beta}$ is drawn boldface on the u-axis).

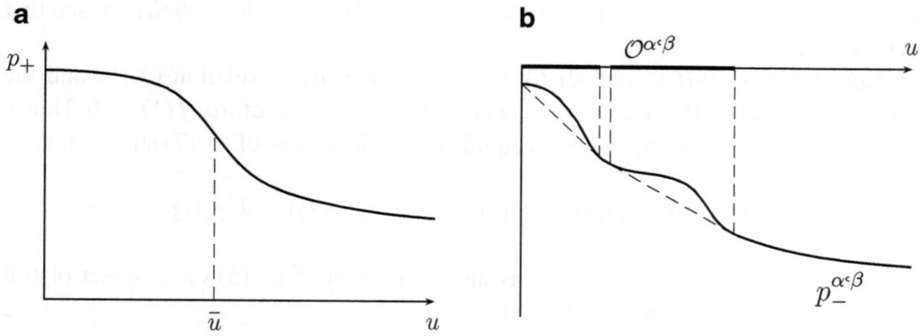

Fig. 3.1 The functions p_+ (**a**) and $p_-^{\alpha,\beta}$ (**b**)

3.2.2 Lemma of Reduction

The following lemma reduces minimization problems (3.14) to a simpler problem of the minimization of a function depending on a parameter.

Let the function $p \in C(\mathbb{R}_+)$ and the values $d \geq 2$, $\lambda > 0$ be fixed.

Lemma 3.3. *Let $f_h \in \mathcal{M}(h)$ be a function such that*

(C_λ) $f_h(1) = 0$ and for almost all t, $u = f_h'(t)$ is a solution of the problem

$$t^{d-2}\, p(u) + \lambda\, u \to \min\,. \tag{3.15}$$

Then f_h is a solution of the minimization problem

$$\inf_{f \in \mathcal{M}(h)} \mathcal{R}(f), \quad \mathcal{R}(f) = \int_0^1 p(f'(t))\, dt^{d-1}. \tag{3.16}$$

Moreover, all other solutions of (3.16) satisfy condition (C_λ) with the same λ.

Proof. Problem (3.16) can in fact be regarded as a degenerate case of the classical optimal control problem [61], and the statement of the lemma is a consequence of Pontryagin's maximum principle. However, we give here an elementary proof not appealing to the maximum principle (cf. [69]).

For each $f \in \mathcal{M}(h)$ one has

$$t^{d-2}\, p(f'(t)) + \lambda\, f'(t) \geq t^{d-2}\, p(f_h'(t)) + \lambda\, f_h'(t) \tag{3.17}$$

for almost all t. Integrating both sides of (3.17) with respect to t one obtains

$$\frac{1}{d-1} \int_0^1 p(f'(t))\, dt^{d-1} + \lambda\, (f(1) - f(0)) \geq$$

$$\geq \frac{1}{d-1} \int_0^1 p(f_h'(t))\, dt^{d-1} + \lambda\, (f_h(1) - f_h(0)), \tag{3.18}$$

and bearing in mind that $f(1) \le 0 = f_h(1)$ and $f(0) = -h = f_h(0)$ we see that $\mathcal{R}(f) \ge \mathcal{R}(f_h)$.

Now let $f \in \mathcal{M}(h)$ and $\mathcal{R}(f) = \mathcal{R}(f_h)$; then using relation (3.18) and the equality $f(0) = f_h(0)$ one obtains $f(1) \ge f_h(1) = 0$; therefore, $f(1) = 0$. Hence the inequality in (3.18) becomes an equality, which in view of (3.17) shows that

$$t^{d-2}\, p(f'(t)) + \lambda\, f'(t) = t^{d-2}\, p(f_h'(t)) + \lambda\, f_h'(t)$$

for almost all t. Hence $u = f'(t)$ is also a solution of (3.15) on a subset of full measure. Thus, f satisfies condition C_λ. □

3.2.3 *The Minimizing Function for d = 2*

Assume that $p \in C^1(\mathbb{R}_+)$ is bounded below. Let $\bar{p}$ be the largest convex function on $\mathbb{R}_+$ not exceeding p. The function $\bar{p}$ also has a continuous derivative, and for arbitrary $h \ge 0$ and $u \ge 0$ one has

$$p(u) \ge \bar{p}(u) \ge \bar{p}(h) + \bar{p}'(h) \cdot (u - h). \tag{3.19}$$

Consider $\mathcal{O}_p = \{u : p(u) > \bar{p}(u)\}$. Obviously, $\mathcal{O}_p$ is open and therefore is the union of a countable (perhaps empty) set of disjoint open intervals.

Assume additionally that $h \ge 0$, $\bar{p}'(h) < 0$, and let $(h^{(-)}, h^{(+)})$ be the largest subinterval in $\mathcal{O}_p$ such that $h^{(-)} \le h \le h^{(+)}$ [it may happen that $h^{(-)} = h = h^{(+)}$, that is, the interval is empty]. Since p is bounded below, there exists a quantity $u \ge h$ such that $\bar{p}(u) = p(u)$; hence $h^{(+)} \le u < +\infty$.

The following lemma specifies the solution f_h of the minimization problem (3.16) in the case $d = 2$.

Lemma 3.4. *(a) The function f_h defined by the formula*

$$f_h(t) = -h + ht,$$

if $h^{(-)} = h = h^{(+)}$, and by the formula

$$f_h(t) = \begin{cases} -h + h^{(-)}t, & t \le t_0\,; \\ -h + h^{(-)}t_0 + h^{(+)}(t - t_0), & t \ge t_0\,, \end{cases} \quad \text{where } t_0 = \frac{h^{(+)} - h}{h^{(+)} - h^{(-)}},$$

if $h^{(-)} < h^{(+)}$, is a solution of the minimization problem

$$\inf_{f \in \mathcal{M}(h)} \mathcal{R}(f), \qquad \mathcal{R}(f) = \int_0^1 p(f'(t))\, dt\,. \tag{3.20}$$

(b) The following equality holds:

$$\inf_{f \in \mathcal{M}(h)} \mathcal{R}(f) = \mathcal{R}(f_h) = \bar{p}(h). \tag{3.21}$$

(c) If f is a solution of (3.20), then for almost all t the quantity $u = f'(t)$ satisfies the relation $p(u) = \bar{p}(h) + \bar{p}'(h) \cdot (u - h)$ and $u \notin \mathcal{O}_p$.

Proof. We have $\bar{p}(h^{(-)}) = p(h^{(-)})$, $\bar{p}(h^{(+)}) = p(h^{(+)})$, and

$$p(h^{(\pm)}) = \bar{p}(h) + \bar{p}'(h) \cdot (h^{(\pm)} - h). \tag{3.22}$$

It follows from (3.19) and (3.22) that for each u,

$$p(u) - p(h^{(\pm)}) \geq \bar{p}'(h) \cdot (u - h^{(\pm)}),$$

so that setting $\lambda = -\bar{p}'(h)$ one obtains

$$p(u) + \lambda u \geq p(h^{(\pm)}) + \lambda h^{(\pm)}.$$

This means that the quantities $u = h^{(-)}$ and $h^{(+)}$ minimize the function $p(u) + \lambda u$.

Further, one easily sees that $f_h \in \mathcal{M}(h)$, $f_h(1) = 0$, and the function f_h' takes the values $h^{(-)}$ and $h^{(+)}$ (which may be equal). Therefore, f_h satisfies condition C_λ. Using Lemma 3.3 one sees that f_h is a solution of problem (3.20). Claim (a) is proved.

If $h^{(-)} = h^{(+)} = h$, then $\mathcal{R}(f_h) = p(h) = \bar{p}(h)$. If $h^{(-)} \neq h^{(+)}$, then

$$\mathcal{R}(f_h) = \int_0^{t_0} p(h^{(-)})\, dt + \int_{t_0}^1 p(h^{(+)})\, dt =$$

$$= \frac{h^{(+)} - h}{h^{(+)} - h^{(-)}}\, p(h^{(-)}) + \frac{h - h^{(-)}}{h^{(+)} - h^{(-)}}\, p(h^{(+)}).$$

On the other hand, excluding $\bar{p}'(h)$ from (3.22) one obtains

$$\frac{h^{(+)} - h}{h^{(+)} - h^{(-)}}\, p(h^{(-)}) + \frac{h - h^{(-)}}{h^{(+)} - h^{(-)}}\, p(h^{(+)}) = \bar{p}(h);$$

therefore, $\mathcal{R}(f_h) = \bar{p}(h)$. Thus, the proof of (3.21), and therefore claim (b), is complete.

Now let f be a solution of (3.20). By Lemma 3.3 for almost every t the quantity $\hat{u} = f'(t)$ minimizes the function $p(u) + \lambda u$; therefore, $p(\hat{u}) + \lambda \hat{u} = p(h^{(+)}) + \lambda h^{(+)}$, and setting $\lambda = -\bar{p}'(h)$ while bearing in mind that $p(h^{(+)}) = \bar{p}(h) + \bar{p}'(h) \cdot (h^{(+)} - h)$ one obtains

$$p(\hat{u}) = \bar{p}(h) + \bar{p}'(h) \cdot (\hat{u} - h).$$

Taking account of (3.19) one sees that

$$p(\hat{u}) = \bar{p}(\hat{u}) = \bar{p}(h) + \bar{p}'(h) \cdot (\hat{u} - h);$$

therefore, $\hat{u} \notin \mathcal{O}_p$. The proof of claim (c) is complete. □

3.2.4 The Minimizing Function for $d \geq 3$

In addition to our previous assumptions, assume that there exists the limit $p(+\infty) = \lim_{u \to +\infty} p(u)$ and $p(u) > p(+\infty)$ for each $u \in \mathbb{R}_+$. Then $\bar{p}'(u) < 0$ for each u. We set $b = -\bar{p}'(0)$; then $b > 0$; the function $\bar{p}'$ is continuous and varies nondecreasingly from b to 0.

We now assume that $d \geq 3$ and introduce auxiliary notation: $q(u) = |\bar{p}'(u)|^{-1/(d-2)}$ and $Q(u) = \int_0^u q(v)\,dv$. Both functions q and Q are continuous and nondecreasing on $\mathbb{R}_+$; q ranges from $b^{-1/(d-2)}$ to $+\infty$, and Q from 0 to $+\infty$.

Fix $h > 0$.

Lemma 3.5. *(a) The set of quantities $U \geq 0$ satisfying the equation*

$$U - \frac{Q(U)}{q(U)} = h \tag{3.23}$$

is nonempty.

(b) Let U be a solution of (3.23) and $t_0 = \frac{q(0)}{q(U)}$. Consider the function $f_h(t)$, $t \in [0, 1]$, defined as follows:

$f_h(t) = -h$ for $t \in [0, t_0]$;
$f_h(t)$ is defined parametrically for $t \in [t_0, 1]$:

$$f_h = -h + \frac{u\,q(u) - Q(u)}{q(U)}, \tag{3.24}$$

$$t = \frac{q(u)}{q(U)}, \quad u \in [0, U]. \tag{3.25}$$

Then the function f_h is well defined, strictly convex on $[t_0, 1]$, and is the unique solution of the minimization problem (3.16).

(c) The minimum value of $\mathcal{R}$ is

$$\inf_{f \in \mathcal{M}(h)} \mathcal{R}(f) = \mathcal{R}(f_h) = \bar{p}(U) + \frac{Q(U)}{q(U)^{d-1}}. \tag{3.26}$$

Proof. (a) For arbitrary $c > 0$ and $U \geq c$ we have

$$U - \frac{Q(U)}{q(U)} = \int_0^U \left(1 - \frac{q(v)}{q(U)}\right) dv \geq \int_0^c \left(1 - \frac{q(c)}{q(U)}\right) dv = c\left(1 - \frac{q(c)}{q(U)}\right),$$

and we conclude from $\lim_{U\to+\infty} q(U) = +\infty$ that $U - \frac{Q(U)}{q(U)} > c/2$ for sufficiently large U. Hence the continuous function $U - \frac{Q(U)}{q(U)}$ approaches $+\infty$ as $U \to +\infty$; furthermore, it vanishes at $U = 0$; therefore, the solution set of (3.23) is nonempty. Thus, the proof of (a) is complete.

Let $S(h)$ be the set of points u such that $\bar{p}(u) = \bar{p}(h) + \bar{p}'(h) \cdot (u - h)$. Note that $S(h)$ coincides with the connected component of $\bar{\mathcal{O}}_p$ containing h if $h \in \bar{\mathcal{O}}_p$, and $S(h) = \{h\}$ otherwise. Thus, $S(h)$ is always a closed interval containing h; two intervals $S(h_1)$ and $S(h_2)$ either coincide or are disjoint. It follows by the condition $p(u) > p(+\infty)$, which holds for each u, that all the intervals $S(u)$ are bounded. Clearly, a family of nondegenerate (that is, distinct from singletons) segments is countable. Let $\mathcal{S}$ be the union of nondegenerate closed intervals. The function $\bar{p}'(u)$ increases on $\mathbb{R}_+ \setminus \mathcal{S}$ and is constant on each nondegenerate interval $S(u) \subset \mathcal{S}$; therefore, the set $\{\bar{p}'(u),\ u \in \mathcal{S}\}$ is countable.

Assume that $0 \le u_1 \le u_2$. After some simple algebra one obtains

$$\left[u_2 - \frac{Q(u_2)}{q(u_2)}\right] - \left[u_1 - \frac{Q(u_1)}{q(u_1)}\right] =$$
$$= Q(u_1)\left[\frac{1}{q(u_1)} - \frac{1}{q(u_2)}\right] + \frac{1}{q(u_2)} \int_{u_1}^{u_2} (q(u_2) - q(v))\, dv. \tag{3.27}$$

Both terms on the right-hand side of (3.27) are nonnegative; therefore, the function $U - \frac{Q(U)}{q(U)}$ is nondecreasing. If both u_1 and u_2 are solutions of (3.23), then both terms in (3.27) vanish, so that the function q is constant on $[u_1, u_2]$; that is, $\bar{p}'$ is constant on $[u_1, u_2]$ or, equivalently, $u_1 \in S(u_2)$. Hence the solution set of (3.23) coincides with some closed interval $S(u)$.

(b) Relations (3.24) and (3.25) define the continuous functions f_h and t of u, ranging from $-h$ to 0 and from t_0 to 1, respectively, as u ranges over $[0, U]$. Moreover, the function $t = q(u)/q(U)$ is nondecreasing, and each interval of constancy $\{u : q(u)/q(U) = t\}$ coincides with some interval $S(u)$, and f_h is constant on each such interval. This means that the function $f_h(t)$ is well defined.

We now calculate the left-hand and right-hand derivatives $f_h'(t^-)$ and $f_h'(t^+)$ of the function f_h at t. Let $[u^-(t), u^+(t)]$ be the interval $\{u : q(u)/q(U) = t\}$, and let $u = u^+(t)$. Assume that the values of the functions $f_h + \Delta f_h$ and $t + \Delta t$, where $\Delta t > 0$, correspond to the value of the independent variable $u + \Delta u$. Then we have

$$\Delta t = \frac{q(u + \Delta u) - q(u)}{q(U)},$$

$$\Delta f_h = \frac{(u + \Delta u)\, q(u + \Delta u) - Q(u + \Delta u)}{q(U)} - \frac{u\, q(u) - Q(u)}{q(U)} Q(u) q(U)$$

$$= \frac{u\,(q(u + \Delta u) - q(u)) + \int_u^{u+\Delta u} (q(u + \Delta u) - q(v)) dv}{q(U)};$$

therefore,

$$\frac{\Delta f_h}{\Delta t} = u + \int_u^{u+\Delta u} \frac{q(u+\Delta u) - q(v)}{q(u+\Delta u) - q(u)}\, dv\,. \tag{3.28}$$

The integrand on the right-hand side of (3.28) is less than 1; by the definition of $u = u^+(t)$, we have $\Delta u \to 0^+$ as $\Delta t \to 0^+$; therefore,

$$f_h'(t^+) = \lim_{\Delta t \to 0+} \frac{\Delta f_h}{\Delta t} = u^+(t).$$

In a similar fashion,

$$f_h'(t^-) = u^-(t).$$

Both functions $u^-(t)$ and $u^+(t)$ are positive and for arbitrary t_1 and t_2, $t_1 < t_2$, we have $u^-(t_1) \le u^+(t_1) < u^-(t_2) \le u^+(t_2)$. Hence the function f_h is increasing and strictly convex for $t \in [t_0, 1]$; moreover, it is constant on $[0, t_0]$, $f_h(0) = -h$ and $f_h(1) = 0$. We have thus proved that $f_h \in \mathcal{M}(h)$.

For each $t \in [t_0, 1]$, with the possible exception of countably many values, we have $u^-(t) = u^+(t) := \tilde{u} \in \mathbb{R}_+ \setminus \mathcal{S}$; hence there exists the derivative $f_h'(t) = \tilde{u}$. For $u \ne \tilde{u}$ we have

$$\bar{p}(u) > \bar{p}(\tilde{u}) + \bar{p}'(\tilde{u}) \cdot (u - \tilde{u}),$$

and bearing in mind that $p(u) \ge \bar{p}(u)$ and $p(\tilde{u}) = \bar{p}(\tilde{u})$ we obtain

$$p(u) > p(\tilde{u}) + \bar{p}'(\tilde{u}) \cdot (u - \tilde{u});$$

therefore,

$$p(u) - \bar{p}'(\tilde{u}) \cdot u > p(\tilde{u}) - \bar{p}'(\tilde{u}) \cdot \tilde{u}. \tag{3.29}$$

Recall that

$$t = \frac{q(\tilde{u})}{q(U)} = \frac{|\bar{p}'(U)|^{1/(d-2)}}{|\bar{p}'(\tilde{u})|^{1/(d-2)}}.$$

We have $t^{d-2} = \frac{\bar{p}'(U)}{\bar{p}'(\tilde{u})}$, and multiplying both sides of (3.29) by t^{d-2} and setting $-\bar{p}'(U) = \lambda$, we obtain

$$t^{d-2} p(u) + \lambda u > t^{d-2} p(\tilde{u}) + \lambda \tilde{u}$$

for each $u \ne \tilde{u}$. Thus, $\tilde{u} = f_h'(t)$ is the unique quantity minimizing the function $t^{d-2} p(u) + \lambda u$.

Now let $t \in (0, t_0)$. For $u > 0$ we have

$$\bar{p}(u) \ge \bar{p}(0) + \bar{p}'(0) \cdot u.$$

Using the relations $p(u) \ge \bar{p}(u)$, $p(0) = \bar{p}(0)$, $t_0^{2-d} = \bar{p}'(0)/\bar{p}'(U) = -\bar{p}'(0)/\lambda$, we obtain

$$p(u) \ge p(0) + \bar{p}'(0) \cdot u = p(0) - \lambda t_0^{2-d} u;$$

hence

$$p(u) + \lambda t^{2-d} u > p(0)$$

for each $t \in (0, t_0)$, and $f_h'(t) = 0$ is the unique minimum of the function $t^{d-2} p(u) + \lambda u$. Applying Lemma 3.3 we conclude that f_h is the unique solution of (3.16).

(c) We have

$$\mathcal{R}(f_h) = \int_0^{t_0} p(0)\, dt^{d-1} + \int_{t_0}^1 p(f_h'(t))\, dt^{d-1}. \tag{3.30}$$

The first integral on the right-hand side of (3.30) is

$$\int_0^{t_0} (\ldots) = p(0) \left(\frac{q(0)}{q(U)} \right)^{d-1}.$$

We set $\tilde{U} = \inf S(U)$. Making the change of variable $t = q(u)/q(U)$, $u \in [0, \tilde{U}]$, in the second integral and bearing in mind that $f_h'(t) = u$ for almost all t, we see that the second integral is equal to

$$\int_{t_0}^1 (\ldots) = \int_0^{\tilde{U}} p(u)\, d\left(\frac{q(u)}{q(U)} \right)^{d-1} = p(u) \left(\frac{q(u)}{q(U)} \right)^{d-1} \Bigg|_0^{\tilde{U}} - \int_0^{\tilde{U}} \left(\frac{q(u)}{q(U)} \right)^{d-1} dp(u).$$

Summing the first and second integrals and taking account of the equality $q(\tilde{U}) = q(U)$, we obtain

$$\mathcal{R}(f_h) = p(\tilde{U}) - \int_0^{\tilde{U}} \left(\frac{q(u)}{q(U)} \right)^{d-1} dp(u). \tag{3.31}$$

The integral in (3.31) can be represented as the sum

$$\int_0^{\tilde{U}} (\ldots) = \int_{[0,\tilde{U}] \setminus \mathcal{S}} (\ldots) + \sum_i \int_{S_i} (\ldots),$$

where $S_i = S(u_i)$ are closed nondegenerate intervals whose union is $[0, \tilde{U}] \cap \mathcal{S}$. If $u \in [0, \tilde{U}] \setminus \mathcal{S}$, then we have $p'(u) = \bar{p}'(u) = -q(u)^{2-d}$; therefore,

$$\int_{[0,\tilde{U}] \setminus \mathcal{S}} (\ldots) = -\int_{[0,\tilde{U}] \setminus \mathcal{S}} \frac{q(u)}{q(U)^{d-1}}\, du.$$

Next, bearing in mind that the function q is constant on $u \in S_i$ and that p and $\bar{p}$ are equal at the endpoints of this interval we obtain

$$\int_{S_i} (\ldots) = \int_{S_i} \left(\frac{q(u)}{q(U)} \right)^{d-1} d\,\bar{p}(u) = -\int_{S_i} \frac{q(u)}{q(U)^{d-1}}\, du.$$

Summing these integrals we see that

$$\int_0^{\tilde U}\left(\frac{q(u)}{q(U)}\right)^{d-1} dp(u) = -\int_0^{\tilde U}\frac{q(u)}{q(U)^{d-1}}\,du. \tag{3.32}$$

Now the function $q(u) = |\bar p'(u)|^{-1/(d-2)}$ is constant on $[\tilde U, U]$; therefore,

$$-\int_{\tilde U}^{U}\frac{q(u)\,du}{q(U)^{d-1}} = -q(U)^{2-d}(U-\tilde U) = \bar p'(U)\,(U-\tilde U) = \bar p(U) - \bar p(\tilde U). \tag{3.33}$$

Using the equality $\bar p(\tilde U) = p(\tilde U)$ and relations (3.31)–(3.33) we obtain

$$\mathcal{R}(f_h) = \bar p(U) + \int_0^U \frac{q(u)}{q(U)^{d-1}}\,du.$$

Recalling that Q is a primitive of q, we arrive at (3.26). □

3.3 Solution of Minimum Resistance Problem

3.3.1 *Two-Dimensional Problem*

Minimization of $\mathcal{R}_+$

It follows from part (a) of Lemma 3.2 that there exist quantities $u_+^0 > 0$ and $b_+ > 0$ such that

$$\frac{p_+(u_+^0) - p_+(0)}{u_+^0} = p_+'(u_+^0) = -b_+$$

and

$$\bar p_+(u) = \begin{cases} p_+(0) - b_+u, & \text{if } 0 \le u \le u_+^0, \\ p_+(u), & \text{if } u \ge u_+^0. \end{cases}$$

Hence $\mathcal{O}_{p_+} = (0, u_+^0)$. Using Lemma 3.4 one sees that there exists a unique solution f_h^+ of the minimization problem

$$\inf_{f\in\mathcal{M}(h)} \mathcal{R}_+(f), \quad \mathcal{R}_+(f) = \int_0^1 p_+(f'(t))\,dt,$$

which is defined by the relation

$$f_h^+(t) = \begin{cases} -h & \text{for } t \le t_0, \\ -h + u_+^0\cdot(t-t_0) & \text{for } t \ge t_0, \end{cases} \tag{3.34}$$

$$t_0 = 1 - h/u_+^0,$$

if $0 \le h < u_+^0$, and by the relation

$$f_h^+(t) = -h + ht,$$

if $h \ge u_+^0$. The minimum resistance is

$$\inf_{f \in \mathcal{M}(h)} \mathcal{R}_+(f) = \mathcal{R}_+(f_h^+) = \bar{p}_+(h).$$

Minimization of $\mathcal{R}_-$

One has $p'_-(0) = 0$ and $\bar{p}'_-(0) < 0$; therefore, $\mathcal{O}_{p_-}$ contains an interval $(0, u_-^0)$, $u_-^0 > 0$; in addition, $p_-(u_-^0) = p_-(0) + \bar{p}'_-(0) \cdot u_-^0$. Let $b_- = -\bar{p}'_-(0)$; we represent the open set $\mathcal{O}_{p_-}$ as the union of its connected components $\mathcal{O}_i = (u_i^-, u_i^+)$: $\mathcal{O}_{p_-} = \cup_i \mathcal{O}_i$. We shall assume that the index set $\{i\}$ contains 1 and that $\mathcal{O}_1 = (0, u_-^0)$. Statement (b) of Lemma 3.2 and Example 3.2 show that in some cases (for example, when one considers the distribution of pressure of a mixture of two homogeneous rarefied gases on the rear part of the surface of a moving body), $\mathcal{O}_{p_-}$ has at least two connected components.

Consider the minimization problem

$$\inf_{f \in \mathcal{M}(h)} \mathcal{R}_-(f), \quad \mathcal{R}_-(f) = \int_0^1 p_-(f'(t))\, dt. \tag{3.35}$$

Using Lemma 3.4 one sees that this problem has a solution f_h^-; moreover,

$$\inf_{f \in \mathcal{M}(h)} \mathcal{R}_-(f) = \mathcal{R}_-(f_h^-) = \bar{p}_-(h).$$

For $0 \le h < u_-^0$ one has

$$f_h^-(t) = \begin{cases} -h & \text{for } t \le t_0, \\ -h + u_-^0 \cdot (t - t_0) & \text{for } t \ge t_0, \end{cases} \tag{3.36}$$

$$t_0 = 1 - h/u_-^0.$$

For $h \in \mathbb{R} \setminus \mathcal{O}_{p_-}$ one has

$$f_h^-(t) = -h + ht.$$

Finally, for $h \in \mathcal{O}_i$, $i \ne 1$, one has

$$f_h^-(t) = \begin{cases} -h + u_i^- t & \text{for } t \le t_i, \\ -h + u_i^- t_i + u_i^+(t - t_i) & \text{for } t \ge t_i, \end{cases} \tag{3.37}$$

$$t_i = \frac{u_i^+ - h}{u_i^+ - u_i^-}.$$

Note that f_h is not necessarily the unique solution of (3.35). In some degenerate cases, it can occur that the right endpoint of some interval $\mathcal{O}_i$ coincides with the left endpoint of another interval, $u_i^+ = u_j^-$, $i \neq j$; then there exists a continuous family of functions solving (3.35); the derivative of each function in this family takes the values u_i^-, u_i^+, and u_j^+.

Solution of Two-Dimensional Problem

We see that the problem of finding

$$\mathrm{R}(h) = \inf_{h_+ + h_- = h} \left(\mathcal{R}_+(f_{h_+}^+) + \mathcal{R}_-(f_{h_-}^-) \right)$$

reduces to finding

$$\min_{0 \le z \le h} p_h(z), \quad \text{where} \quad p_h(z) = \bar{p}_+(z) + \bar{p}_-(h - z). \tag{3.38}$$

The functions $\bar{p}'_-$ and $\bar{p}'_+$ are continuous and nondecreasing; therefore, $p'_h(z)$ for $0 \le z \le h$ is also nondecreasing.

From statement (c) of Lemma 3.1 one concludes that $b_+ > b_-$. Indeed, if $u_-^0 \le u_+^0$, then

$$-b_- = \frac{p_-(u_-^0) - p_-(0)}{u_-^0} > \frac{p_+(u_-^0) - p_+(0)}{u_-^0} \ge \frac{\bar{p}_+(u_-^0) - p_+(0)}{u_-^0} = -b_+,$$

whereas if $u_-^0 > u_+^0$, then

$$-b_- = p'_-(u_-^0) > p'_+(u_-^0) > p'_+(u_+^0) = -b_+.$$

Hence there exists a unique quantity $u_* > u_+^0$ such that $\bar{p}'_+(u_*) = p'_+(u_*) = -b_-$. Consider now four cases:

1. $0 < h < u_+^0$.
2. $u_+^0 \le h \le u_*$.
3. $u_* < h < u_* + u_-^0$.
4. $h \ge u_* + u_-^0$.

In cases (1) and (2) one has $p'_h(z) < \bar{p}'_+(u_*) + b_- = 0$ for $0 \le z < h$; therefore, $z = h$ is the unique value of the independent variable minimizing p_h. Hence the optimal values h_+ and h_- are $h_+ = h$, $h_- = 0$, and $f_{h_-=0}^- \equiv 0$.

1. $0 < h < u_+^0$. The function $f_{h_+=h}^+$ is defined by (3.34). The least-resistance body is a trapezium; the slope of its lateral sides is u_+^0 (Fig. 1.4a). The minimum resistance is

$$\mathrm{R}(h) = \mathcal{R}_+(f_{h_+}^+) + \mathcal{R}_-(f_{h_-}^-) = p_+(0) - b_+\, h + p_-(0).$$

2. $u_+^0 \leq h \leq u_*$. Here one has $f_{h_+=h}^+(t) = -h + h\,t$; therefore, the optimal body is an isosceles triangle (Fig. 1.4b) and

$$\mathrm{R}(h) = p_+(h) + p_-(0).$$

In cases (3) and (4), one has $\bar{p}'_+(h) > -b_-$; therefore, $p'_h(h) = \bar{p}'_+(h) - \bar{p}'_-(0) > 0$. On the other hand, $p'_h(u_+^0) = \bar{p}'_+(u_+^0) - \bar{p}'_-(h - u_+^0) \leq -b_+ + b_- < 0$. Moreover, using statement (a) of Lemma 3.2 one sees that $\bar{p}'_+(u) = p'_+(u)$, $u \in [u_+^0, h]$, is an increasing function, so that p'_h is also increasing on this interval. Hence the function p_h has a unique minimum $z \in (u_+^0, h)$ and $f_{h_+=z}^+(t) = -z + zt$.

3. $u_* < h < u_* + u_-^0$. We have $p'_h(u_*) = \bar{p}'_+(u_*) - \bar{p}'_-(h - u_*) = -b_- + b_- = 0$; therefore, p_h reaches its minimum value at $z = u_*$; thus, the optimal values h_+ and h_- are $h_+ = u_*$ and $h_- = h - u_*$, respectively. The function $f_{h_-=h-u_*}^-$ is defined by (3.36). Here the optimal body is the union of a triangle and a trapezium, as shown in Fig. 1.4c. The slope of the lateral sides of the trapezium is $-u_-^0$. The minimum resistance is

$$\mathrm{R}(h) = p_+(u_*) + p_-(0) - b_-\,(h - u_*).$$

4. $h \geq u_* + u_-^0$. We have $p'_h(h - u_-^0) = \bar{p}'_+(h - u_-^0) + b_- \geq 0$; therefore, p_h reaches its minimum at a point $z \in (u_+^0, h - u_-^0]$, and the optimal values $h_+ = z$ and $h_- = h - z$ and the minimum resistance can be found from the relations

$$\begin{aligned}
&h_+ + h_- = h;\\
&p'_+(h_+) = \bar{p}'_-(h_-);\\
&h_+ \geq u_+^0,\ h_- \geq u_-^0;\\
&\mathrm{R}(h) = p_+(h_+) + \bar{p}_-(h_-).
\end{aligned}$$

Here one must distinguish between two cases.

(4a) If $h_- \in \mathbb{R}\backslash\mathcal{O}_{p_-}$, then $f_{h_-}^-(t) = -h_- + h_- t$, and the optimal body is a union of two isosceles triangles with a common base, of heights h_+ and h_- (Fig. 1.4d).

(4b) If h_- belongs to an interval $\mathcal{O}_i = (u_i^-, u_i^+)$, $i \neq 1$, then $f_{h_-}^-$ is given by formula (3.37), and the optimal body is the union of two isosceles triangles and a trapezium (Fig. 1.4e).

Note that case (4b) is realized for h in some open (perhaps empty) set contained in $(u_* + u_-^0, +\infty)$. This set is defined by the parameters σ and V. Case (4a) is realized for h in the complement of this set in $(u_* + u_-^0, +\infty)$, which is always nonempty.

3.3.2 *The Problem in Three and More Dimensions*

Assume that $d \geq 3$. From Lemma 3.5 one concludes that there exists a unique solution $f_h^{\pm}$ of the problem

$$\inf_{f \in \mathcal{M}(h)} \mathcal{R}_{\pm}(f), \quad \mathcal{R}_{\pm}(f) = \int_0^1 p_{\pm}(f'(t))\, dt^{d-1};$$

furthermore,

$$\mathcal{R}_{\pm}(f_h^{\pm}) = \bar{p}_{\pm}(U) + \frac{Q_{\pm}(U)}{q_{\pm}(U)^{d-1}},$$

where U is defined (not necessarily uniquely) by the relation

$$U - \frac{Q_{\pm}(U)}{q_{\pm}(U)} = h;$$

here $q_{\pm}(U) = |\bar{p}'_{\pm}(U)|^{-1/(d-2)}$, $Q_{\pm}(U) = \int_0^U q_{\pm}(u)\, du$.

Thus, the problem

$$\inf_{h_+ + h_- = h} \left(\mathcal{R}_+(f_{h_+}^+) + \mathcal{R}_-(f_{h_-}^-) \right)$$

amounts to finding

$$\inf_{\mathrm{h}_+(u_+) + \mathrm{h}_-(u_-) = h} \left(\mathrm{r}_+(u_+) + \mathrm{r}_-(u_-) \right),$$

where

$$\mathrm{r}_+(u) = \bar{p}_+(u) + \frac{Q_+(u)}{q_+(u)^{d-1}}, \quad \mathrm{r}_-(u) = \bar{p}_-(u) + \frac{Q_-(u)}{q_-(u)^{d-1}}$$

and

$$\mathrm{h}_+(u) = u - \frac{Q_+(u)}{q_+(u)}, \quad \mathrm{h}_-(u) = u - \frac{Q_-(u)}{q_-(u)}, \quad u \geq 0.$$

The functions $\mathrm{r}_{\pm}$ and $\bar{p}_{\pm}$ are nonincreasing and $\mathrm{h}_{\pm}$ vary nondecreasingly from 0 to $+\infty$ for $u \in \mathbb{R}_+$; each constancy interval of one of these functions is a constancy interval of the others. For each $z \geq 0$ we choose u such that $\mathrm{h}_{\pm}(u) = z$ and set

$\mathrm{r}^{(\pm)}(z) := \mathrm{r}_\pm(u)$, $\pi^{(\pm)}(z) := \bar{p}'_\pm(u)$. It follows from the preceding dicussion that the functions $\mathrm{r}^{(\pm)}$ and $\pi^{(\pm)}$ are well defined on $\mathbb{R}_+$ and decrease there. We set

$$\mathrm{r}_h(z) = \mathrm{r}^{(+)}(z) + \mathrm{r}^{(-)}(h - z).$$

After some algebra one sees that the function r_h is differentiable and

$$\mathrm{r}'_h(z) = (d-1)\,(\bar{p}'_+(u_+) - \bar{p}'_-(u_-)), \tag{3.39}$$

where u_+ and u_- are selected from the relations $\mathrm{h}_+(u_+) = z$ and $\mathrm{h}_-(u_-) = h - z$. Both $\bar{p}'_+(u_+) = \pi^{(+)}(z)$ and $-\bar{p}'_-(u_-) = -\pi^{(-)}(h-z)$ on the right-hand side of (3.39) are increasing functions of z; therefore, $\mathrm{r}'_h(z)$ also increases monotonically from $\mathrm{r}'_h(0) = (d-1)\,(\bar{p}'_+(0) - \bar{p}'_-(U_-))$ to $\mathrm{r}'_h(h) = (d-1)\,(\bar{p}'_+(U_+) - \bar{p}'_-(0))$, where U_+ and U_- are defined by the relations $\mathrm{h}_+(U_+) = h$ and $\mathrm{h}_-(U_-) = h$. Note that $\bar{p}'_+(0) = -b_+$ and $\bar{p}'_-(U_-) \geq -b_-$. From statement (c) of Lemma 3.1 we have that $b_+ > b_-$; therefore, $\mathrm{r}'_h(0) < 0$.

Recall that u_* is defined by $\bar{p}'_+(u_*) = -b_-$. We set

$$h_* := \mathrm{h}_+(u_*) = u_* - b_-^{\frac{1}{d-2}}\, Q_+(u_*) \tag{3.40}$$

and consider two cases.

1. $h \leq h_*$. Take U_+ such that $h_+(U_+) = h$. Then we have $\mathrm{h}_+(U_+) \leq \mathrm{h}_+(u_*)$; consequently $U_+ \leq u_*$, and therefore $\mathrm{r}'_h(h) = (d-1)\,(\bar{p}'_+(U_+) + b_-) \leq (d-1)\,(\bar{p}'_+(u_*) + b_-) = 0$. It now follows that $\mathrm{r}'_h(z) < 0$ for $z \in [0, h)$; therefore, r_h has a unique minimum at the point $z = h$, which corresponds to the values $h_+ = h$ and $h_- = 0$. The minimum resistance is

$$\mathrm{R}(h) = \bar{p}_+(u_+) + Q_+(u_+)\, q_+(u_+)^{-d+1} + \bar{p}_-(0), \quad \text{where} \quad h_+(U_+) = h.$$

2. $h > h_*$. Take $\check{U}_+$ such that $h_+(\check{U}_+) = h$; then we have $\check{U}_+ > u_*$; therefore, $\mathrm{r}'_h(h) = (d-1)\,(\bar{p}'_+(\check{U}_+) + b_-) > 0$. On the other hand, $\mathrm{r}'_h(0) < 0$. Hence there exists a unique value $z \in (0, h)$ such that $\mathrm{r}'_h(z) = 0$. Thus, the function r_h has a unique minimum at z; the optimal values h_+ and h_- are $h_+ = z > 0$ and $h_- = h - z > 0$, respectively. The quantities h_-, h_+ and the related auxiliary quantities u_- and u_+ are uniquely defined by the equations

$$h_+ = u_+ - \frac{Q_+(u_+)}{q_+(u_+)},$$
$$h_- = u_- - \frac{Q_-(u_-)}{q_-(u_-)},$$
$$h_+ + h_- = h,$$
$$\bar{p}'_+(u_+) = \bar{p}'_-(u_-),$$

and the minimum resistance is

$$\mathrm{R}(h) = \bar{p}_+(u_+) + Q_+(u_+)\, q_+(u_+)^{-d+1} + \bar{p}_-(u_-) + Q_-(u_-)\, q_-(u_-)^{-d+1}.$$

3.3.3 Limiting Cases

We consider the limiting behavior of solutions for high and low velocities, that is, as $V \to +\infty$ and $V \to 0$, for fixed h and a fixed distribution function σ. We shall denote the pressure and the resistance functions by $p_\pm(u, V)$ and $\mathrm{R}(h, V)$, indicating explicitly in this way the dependence of these functions on the parameter V. Our arguments will be heuristic here.

The case. $\mathbf{V \to +\infty}$.

Let $\tilde{p}_\pm(u, V) = V^{-2} p_\pm(u, V)$ be the reduced pressure and $\tilde{\mathrm{R}}(h, V) = V^{-2}\mathrm{R}(h, V)$ the minimum reduced resistance. Then one has

$$\begin{aligned} &\tilde{p}_+(u, V) = 1/(1 + u^2) + o(1), \\ &\tilde{p}_-(u, V) = o(1), \quad \tilde{p}'_-(u, V) = o(1), \quad V \to +\infty. \end{aligned}$$

In other words, as $V \to +\infty$, the functions $\tilde{p}_+(u, V)$ and $\tilde{p}_-(u, V)$ approach $1/(1 + u^2)$ and 0, respectively, that is, the functions describing the pressure distribution on the front and rear parts of the body surface in the classical Newton problem, respectively.

We consider separately the cases $d = 2$ and $d = 3$.

Let $d = 2$. If $h < 1$, then for sufficiently large V the figure of least resistance is a trapezium and the inclination angle of its lateral sides approaches 45° as $V \to +\infty$. If $h > 1$, then for sufficiently large V the figure of least resistance is an isosceles triangle coinciding with the solution of the two-dimensional analog of the classical Newton problem.

Let $d = 3$. Then for sufficiently large V the body of least resistance has a similar shape to the solution of the classical Newton problem and approaches the latter as $V \to +\infty$: the rear part of the surface is a fixed flat disc of radius 1 and the front part is the union of a flat disc of smaller radius parallel to the first one and a strictly convex lateral surface. The front part of the surface can be described as the graph of a radial function in the unit disc; as $V \to +\infty$, this function converges uniformly to the function describing the classical solution.

The case $d > 3$ is similar to the three-dimensional one.

Finally, the limiting value $\tilde{\mathrm{R}}(h, \infty) = \lim_{V \to +\infty} \tilde{\mathrm{R}}(h, V)$ of the reduced resistance is proportional to the resistance of Newton's optimal solution with coefficient equal to the density of the flow of particles

$$\nu = \int_{\mathbb{R}^d} \sigma(|v|)\, dv.$$

The case. $\mathbf{V \to 0}$.

In this limiting case, one has

$$p_{\pm}(u, V) = \pm B^{(d)} + V \frac{c^{(d)}}{\sqrt{1+u^2}} + o(V), \tag{3.41}$$

where

$$B^{(2)} = \frac{\pi}{2} \int_0^{+\infty} \sigma(r) r^3 \, dr, \quad c^{(2)} = 4 \int_0^{+\infty} \sigma(r) r^2 \, dr, \tag{3.42}$$

$$B^{(3)} = \frac{2\pi}{3} \int_0^{+\infty} \sigma(r) r^4 \, dr, \quad c^{(3)} = 2\pi \int_0^{+\infty} \sigma(r) r^3 \, dr. \tag{3.43}$$

We shall deduce this formula in Sect. 3.5. One readily sees that u_-^0 and u_+^0 approach, respectively, the quantities $\sqrt{\tau} \approx 1.272$ and $b_{\pm} = V \cdot \tau^{-5/2} + o(V)$. Here $\tau = (1+\sqrt{5})/2 \approx 1.618$ is the golden section. Taking into account the inequality $\bar{p}_+'(u) < \bar{p}_-'(u) < 0$, one concludes that u_* approaches the same value $\sqrt{\tau}$ and $u_+^0 + u_*$ approaches $2\sqrt{\tau}$.

We now describe the shape of the least-resistance body, determine the minimum reduced resistance $\hat{\mathrm{R}}(h, V) = V^{-1}\mathrm{R}(h, V)$, and find its limiting value as $V \to 0^+$ in the cases $d = 2$ and $d = 3$.

Let $d = 2$. Then the following results hold:

(a) $0 < h < \sqrt{\tau}$: the optimal figure is a trapezium.
(b) $h = \sqrt{\tau}$: an isosceles triangle.
(c) $a < h < 2\sqrt{\tau}$: the union of a triangle and a trapezium.
(d) $h \geq 2\sqrt{\tau}$: a rhombus.

In the first three cases, the inclination of the lateral sides in the horizontal direction is $\arctan \sqrt{\tau} \approx 51.8°$, and in the last case it is larger. Examples of optimal figures are presented in Fig. 3.2; we also present there for comparison the solutions of the classical two-dimensional Newton problem for the same values of h.

The minimum reduced resistance is

$$\hat{\mathrm{R}}(h, V) = 2c^{(2)} \bar{p}(h/2) + o(1) \quad \text{as } V \to 0^+, \tag{3.44}$$

where

$$\bar{p}(u) = \begin{cases} 1 - \tau^{-5/2} u, & u \leq \sqrt{\tau}, \\ 1/\sqrt{1+u^2}, & u \geq \sqrt{\tau}. \end{cases} \tag{3.45}$$

Let $d = 3$. Let

$$q(u) = \begin{cases} \tau^{5/2}, & u \leq \sqrt{\tau}, \\ \frac{(1+u^2)^{3/2}}{u}, & u \geq \sqrt{\tau}, \end{cases}$$

$$Q(u) = \begin{cases} \tau^{5/2} u, & u \leq \sqrt{\tau}, \\ \sqrt{1+u^2}\, \frac{4+u^2}{3} + \frac{2+\tau}{3} + \ln \frac{\sqrt{1+u^2}-1}{u} - \ln \frac{\tau-1}{\sqrt{\tau}}, & u \geq \sqrt{\tau}, \end{cases}$$

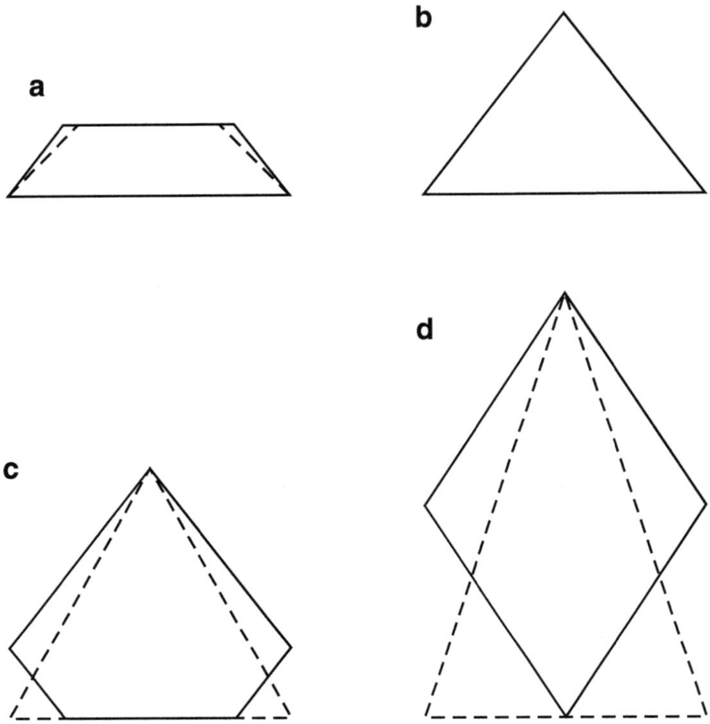

Fig. 3.2 Two-dimensional case; solutions as $V \to 0^+$ are drawn by *continuous lines*, solutions of the classical Newton problem by *dashed lines*; (**b**) is the only case where these solutions coincide

and let U be the solution of (3.23) (it is unique for $h > 0$). Define the function f_h as in statement (b) of Lemma 3.5. Then the body of least resistance is

$$\left\{(x', x_3) \in \mathbb{R}^3 : |x'| \le 1,\ |x_3| \le -f_h(|x'|)\right\},$$

where $x' = (x_1, x_2)$. One sees that this body is symmetric with respect to the horizontal plane $\{x_3 = 0\}$. The front and rear parts of its surface contain flat discs of the same size, and the inclination of the lateral surface to these discs is $\arctan\sqrt{\tau} \approx 51.8°$.

In Fig. 3.3, we present the optimal body as $V \to 0$ and the optimal body of the classical Newton problem for $h = 1$. We display the projections of these bodies onto the plane Ox_1x_3.

The minimum reduced resistance is

$$\hat{\mathrm{R}}(h, V) = 2c^{(3)}\left(\bar{p}(U) + \frac{Q(U)}{q^2(U)}\right),$$

where $\bar{p}$ is defined in (3.45) and U is as in (3.23).

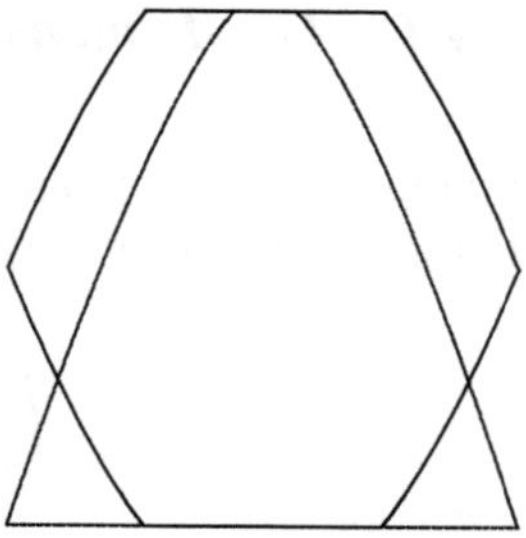

Fig. 3.3 Three-dimensional case, $h = 1$; the solution as $V \to 0^+$ (*solid line*) and the solution of the classical Newton problem (*dashed line*)

It is interesting to note that in these limiting cases the optimal body does not depend on the distribution σ; moreover, the reduced minimum resistance is proportional to ν in the limit $V \to +\infty$ and to the coefficient $c^{(d)}$ defined by (3.42) and (3.43) in the limit $V \to 0^+$. This coefficient can be interpreted as the sum of the absolute values of the momenta of the medium particles per unit volume in the frame of reference associated with the medium.

3.4 Gaussian Distribution of Velocities: Exact Solutions

Let $\rho = \rho_V$ be the density of the circular Gaussian distribution with mean $-Ve_d$ and variance 1, that is,

$$\rho_V(v) = \sigma(|v + Ve_d|), \quad \text{where} \quad \sigma(r) = (2\pi)^{-d/2}\, e^{-r^2/2}. \tag{3.46}$$

This function describes the velocity distribution of particles in a frame of reference moving in a homogeneous monatomic ideal gas, where the velocity of the motion is V times the mean square velocity of the molecules (Example 3.1).

Here we shall analytically calculate the pressure functions $p_\pm(u, V)$ in the cases $d = 2$ and 3 using the results of the previous section; by means of numerical simulation we obtain the following results.

(a) The set of parameters V, h is partitioned into subsets: distinct subsets correspond to distinct types of solution. This partitioning is plotted in Fig. 3.4 for $d = 2$ and in Fig. 3.7 for $d = 3$.
(b) We calculate the least resistance $\mathrm{R}(h, V)$ for various values of h, V. The results are presented in Figs. 3.5 and 3.8.
(c) For several values of the parameters h and V we construct the body of least resistance. Several such bodies are displayed in Figs. 1.3 and 1.4a–d (the central cross sections of the bodies in Fig. 1.3 are shown in Fig. 3.6).

Here V is a variable regarded as an argument of the functions p_+, p_-, and R: we use the notation $p_\pm(u, V)$ and $\mathrm{R}(h, V)$.

3.4.1 Two-Dimensional Case

Fixing the "+" sign and going over to polar coordinates $v = (-r \sin\varphi, -r\cos\varphi)$ in formula (3.11) one obtains

$$p_+(u, V) = \iint \frac{r^2(\cos\varphi + u\sin\varphi)_+^2}{1+u^2}\, \rho_+(r, \varphi, V)\, r dr d\varphi, \tag{3.47}$$

where $z_+ := \max\{0, z\}$ and $\rho_+(r, \varphi, V)$ is the density function ρ_V (3.46) expressed in the preceding polar coordinates:

$$\rho_+(r, \varphi, V) = \frac{1}{\pi}\, e^{-\frac{1}{2}(r^2 - 2Vr\cos\varphi + V^2)}. \tag{3.48}$$

Next, fixing the "–" sign and introducing polar coordinates in a slightly different fashion, $v = (-r \sin\varphi, r\cos\varphi)$, one obtains

$$p_-(u, V) = -\iint \frac{r^2(\cos\varphi + u\sin\varphi)_+^2}{1+u^2}\, \rho_-(r, \varphi, V)\, r dr d\varphi. \tag{3.49}$$

Here $\rho_-(r, \varphi, V)$ is the same density function ρ_V (3.46), expressed in these coordinates:

$$\rho_-(r, \varphi, V) = \frac{1}{\pi}\, e^{-\frac{1}{2}(r^2 + 2rV\cos\varphi + V^2)}. \tag{3.50}$$

Combining formulas (3.47)–(3.50) one arrives at a more general expression:

$$p_\pm(u, V) = \pm\frac{e^{-V^2/2}}{\pi} \iint\limits_{\cos\varphi + u\sin\varphi > 0} \frac{(\cos\varphi + u\sin\varphi)^2}{1+u^2}\, e^{-\frac{1}{2}r^2 \pm rV\cos\varphi}\, r^3\, dr d\varphi\,.$$

Integrating with respect to r one obtains

$$p_\pm(u, V) = \pm 2\frac{e^{-V^2/2}}{\pi} \int\limits_{\cos\varphi + u\sin\varphi > 0} \frac{(\cos\varphi + u\sin\varphi)^2}{1+u^2}\, l(\pm V\cos\varphi)\, d\varphi\,, \tag{3.51}$$

where

$$l(z) = 1 + \frac{z^2}{2} + \frac{\sqrt{\pi}}{2\sqrt{2}}\, e^{z^2/2}\left(3z + z^3\right)\left(1 + \operatorname{erf}\left(z/\sqrt{2}\right)\right),$$

$$\operatorname{erf}(x) = \frac{2}{\sqrt{\pi}} \int_0^x e^{-t^2} dt.$$

Making the change of variable $\tau = \varphi - \arcsin(u/\sqrt{1+u^2})$ one finally arrives at

$$p_\pm(u, V) = \pm 2\frac{e^{-V^2/2}}{\pi} \int_{-\pi/2}^{\pi/2} \cos^2\tau\, I\left(\pm V\,\frac{\cos\tau - u\sin\tau}{\sqrt{1+u^2}}\right) d\tau. \tag{3.52}$$

Further constructions were carried out with the use of Maple and verified with Matlab.

We plotted the graphs of the functions

$$h = u_+^0(V), \quad h = u_*(V), \quad h = u_*(V) + u_-^0(V),$$

where the quantities u_+^0, u_-^0, and u_* (which are functions of V) are defined in the previous sections. These graphs (Fig. 3.4) partition the parameter space $\mathbb{R}_+^2$ into four regions corresponding to the four distinct kinds of solution. For the smaller function $h = u_+^0(V)$ one has $\lim_{V\to\infty} u_+^0(V) = 1$. In the limit $V \to 0$, the smallest, middle, and largest functions take the values $\sqrt{\tau}$, $\sqrt{\tau}$, and $2\sqrt{\tau}$, respectively (recall that $\tau = (1+\sqrt{5})/2 \approx 1.618$ is the golden section):

$$\lim_{V\to 0} u_+^0(V) = \sqrt{\tau}, \quad \lim_{V\to 0} u_*(V) = \sqrt{\tau}, \quad \lim_{V\to 0}(u_*(V) + u_-^0(V)) = 2\sqrt{\tau}.$$

Solutions of the fifth kind (the union of two triangles and a trapezium) were not discovered in numerical simulation. (These solutions correspond to the case where the set $\mathcal{O}_{p_-,V} = \{u : \bar{p}_-(u, V) < p_-(u, V)\}$ contains at leat two connected components.) We believe, although we cannot prove it, that in the case of the Gaussian distribution ρ_V under consideration, this kind of solution does not occur at all. [According to statement (b) of Lemma 3.2, such solutions do occur for some distributions corresponding to mixtures of homogeneous gases.]

Further, using formulae from Sect. 3.3.1 we calculate the functions f_+ and f_-, which enable one to construct the optimal figures (Fig. 1.4a–d), and the minimum resistance $\mathrm{R}(h, V)$. The graphs of the reduced minimum resistance $\tilde{\mathrm{R}}(h, V) = V^{-2}\mathrm{R}(h, V)$ versus h are plotted in Fig. 3.5a, b for several values of V.

3.4.2 Three-Dimensional Case

Fix the "+" sign. In the spherical coordinates $v = (-r\sin\varphi\cos\theta, -r\sin\varphi\sin\theta, -r\cos\varphi)$, $r \ge 0$, $0 \le \varphi \le \pi$, $-\pi \le \theta \le \pi$, formula (3.11) takes the following form:

$$p_+(u, V) = 2\iiint \frac{r^2(\cos\varphi + u\sin\varphi\cos\theta)_+^2}{1+u^2}\, \rho_+(r, \varphi, \theta, V)\, r^2 \sin\varphi\, dr d\varphi d\theta,$$

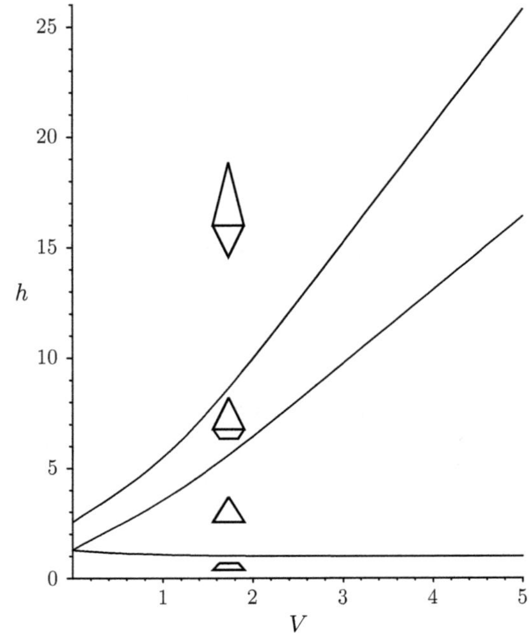

Fig. 3.4 Two-dimensional case: the four subregions of the parameter space corresponding to the four kinds of solution

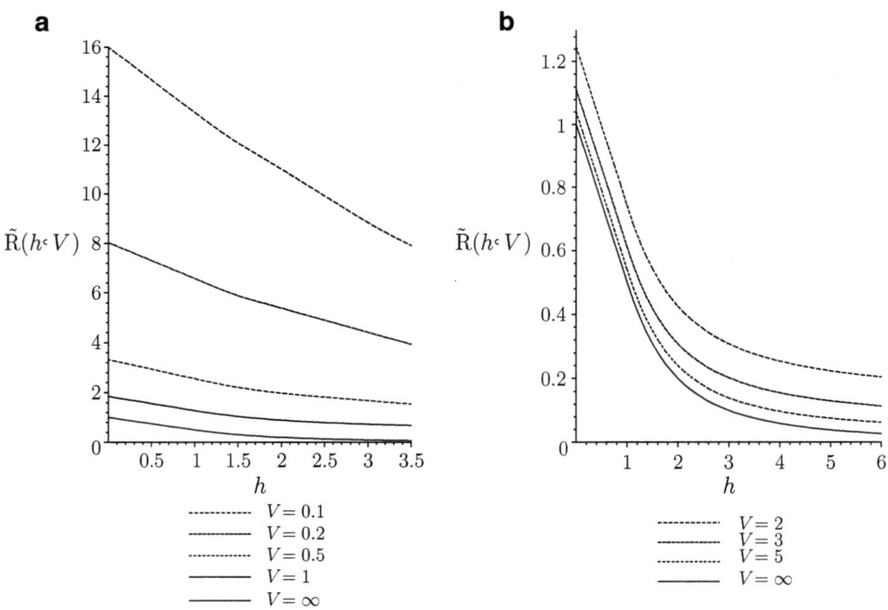

Fig. 3.5 Two-dimensional case: the least reduced resistance $\tilde{R}(h, V)$ versus the height h of the body

where

$$\rho_+(r,\varphi,\theta,V) = \frac{1}{(2\pi)^{3/2}}\, e^{-\frac{1}{2}(r^2-2Vr\cos\varphi+V^2)}.$$

Now fix the "–" sign. In the coordinates $v = (-r\sin\varphi\cos\theta, -r\sin\varphi\sin\theta,\ r\cos\varphi)$, one has

$$p_-(u,V) = -2\iiint \frac{r^2(\cos\varphi + u\sin\varphi\cos\theta)_+^{\,2}}{1+u^2}\,\rho_-(r,\varphi,\theta,V)\,r^2\sin\varphi\,dr d\varphi d\theta,$$

where

$$\rho_-(r,\varphi,\theta,V) = \frac{1}{(2\pi)^{3/2}}\, e^{-\frac{1}{2}(r^2+2Vr\cos\varphi+V^2)}.$$

Combining these cases one arrives at the formula

$$p_\pm(u,V) = \pm 2\frac{e^{-V^2/2}}{(2\pi)^{3/2}} \iiint\limits_{\cos\varphi+u\sin\varphi\cos\theta>0} \frac{(\cos\varphi + u\sin\varphi\,cos\theta)^2}{1+u^2}\cdot$$

$$\cdot\, e^{-\frac{1}{2}r^2\pm Vr\cos\varphi}r^4\sin\varphi\,dr d\varphi d\theta. \tag{3.53}$$

The following two equalities (3.54) and (3.55) are easy to verify. First,

$$\int_0^{+\infty} e^{-\frac{1}{2}r^2\pm Vr\cos\varphi}r^4\,dr = I(\pm V\cos\varphi), \tag{3.54}$$

where

$$I(z) = \sqrt{\pi/2}\,e^{z^2/2}\,(3+6z^2+z^4)(1+\mathrm{erf}\,(z/\sqrt{2})) + 5z + z^3.$$

Second,

$$\int_{\cos\varphi+u\sin\varphi\cos\theta>0} \frac{(\cos\varphi + u\sin\varphi\cos\theta)^2}{1+u^2}\,d\theta = J(u,\cos\varphi), \tag{3.55}$$

where

$$J(u,\zeta) = \begin{cases} 0, & \text{if } -1 \le \zeta \le -u/\sqrt{1+u^2} \\ J_1(u,\zeta), & \text{if } |\zeta| < u/\sqrt{1+u^2} \\ J_2(u,\zeta), & \text{if } u/\sqrt{1+u^2} \le \zeta \le 1 \end{cases}$$

and

$$J_1(u,\zeta) = \frac{1}{1+u^2}\left[\theta_0\,(2\zeta^2 + u^2(1-\zeta^2)) + 3\zeta\sqrt{u^2-\zeta^2(1+u^2)}\right],$$

$$J_2(u,\zeta) = \frac{\pi}{1+u^2}\,[2\zeta^2 + u^2(1-\zeta^2)], \qquad \theta_0 = \arccos\left(-\frac{\zeta}{u\sqrt{1-\zeta^2}}\right).$$

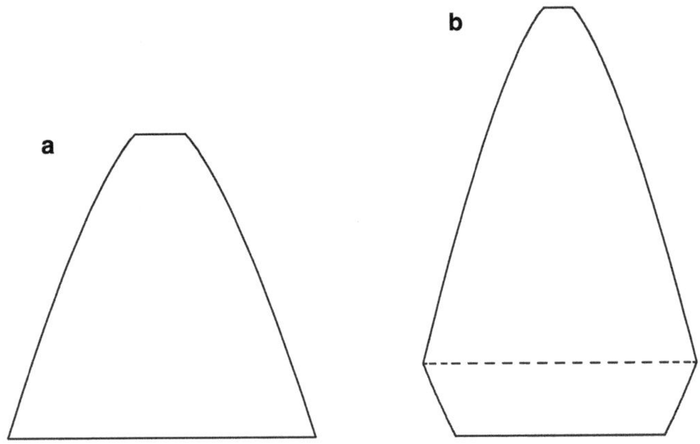

Fig. 3.6 Solutions in three-dimensional case for distribution function ρ_V (3.46)

Taking these formulae into account and making the change of variable $\zeta = \cos\varphi$ in (3.53) one obtains

$$p_{\pm}(u,V) = \pm 2\frac{e^{-V^2/2}}{(2\pi)^{3/2}} \int_{-1}^{1} I(\pm V\zeta)\, J(u,\zeta)\, d\zeta =$$

$$= \pm 2\frac{e^{-V^2/2}}{(2\pi)^{3/2}} \left(\int_{-u/\sqrt{1+u^2}}^{u/\sqrt{1+u^2}} I(\pm V\zeta)\, J_1(u,\zeta)\, d\zeta + \int_{u/\sqrt{1+u^2}}^{1} I(\pm V\zeta)\, J_2(u,\zeta)\, d\zeta \right).$$

Next one calculates numerically the function $h_*(V)$ by formula (3.40). This function is presented in Fig. 3.7a; it partitions the parameter space $\mathbb{R}^2_+$ into two subsets corresponding to two distinct kinds of solution. The function $h_*(V)$ appears linear but is not: we plot the graph of its derivative in Fig. 3.7b. The function $\mathrm{R}(h, V)$ is calculated by formulae from the previous section; the graphs of $\tilde{\mathrm{R}}(h, V)$ versus h are plotted in Fig. 3.8 for several values of V. In Fig. 3.6, we present examples of solutions of the first and second kinds for the indicated values of the parameters.

3.5 Proof of Auxiliary Statements

3.5.1 Proof of Lemma 3.1

Making the change of variable $v = r\nu,\ r \ge 0,\ \nu \in S^{d-1}$, in the integral (3.11), one obtains

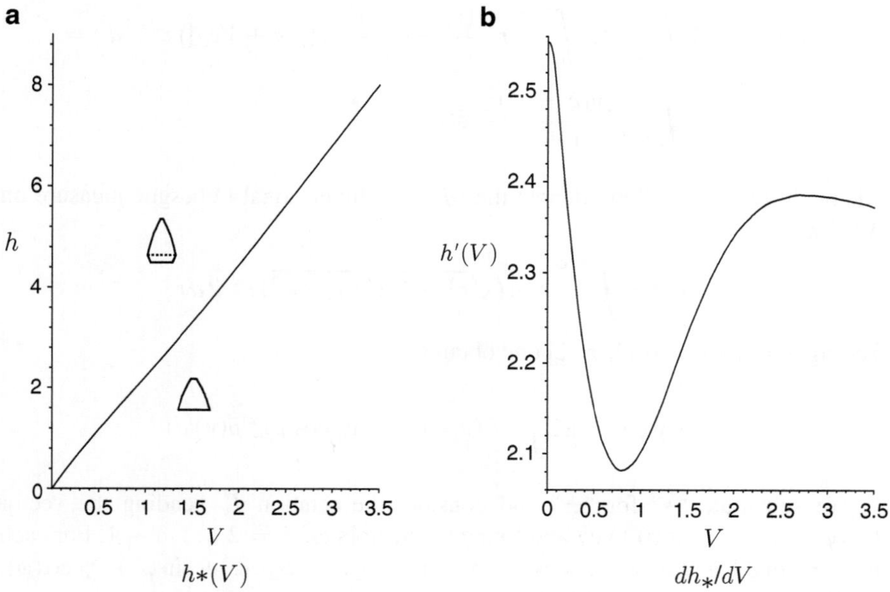

Fig. 3.7 Three-dimensional case: the function $h_*(V)$ partitions the plane of the parameters $\{V, h\}$ into two subsets corresponding to two kinds of solution

Fig. 3.8 Three-dimensional case: the minimum reduced resistance $\tilde{R}(h, V)$ versus the height h of the body

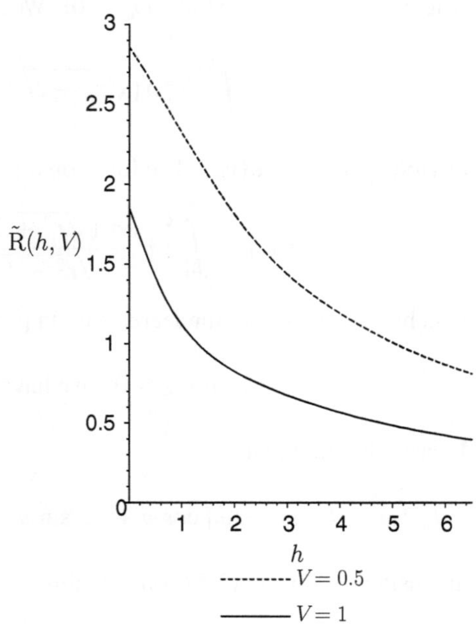

$$p_\pm(u) = \pm 2 \int_{S^{d-1}} d\nu \int_0^{+\infty} r^2 \, \frac{(\nu_1 u \pm \nu_d)_-^2}{1+u^2} \, \sigma(|r\nu + V e_d|) \, r^{d-1} dr =$$

$$= \pm 2 \int_{S^{d-1}} \frac{(\nu_1 u \pm \nu_d)_-^2}{1+u^2} \, \bar{\rho}(\nu) \, d\nu,$$

where $d\nu$ (and $d\omega$ below) denote the $(d-1)$-dimensional Lebesgue measure on S^{d-1} and

$$\bar{\rho}(\nu) := \int_0^{+\infty} r^2 \, \sigma(\sqrt{r^2 + 2rV\,\nu_d + V^2}) \, r^{d-1} dr.$$

Setting $u = \tan\varphi$, $\varphi \in [0, \pi/2]$ one obtains

$$p_\pm(\tan\varphi) = \pm 2 \int_{S^{d-1}} (\nu_1 \sin\varphi \pm \nu_d \cos\varphi)_-^2 \, \bar{\rho}(\nu) \, d\nu.$$

We substitute "+" for "±" and consider the rotation T_φ sending the vector $(\sin\varphi, 0, \ldots, 0, \cos\varphi)$ to e_d and fixing the vectors e_i, $i = 2, \ldots, d-1$. For each $\nu \in \mathbb{R}^d$ one has $T_\varphi \nu = (\nu_1 \cos\varphi - \nu_d \sin\varphi, \ \nu_2, \ldots, \nu_{d-1}, \ \nu_1 \sin\varphi + \nu_d \cos\varphi)$. Making the change of variable $T_\varphi \nu = \omega$ one obtains

$$p_+(\tan\varphi) = 2 \int_{S_-^{d-1}} \omega_d^2 \, \bar{\rho}(T_\varphi^{-1}\omega) \, d\omega, \tag{3.56}$$

where $S_-^{d-1} := \{\omega \in S^{d-1} : \omega_d < 0\}$. We set

$$\varrho(z) := \int_0^{\infty} r^2 \, \sigma(\sqrt{r^2 + 2rVz + V^2}) \, r^{d-1} dr, \quad |z| \le 1; \tag{3.57}$$

obviously, $\bar{\rho}(\nu) = \varrho(\nu_d)$. The function ϱ has the continuous derivative

$$\varrho'(z) = \int_0^{\infty} r^2 \, \frac{\sigma'(\sqrt{r^2 + 2rVz + V^2})}{\sqrt{r^2 + 2rVz + V^2}} \, rV \, r^{d-1} dr,$$

which is negative and nondecreasing; in particular,

$$\text{for } z > 0, \ \text{ we have } \varrho'(z) > \varrho'(-z). \tag{3.58}$$

Bearing in mind that

$$T_\varphi^{-1}\omega = T_{-\varphi}\omega = (\omega_1 \cos\varphi + \omega_d \sin\varphi, \ \omega_2, \ldots, \omega_{d-1}, \ -\omega_1 \sin\varphi + \omega_d \cos\varphi)$$

and using (3.56) and (3.57) one obtains

$$p_+(\tan\varphi) = 2 \int_{S_-^{d-1}} \omega_d^2 \, \varrho(-\omega_1 \sin\varphi + \omega_d \cos\varphi) \, d\omega. \tag{3.59}$$

We now substitute "−" for "±" and consider the orthogonal reflection U_φ with respect to the hyperplane $\{\omega_1 \sin \frac{\varphi}{2} = \omega_d \cos \frac{\varphi}{2}\}$; for each $\nu \in \mathbb{R}^d$ we have

$$U_\varphi \nu = (\nu_1 \cos \varphi + \nu_d \sin \varphi,\ \nu_2, \ldots, \nu_{d-1},\ \nu_1 \sin \varphi - \nu_d \cos \varphi).$$

Making the change of variable $U_\varphi \nu = \omega$ we obtain

$$p_-(\tan \varphi) = -2 \int_{S_-^{d-1}} \omega_d^2\, \bar{\rho}(U_\varphi^{-1}\omega)\, d\omega.$$

Taking account of the equality $U_\varphi^{-1} = U_\varphi$ we see that

$$p_-(\tan \varphi) = -2 \int_{S_-^{d-1}} \omega_d^2\, \varrho(\omega_1 \sin \varphi - \omega_d \cos \varphi)\, d\omega. \tag{3.60}$$

Formulae (3.59) and (3.60) can be written in unified form:

$$p_\pm(\tan \varphi) = \pm 2 \int_{S_-^{d-1}} \omega_d^2\, \varrho(\pm(-\omega_1 \sin \varphi + \omega_d \cos \varphi))\, d\omega. \tag{3.61}$$

Setting $\varphi = \pi/2$ in (3.61) we obtain

$$\lim_{u \to +\infty} p_\pm(u) = p_\pm(+\infty) = \pm 2 \int_{S_-^{d-1}} \omega_d^2\, \varrho(\mp\omega_1)\, d\omega.$$

Since the map $\omega_1 \mapsto -\omega_1$ preserves the Lebesgue measure on S_-^{d-1}, we conclude that $p_+(+\infty) = -p_-(+\infty)$; the proof of part (a) of Lemma 3.1 is thus complete.

From (3.61) one concludes that the functions $p_\pm$ have continuous derivatives and

$$p'_\pm(\tan \varphi) = \pm \cos^2 \varphi \cdot 2 \int_{S_-^{d-1}} \omega_d^2 \cdot \frac{d}{d\varphi}\, \varrho(\pm(-\omega_1 \sin \varphi + \omega_d \cos \varphi))\, d\omega =$$

$$= -\cos^2 \varphi \cdot 2 \int_{S_-^{d-1}} \omega_d^2\, (\omega_1 \cos \varphi + \omega_d \sin \varphi)\, \varrho'(\pm(-\omega_1 \sin \varphi + \omega_d \cos \varphi))\, d\omega. \tag{3.62}$$

Setting $\varphi = 0$ in (3.62) we obtain

$$p'_\pm(0) = -2 \int_{S_-^{d-1}} \omega_d^2\, \omega_1\, \varrho'(\pm\omega_d)\, d\omega.$$

Since S_-^{d-1} is invariant and the integrand is antisymmetric with respect to the reflection $\omega_1 \mapsto -\omega_1$, it follows that $p'_\pm(0) = 0$. Next, setting $\varphi = \pi/2$ in (3.62) we obtain

$$\lim_{u \to +\infty} p'_\pm(u) = p'_\pm(+\infty) = 0.$$

The proof of (b) is thus complete.

Further, we have

$$p'_+(\tan\varphi) - p'_-(\tan\varphi) = \cos^2\varphi \cdot 2\int_{S_-^{d-1}} \omega_d^2 \, \frac{\partial}{\partial\varphi}\, \Phi(\varphi, \omega_1, \omega_d)\, d\omega, \quad (3.63)$$

$$p'_+(\tan\varphi) = \cos^2\varphi \cdot 2\int_{S_-^{d-1}} \omega_d^2 \, \frac{\partial}{\partial\varphi}\, \Phi_+(\varphi, \omega_1, \omega_d)\, d\omega, \quad (3.64)$$

where

$$\Phi(\varphi, \omega_1, \omega_d) = \varrho(-\omega_1 \sin\varphi + \omega_d \cos\varphi) + \varrho(\omega_1 \sin\varphi - \omega_d \cos\varphi),$$

$$\Phi_+(\varphi, \omega_1, \omega_d) = \varrho(-\omega_1 \sin\varphi + \omega_d \cos\varphi).$$

We set

$$I(c,\varphi) = \int_{\Gamma_c} \omega_d^2 \, \frac{\partial}{\partial\varphi}\, \Phi(\varphi, \omega_1, \omega_1, \omega_d)\, ds,$$

$$I_+(c,\varphi) = \int_{\Gamma_c} \omega_d^2 \, \frac{\partial}{\partial\varphi}\, \Phi_+(\varphi, \omega_1, \omega_d)\, ds,$$

where $\Gamma_c = \{(\omega_1, \omega_d) : \omega_1^2 + \omega_d^2 = c^2,\ \omega_d < 0\}$ is the lower half of a circumference of radius c and ds is a Lebesgue measure on Γ_c. Our immediate aim is to prove that

$$I(c,\varphi) < 0, \quad I_+(c,\varphi) < 0 \quad \text{for all } c \in (0,\,1) \text{ and } \varphi \in (0,\,\pi/2); \quad (3.65)$$

then, integrating $I(1 - |\tilde\omega|^2, \varphi)$ and $I_+(1 - |\tilde\omega|^2, \varphi)$ with respect to $\tilde\omega = (\omega_2, \ldots, \omega_{d-1})$ and multiplying by $2\cos^2\varphi$, we shall be able to conclude that the right-hand sides of (3.63) and (3.63) are negative, which proves statement (c) and the first inequality in (d) of Lemma 3.1.

We parameterize the curve Γ_c in accordance with the formulae $\omega_1 = c\cos\theta$, $\omega_d = -c\sin\theta$, $\theta \in [0, \pi]$; then

$$I(c,\varphi) = \int_0^{\pi} c^2 \sin^2\theta \, \frac{\partial}{\partial\varphi} \big[\varrho(c\sin(\varphi+\theta)) + \varrho(-c\sin(\varphi+\theta))\big]\, c\, d\theta = c^3\mathcal{I}_1 + c^3\mathcal{I}_2,$$

where

$$\mathcal{I}_1 = \int_0^{\pi-2\varphi} \sin^2\theta \, \frac{\partial}{\partial\varphi} \big[\varrho(c\sin(\varphi+\theta)) + \varrho(-c\sin(\varphi+\theta))\big]\, d\theta, \quad (3.66)$$

$$\mathcal{I}_2 = \int_{\pi-2\varphi}^{\pi} \sin^2\theta \, \frac{\partial}{\partial\varphi} \big[\varrho(c\sin(\varphi+\theta)) + \varrho(-c\sin(\varphi+\theta))\big]\, d\theta, \quad (3.67)$$

and

$$I_+(c,\varphi) = \int_0^\pi c^2 \sin^2\theta \, \frac{\partial}{\partial\varphi}\, \varrho(-c\sin(\varphi+\theta))\, c\, d\theta = c^3\mathcal{I}_1^+ + c^3\mathcal{I}_2^+,$$

where

$$\mathcal{I}_1^+ = \int_0^{\pi-2\varphi} \sin^2\theta \, \frac{\partial}{\partial\varphi}\, \varrho(-c\sin(\varphi+\theta))\, d\theta, \tag{3.68}$$

$$\mathcal{I}_2^+ = \int_{\pi-2\varphi}^{\pi} \sin^2\theta \, \frac{\partial}{\partial\varphi}\, \varrho(-c\sin(\varphi+\theta))\, d\theta. \tag{3.69}$$

The change of variable $\psi = \theta + \varphi - \pi/2$ in (3.66) yields

$$\mathcal{I}_1 = \int_{-\pi/2+\varphi}^{\pi/2-\varphi} \cos^2(\varphi-\psi)\, \frac{d}{d\psi}\left[\varrho(c\cos\psi) + \varrho(-c\cos\psi)\right] d\psi,$$

and using the fact that the function $\frac{d}{d\psi}[\cdots]$ under the integral sign is odd one obtains

$$\mathcal{I}_1 = \int_0^{\pi/2-\varphi} (\cos^2(\varphi-\psi) - \cos^2(\varphi+\psi))\, \frac{d}{d\psi}\left[\varrho(c\cos\psi) + \varrho(-c\cos\psi)\right] d\psi.$$

One has $\cos^2(\varphi-\psi) - \cos^2(\varphi+\psi) = \sin 2\varphi\, \sin 2\psi > 0$. Taking account of (3.58) one also obtains

$$\frac{d}{d\psi}\left[\varrho(c\cos\psi) + \varrho(-c\cos\psi)\right] = -c\sin\psi\, (\varrho'(c\cos\psi) - \varrho'(-c\cos\psi)) < 0.$$

Hence $\mathcal{I}_1 < 0$.

Making the same change of variable in (3.68) one obtains

$$\mathcal{I}_1^+ = \int_{-\pi/2+\varphi}^{\pi/2-\varphi} \cos^2(\varphi-\psi)\, \frac{d}{d\psi}\, \varrho(-c\cos\psi)\, d\psi =$$

$$= \int_0^{\pi/2-\varphi} (\cos^2(\varphi-\psi) - \cos^2(\varphi+\psi))\, \frac{d}{d\psi}\, \varrho(-c\cos\psi)\, d\psi.$$

One has $\cos^2(\varphi-\psi) - \cos^2(\varphi+\psi) > 0$ and

$$\frac{d}{d\psi}\, \varrho(-c\cos\psi) = c\sin\psi\, \varrho'(-c\cos\psi)) < 0;$$

therefore, $\mathcal{I}_1^+ < 0$.

On the other hand, the change of variable $\chi = \theta + \varphi - \pi$ in (3.67) yields

$$\mathcal{I}_2 = \int_{-\varphi}^{\varphi} \sin^2(\varphi - \chi)\, \frac{d}{d\chi} \left[\varrho(c \sin \chi) + \varrho(-c \sin \chi)\right] d\chi.$$

The function $\frac{d}{d\chi}[\ldots]$ is odd; therefore,

$$\mathcal{I}_2 = \int_0^{\varphi} (\sin^2(\varphi - \chi) - \sin^2(\varphi + \chi))\, \frac{d}{d\chi} \left[\varrho(c \sin \chi) + \varrho(-c \sin \chi)\right] d\chi.$$

One has $\sin^2(\varphi - \chi) - \sin^2(\varphi + \chi) = -\sin 2\varphi\, \sin 2\chi < 0$ and

$$\frac{d}{d\chi}\left[\ldots\right] = c \cos \chi\, (\varrho'(c \sin \chi) - \varrho'(-c \sin \chi)) > 0.$$

Hence $\mathcal{I}_2 < 0$.

Further, for $\pi - 2\varphi \le \theta \le \pi$ one has

$$\frac{\partial}{\partial \varphi}\, \varrho(-c \sin(\varphi + \theta)) = -c \cos(\varphi + \theta)\, \varrho'(-c \sin(\varphi + \theta)) < 0,$$

and from (3.69) one concludes that $\mathcal{I}_2^+ < 0$.

We have thus proved the inequalities in (3.65), which completes the proof of (c) and the first inequality in (d).

Passing to the limit as $\varphi \to \pi/2$ in (3.60) one obtains

$$\lim_{\varphi \to \pi/2} p_-(\tan \varphi) = p_-(+\infty) = -2 \int_{S_-^{d-1}} \omega_d^2\, \varrho(\omega_1)\, d\omega.$$

Thus, for the proof of the second inequality in (d) one must verify that

$$\int_{S_-^{d-1}} \omega_d^2\, \varrho(\omega_1)\, d\omega > \int_{S_-^{d-1}} \omega_d^2\, \varrho(\omega_1 \sin \varphi - \omega_d \cos \varphi)\, d\omega.$$

We set

$$J(c, \varphi) = \int_{\Gamma_c} \omega_d^2\, \varrho(\omega_1 \sin \varphi - \omega_d \cos \varphi)\, ds = c^3 \int_0^{\pi} \sin^2 \theta\, \varrho(c \sin(\varphi + \theta))\, d\theta.$$

It is sufficient to prove that

$$J(c, \pi/2) > J(c, \varphi) \quad \text{for all } c \in (0,\, 1) \text{and } \varphi \in (0,\, \pi/2); \tag{3.70}$$

then one establishes inequality (3.70) by integrating $J(1 - |\tilde{\omega}|^2, \pi/2)$ and $J(1 - |\tilde{\omega}|^2, \varphi)$ with respect to $\tilde{\omega} = (\omega_2, \ldots, \omega_{d-1})$.

One has $J(c,\varphi) = c^3(J_1 + J_2)$, $J(c,\pi/2) = c^3(J_1^* + J_2^*)$, where

$$J_1 = \int_0^{\pi/2-\varphi} \sin^2\theta\, \varrho(c\sin(\varphi+\theta))\, d\theta, \quad J_2 = \int_{\pi/2-\varphi}^{\pi} \sin^2\theta\, \varrho(c\sin(\varphi+\theta))\, d\theta,$$

$$J_1^* = \int_0^{\pi/2+\varphi} \sin^2\theta\, \varrho(c\cos\theta)\, d\theta, \quad J_2^* = \int_{\pi/2+\varphi}^{\pi} \sin^2\theta\, \varrho(c\cos\theta)\, d\theta.$$

For $0 < \theta < \pi/2 - \varphi$ one has $-\cos\theta < 0 < \sin(\varphi+\theta)$ and $\varrho(-c\cos\theta) > \varrho(c\sin(\varphi+\theta)$, so that

$$J_2^* = \int_0^{\pi/2-\varphi} \sin^2\theta\, \varrho(-c\cos\theta)\, d\theta > \int_0^{\pi/2-\varphi} \sin^2\theta\, \varrho(c\sin(\varphi+\theta))\, d\theta = J_1. \tag{3.71}$$

Next, one has

$$\begin{aligned} 2J_1^* &= \int_0^{\pi/2+\varphi} \sin^2\theta\, \varrho(c\cos\theta)\, d\theta \\ &\quad + \int_0^{\pi/2+\varphi} \sin^2(\pi/2+\varphi-\theta)\, \varrho(c\cos(\pi/2+\varphi-\theta)\, d\theta, \\ 2J_2 &= \int_0^{\pi/2+\varphi} \sin^2\theta\, \varrho(c\sin(\theta-\varphi))\, d\theta \\ &\quad + \int_0^{\pi/2+\varphi} \sin^2(\pi/2+\varphi-\theta)\, \varrho(c\sin(\pi/2-\theta))\, d\theta; \end{aligned}$$

therefore,

$$2J_1^* - 2J_2 = \int_0^{\pi/2+\varphi} [\sin^2\theta - \cos^2(\theta-\varphi)][\varrho(c\cos\theta) - \varrho(c\sin(\theta-\varphi)]\, d\theta. \tag{3.72}$$

Bearing in mind that the function ϱ is monotone decreasing and the equality $\sin^2\theta - \cos^2(\theta-\varphi) = (\sin(\theta-\varphi) - \cos\theta)(\sin(\theta-\varphi) + \cos\theta)$ one concludes that the integrand in (3.72) is positive; therefore, $J_1^* > J_2$. Hence (3.70) follows by (3.71). The proof of Lemma 3.1 is now complete.

3.5.2 *Proof of Lemma 3.2*

(a) We parameterize the semicircle S_-^1: $v_1 = -\sin\theta$, $v_2 = -\cos\theta$, $\theta \in [-\pi/2, \pi/2]$; then (3.61) takes the following form:

$$p_\pm(\tan\varphi) = \pm 2 \int_{-\pi/2}^{\pi/2} \cos^2\theta\, \varrho(\mp\cos(\varphi+\theta))\, d\theta. \tag{3.73}$$

Differentiating twice both sides of this equality with respect to φ one obtains

$$\frac{p''_\pm(\tan\varphi)}{\cos^3\varphi} = \mp 4 \int_{-\pi/2}^{\pi/2} \cos^2\theta \sin\varphi \,\frac{d}{d\varphi}\,\varrho(\mp\cos(\varphi+\theta))\,d\theta \pm$$

$$\pm 2 \int_{-\pi/2}^{\pi/2} \cos^2\theta \,\cos\varphi \,\frac{d^2}{d\varphi^2}\,\varrho(\mp\cos(\varphi+\theta))\,d\theta.$$

Integrating the second integral by parts and taking account of the equalities

$$\frac{d^k}{d\varphi^k}\,\varrho(\mp\cos(\varphi+\theta)) = \frac{d^k}{d\theta^k}\,\varrho(\mp\cos(\varphi+\theta)), \quad k = 1,\ 2,$$

one sees that

$$\frac{p''_\pm(\tan\varphi)}{\cos^3\varphi} = \pm 4 \int_{-\pi/2}^{\pi/2} \cos\theta \,\sin(\theta-\varphi)\,\frac{d}{d\theta}\,\varrho(\mp\cos(\varphi+\theta))\,d\theta.$$

Integrating by parts again and setting $g_\pm(\varphi) := p''_\pm(\tan\varphi)/(4\cos^3\varphi)$ one obtains

$$g_\pm(\varphi) = \mp \int_{-\pi/2}^{\pi/2} \varrho(\mp\cos(\varphi+\theta))\,\cos(2\theta-\varphi)\,d\theta. \tag{3.74}$$

We fix the sign "+" and prove that

(I) $g_+(\varphi) < 0$ for $0 < \varphi < \pi/6$,
(II) $g_+(\varphi) > 0$ for $\varphi \ge 0.3\pi$,
(III) $g'_+(\varphi) > 0$ for $\pi/6 \le \varphi \le 0.3\pi$.

Relations (I)–(III) demonstrate that there exists $\bar{u}_+ \in (1/\sqrt{3},\ \tan(0.3\pi))$ such that $p''_+(u) < 0$ for $u \in (0,\ \bar{u}_+)$ and $p''_+(u) > 0$ for $u \in (\bar{u}_+,\ +\infty)$.

(I) Making the change of variable $\psi = \theta - \varphi/2 + \pi/4$ one obtains

$$g_+(\varphi) = -\int_{-\pi/4-\varphi/2}^{3\pi/4-\varphi/2} \varrho(-\cos(3\varphi/2 - \pi/4 + \psi))\,\sin 2\psi\,d\psi = \mathcal{L}_1 + \mathcal{L}_2, \tag{3.75}$$

$$\text{where} \quad \mathcal{L}_1 = -\int_{-\pi/4-\varphi/2}^{\pi/4+\varphi/2} (\ldots) \quad \text{and} \quad \mathcal{L}_2 = -\int_{\pi/4+\varphi/2}^{3\pi/4-\varphi/2} (\ldots).$$

One has

$$\mathcal{L}_1 = \int_0^{\pi/4+\varphi/2} [\varrho(-\cos(3\varphi/2 - \pi/4 - \psi))$$
$$-\varrho(-\cos(3\varphi/2 - \pi/4 + \psi))]\,\sin 2\psi\,d\psi. \tag{3.76}$$

Since $0 \leq 2\psi \leq \pi/2 + \varphi \leq \pi$, it follows that $\sin 2\psi \geq 0$. Using the inequality $0 < \varphi < \pi/6$ one obtains

$$-\pi/2 \leq 3\varphi/2 - \pi/4 - \psi < -|3\varphi/2 - \pi/4 + \psi|;$$

therefore,

$$-\cos(3\varphi/2 - \pi/4 - \psi) > -\cos(3\varphi/2 - \pi/4 + \psi).$$

The function ϱ decreases, so that

$$\varrho\,(-\cos(3\varphi/2 - \pi/4 - \psi)) < \varrho\,(-\cos(3\varphi/2 - \pi/4 + \psi)).$$

Thus, the integrand in (3.76) is negative, and therefore $\mathcal{L}_1 < 0$.

We make the change of variable $\chi = \psi - \pi/2$ in the integral $\mathcal{L}_2$. Then

$$\mathcal{L}_2 = \int_{-\pi/4+\varphi/2}^{\pi/4-\varphi/2} \varrho\,(-\cos\,(3\varphi/2 + \pi/4 + \chi))\,\sin 2\chi\, d\chi =$$

$$= \int_0^{\pi/4-\varphi/2} [\varrho\,(-\cos\,(3\varphi/2 + \pi/4 + \chi))$$

$$-\varrho\,(-\cos\,(3\varphi/2 + \pi/4 - \chi))]\,\sin 2\chi\, d\chi. \tag{3.77}$$

One has

$$0 \leq 3\varphi/2 + \pi/4 - \chi \leq 3\varphi/2 + \pi/4 + \chi \leq \pi,$$

hence

$$-\cos(3\varphi/2 + \pi/4 - \chi) \leq -\cos(3\varphi/2 + \pi/4 + \chi),$$

so that

$$\varrho(-\cos(3\varphi/2 + \pi/4 - \chi)) \geq \varrho(-\cos(3\varphi/2 + \pi/4 + \chi)).$$

Therefore, the integrand in (3.77) is negative and $\mathcal{L}_2 \leq 0$. The proof of (I) is complete.

(II) By (3.75) one obtains

$$g_+(\varphi) = \mathcal{I}_1 + \mathcal{I}_2 + \mathcal{I}_3 + \mathcal{I}_4,$$

where

$$\mathcal{I}_1 = -\int_{-\varphi}^{\varphi}(\ldots),\ \ \mathcal{I}_2 = -\int_{-\pi/4-\varphi/2}^{-\varphi}(\ldots),\ \ \mathcal{I}_3 = -\int_{\varphi}^{\pi/4+\varphi/2}(\ldots),$$

$$\mathcal{I}_4 = -\int_{\pi/4+\varphi/2}^{3\pi/4-\varphi/2}(\ldots),$$

and $(\ldots) = \varrho(-\cos(3\varphi/2 - \pi/4 + \psi))\,\sin 2\psi\, d\psi$.

One has

$$\mathcal{I}_1 = \int_0^{\varphi} [\varrho(-\cos(3\varphi/2 - \pi/4 - \psi)) - \varrho(-\cos(3\varphi/2 - \pi/4 + \psi))] \sin 2\psi \, d\psi. \tag{3.78}$$

Using the inequality $\varphi \geq 0.3\pi$ one verifies easily that for $0 < \psi < \varphi$,

$$|3\varphi/2 - \pi/4 - \psi| < 3\varphi/2 - \pi/4 + \psi \leq \pi;$$

therefore,

$$-\cos(3\varphi/2 - \pi/4 - \psi) < -\cos(3\varphi/2 - \pi/4 + \psi).$$

Bearing in mind that ϱ decreases one concludes that the integrand in (3.78) is positive, and therefore $\mathcal{I}_1 > 0$.

Next, for $-\pi/4 - \varphi/2 \leq \psi \leq -\varphi$ one has $\sin 2\psi \leq 0$ and

$$|3\varphi/2 - \pi/4 + \psi| \leq 3\varphi/2 - \pi/4 + \psi + 2\varphi \leq \pi,$$

hence

$$-\cos(3\varphi/2 - \pi/4 + \psi) \leq -\cos(3\varphi/2 - \pi/4 + \psi + 2\varphi),$$

so that

$$\varrho(-\cos(3\varphi/2 - \pi/4 + \psi)) \geq \varrho(-\cos(3\varphi/2 - \pi/4 + \psi + 2\varphi)).$$

Therefore,

$$\mathcal{I}_2 \geq -\int_{-\pi/4-\varphi/2}^{-\varphi} \varrho(-\cos(3\varphi/2 - \pi/4 + \psi + 2\varphi)) \sin 2\psi \, d\psi =$$

$$= \int_{\varphi}^{\pi/4+\varphi/2} \varrho(-\cos(3\varphi/2 - \pi/4 - \chi + 2\varphi)) \sin 2\chi \, d\chi$$

and

$$\mathcal{I}_2 + \mathcal{I}_3 \geq \int_{\varphi}^{\pi/4+\varphi/2} [\varrho\,(-\cos(3\varphi/2 - \pi/4 - \psi + 2\varphi)) - \\ -\varrho\,(-\cos(3\varphi/2 - \pi/4 + \psi))] \sin 2\psi \, d\psi. \tag{3.79}$$

On the other hand one has

$$\mathcal{I}_4 = \int_{\pi/2}^{3\pi/4-\varphi/2} [\varrho\,(-\cos(3\varphi/2 + 3\pi/4 - \psi)) \\ -\varrho\,(-\cos(3\varphi/2 - \pi/4 + \psi))] \sin 2\psi \, d\psi. \tag{3.80}$$

Making the changes of variable $\theta = \psi - \varphi$ in (3.79) and $\theta = \psi - \pi/2$ in (3.80) and summing both sides of these relations one obtains

$$\mathcal{I}_2 + \mathcal{I}_3 + \mathcal{I}_4 \geq \int_0^{\pi/4-\varphi/2} \Psi(\theta)\, d\theta,$$

where

$$\Psi(\theta) = [\varrho\,(-\cos(5\varphi/2-\pi/4-\theta)) - \varrho\,(-\cos(5\varphi/2-\pi/4+\theta))]\, \sin(2\theta+2\varphi)-$$
$$-[\varrho\,(-\cos(3\varphi/2+\pi/4-\theta)) - \varrho\,(-\cos(3\varphi/2+\pi/4+\theta))]\, \sin 2\theta.$$

We claim that $\Psi(\theta) \geq 0$; hence $\mathcal{I}_2 + \mathcal{I}_3 + \mathcal{I}_4 \geq 0$, and the proof of (II) will be complete. One has $0 \leq 2\theta \leq \pi/2 - \varphi,\ \pi/2 - \varphi \leq 2\varphi \leq 2\theta + 2\varphi \leq \pi/2 + \varphi$; therefore,

$$0 \leq \sin 2\theta \leq \sin(2\theta + 2\varphi). \tag{3.81}$$

We set

$$J_1(\theta) = \varrho\,(-\cos(5\varphi/2-\pi/4-\theta)) - \varrho\,(-\cos(5\varphi/2-\pi/4+\theta)),$$
$$J_2(\theta) = \varrho\,(-\cos(3\varphi/2+\pi/4-\theta)) - \varrho\,(-\cos(3\varphi/2+\pi/4+\theta)).$$

One has

$$J_1(\theta) = -\int_{-\theta}^{\theta} \varrho'(-\cos(5\varphi/2-\pi/4+\chi))\, \sin(5\varphi/2-\pi/4+\chi)\, d\chi\,,$$

$$J_2(\theta) = -\int_{-\theta}^{\theta} \varrho'(-\cos(3\varphi/2+\pi/4+\chi))\, \sin(3\varphi/2+\pi/4+\chi)\, d\chi\,.$$

Using the inequality $\varphi \geq 0.3\pi$ one obtains $\pi - 2\varphi \leq 3\varphi - \pi/2 \leq 5\varphi/2 - \pi/4 + \chi \leq 2\varphi$ and $2\varphi \leq 3\varphi/2 + \pi/4 + \chi \leq \varphi + \pi/2 \leq \pi$; hence

$$\sin(5\varphi/2-\pi/4+\chi) \geq \sin 2\varphi \geq \sin(3\varphi/2+\pi/4+\chi) \geq 0 \tag{3.82}$$

and

$$-\cos(5\varphi/2-\pi/4+\chi) \leq -\cos 2\varphi \leq -\cos(3\varphi/2+\pi/4+\chi).$$

Bearing in mind that ϱ' is negative and increasing one obtains

$$\varrho'(-\cos(5\varphi/2-\pi/4+\chi)) \leq \varrho'(-\cos(3\varphi/2+\pi/4+\chi)) \leq 0. \tag{3.83}$$

It follows from (3.82) and (3.83) that $J_1(\theta) \geq J_2(\theta) \geq 0$, and taking (3.81) into account one concludes that $\Psi(\theta) \geq 0$.

(III) One has

$$g'_+(\varphi) = -\frac{d}{d\varphi} \int_{-\pi/4-\varphi/2}^{3\pi/4-\varphi/2} \varrho(-\cos(3\varphi/2-\pi/4+\psi))\, \sin 2\psi\, d\psi =$$

$$= -\int_{-\pi/4-\varphi/2}^{3\pi/4-\varphi/2} \frac{d}{d\varphi}\, \varrho(-\cos(3\varphi/2 - \pi/4 + \psi))\ \sin 2\psi\, d\psi +$$
$$+\frac{1}{2}\ \cos\varphi\, [\varrho(-\sin\varphi) - \varrho(\sin\varphi)].$$

One also has $\cos\varphi > 0$; consequently, $\varrho(-\sin\varphi) - \varrho(\sin\varphi) > 0$, and therefore

$$g'_+(\varphi) > -\int_{-\pi/4-\varphi/2}^{3\pi/4-\varphi/2} \frac{d}{d\varphi}\, \varrho(-\cos(3\varphi/2 - \pi/4 + \psi))$$
$$\sin 2\psi\, d\psi = -(\mathcal{K}_1 + \mathcal{K}_2 + \mathcal{K}_3 + \mathcal{K}_4),$$

where

$$\mathcal{K}_1 = \int_{-\pi/4-\varphi/2}^{\pi/2-3\varphi} (\ldots),\ \ \mathcal{K}_2 = \int_{\pi/2-3\varphi}^{0} (\ldots),$$
$$\mathcal{K}_3 = \int_{0}^{\pi/4+\varphi/2} (\ldots),\ \ \mathcal{K}_4 = \int_{\pi/4+\varphi/2}^{3\pi/4-\varphi/2} (\ldots).$$

To prove (III), it is sufficient to verify that $\mathcal{K}_1 \le 0$, $\mathcal{K}_2 \le 0$, $\mathcal{K}_3 \le 0$, and $\mathcal{K}_4 \le 0$.

One has

$$\mathcal{K}_1 = \frac{3}{2}\int_{-\pi/4-\varphi/2}^{\pi/2-3\varphi} \varrho'(-\cos(3\varphi/2 - \pi/4 + \psi))\ \sin(3\varphi/2 - \pi/4 + \psi)\ \sin 2\psi\, d\psi.$$

Using the inequality $\varphi \le 0.3\pi$ one verifies easily that $-\pi/4 - \varphi/2 \le \pi/2 - 3\varphi$. For $-\pi/4 - \varphi/2 \le \psi \le \pi/2 - 3\varphi$ one has $-\pi/2 \le 3\varphi/2 - \pi/4 + \psi \le 0$, $-\pi \le 2\psi \le 0$; therefore, $\sin(3\varphi/2 - \pi/4 + \psi) \le 0$, $\sin 2\psi \le 0$, and taking into account the inequality $\varrho' \le 0$ one concludes that $\mathcal{K}_1 \le 0$.

Making the change of variable $\chi = 3\varphi/2 - \pi/4 + \psi$ one obtains

$$\mathcal{K}_2 = \frac{3}{2}\int_{\pi/4-3\varphi/2}^{3\varphi/2-\pi/4} \frac{d}{d\chi}\, \varrho(-\cos\chi)\ \cos(2\chi - 3\varphi)\, d\chi =$$
$$= \frac{3}{2}\int_{0}^{3\varphi/2-\pi/4} \frac{d}{d\chi}\, \varrho(-\cos\chi)\, [\cos(2\chi - 3\varphi) - \cos(2\chi + 3\varphi)]\, d\chi.$$

One has $\frac{d}{d\chi}\,\varrho(-\cos\chi) \le 0$, $\cos(2\chi - 3\varphi) - \cos(2\chi + 3\varphi) = 2\sin 2\chi \sin 3\varphi \ge 0$, therefore $\mathcal{K}_2 \le 0$.

Further,

$$\mathcal{K}_3 = \frac{3}{2}\int_{0}^{\pi/4+\varphi/2} \varrho'(-\cos(3\varphi/2 - \pi/4 + \psi))\ \sin(3\varphi/2 - \pi/4 + \psi)\ \sin 2\psi\, d\psi.$$

It is easy to verify that the integrand is negative; hence $\mathcal{K}_3 \le 0$.

Making the change of variable $\theta = \psi - \pi/2$ in the integral $\mathcal{K}_4$ one obtains

$$\mathcal{K}_4 = -\int_{-\pi/4+\varphi/2}^{\pi/4-\varphi/2} \frac{d}{d\varphi}\varrho(-\cos(3\varphi/2+\pi/4+\theta))\sin 2\theta\, d\theta =$$

$$= \frac{3}{2}\int_0^{\pi/4-\varphi/2} [\varrho'(-\cos(3\varphi/2+\pi/4-\theta))\sin(3\varphi/2+\pi/4-\theta) -$$

$$-\varrho'(-\cos(3\varphi/2+\pi/4+\theta))\sin(3\varphi/2+\pi/4+\theta)]\sin 2\theta\, d\theta.$$

Using the inequality $\varphi \ge \pi/6$ one verifies easily that

$$\max\{3\varphi/2+\pi/4-\theta,\ -3\varphi/2+3\pi/4+\theta\} \le 3\varphi/2+\pi/4+\theta \le \pi;$$

hence

$$0 \le \sin(3\varphi/2+\pi/4+\theta) \le \sin(3\varphi/2+\pi/4-\theta),$$
$$\cos(3\varphi/2+\pi/4+\theta) \le \cos(3\varphi/2+\pi/4-\theta),$$

and therefore

$$0 \le -\varrho'(-\cos(3\varphi/2+\pi/4+\theta)) \le -\varrho'(-\cos(3\varphi/2+\pi/4-\theta)).$$

This shows that $\mathcal{K}_4 \le 0$ and completes the proof of statement (a) of Lemma 3.2.

(b) We shall change our notation slightly for the functions defined by the preceding formulae (3.57), (3.73), and (3.74); we shall write

$$\varrho(z,V) = \int_0^\infty r^3\,\sigma(\sqrt{r^2+2rVz+V^2})\,dr,$$

$$p_-(\tan\varphi, V) = -2\int_{-\pi/2}^{\pi/2} \cos^2\theta\,\varrho(\cos(\varphi+\theta),V)\,d\theta,$$

$$g_-(\varphi,V) = \frac{1}{4\cos^3\varphi}\frac{\partial^2}{\partial u^2}\Big|_{u=\tan\varphi} p_-(u,V) = \int_{-\pi/2}^{\pi/2}\varrho(\cos(\varphi+\theta),V) \times\cos(2\theta-\varphi)\,d\theta,$$

explicitly indicating in this way the dependence of the functions on V. In accordance with these formulae, the function $\sigma^{\alpha,\beta}(r) = \sigma(r) + \alpha\,\sigma(\beta r)$ gives rise to the function

$$\varrho^{\alpha,\beta}(z) = \varrho(z,V) + \frac{\alpha}{\beta^4}\varrho(z,\beta V)$$

and the pressure function

$$p_-^{\alpha,\beta}(u) = p_-(u,V) + \frac{\alpha}{\beta^4}p_-(u,\beta V).$$

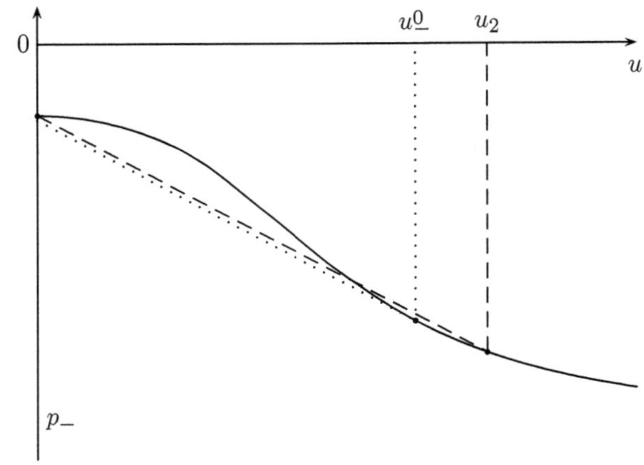

Fig. 3.9 Graph of $p_-(\cdot, V)$ and an auxiliary construction

Assume that the set $\mathcal{O}^0 := \{u : p_-(u, V) > \bar{p}_-(u, V)\}$ is an interval $(0, u_-^0)$; otherwise the result of Lemma 3.2 (b) holds for $\alpha = 0$ and arbitrary $\beta > 0$. This assumption implies that $p_-(u)$ is nondecreasing for $u \geq u_-^0$.

Consider an arbitrary value $u_2 > u_-^0$ and let

$$\Delta_2 = \frac{p_-(u_2, V) - p_-(0, V)}{u_2};$$

one has $p'_-(u_2, V) > \Delta_2(p)$ (Fig. 3.9). Note that the function $g_-(\varphi, V)$ is continuous in $\varphi \in [0, \pi/2)$ and has a finite limit as $\varphi \to \pi/2 - 0$; therefore, the quantity $\mathrm{g} := \sup_{\varphi \in [0, \pi/2)} |g_-(\varphi, V)|$ is finite. We set

$$\omega = \frac{p'_-(u_2, V) - \Delta_2}{2}.$$

It is easy to see that if the functions

$$\check{p}(u) := \frac{\alpha}{\beta^4}\, p_-(u, \beta V), \quad \check{g}(\varphi) := \frac{\alpha}{\beta^4}\, g_-(\varphi, \beta V)$$

satisfy the inequalities

$$|\check{p}'(u)| < \omega \ \text{ for } \ 0 \leq u \leq u_2, \quad \check{p}'(u) > -\omega \ \text{ for } \ u > u_2 \tag{3.84}$$

and

$$\check{g}(\varphi) < -\mathrm{g} \quad \text{for some } \varphi > \arctan u_2, \tag{3.85}$$

then the function $p_-^{\alpha,\beta}(u) = p_-(u) + \check{p}(u)$ possesses the following property: the set $\mathcal{O}^{\alpha,\beta} = \{u : p_-^{\alpha,\beta}(u) > \bar{p}_-^{\alpha,\beta}(u)\}$ has at least two connected components; one component lies in $(0, u_2)$ and the second contains the point $\tan\varphi$.

Let

$$\Delta_2^{\alpha,\beta} = \frac{p_-^{\alpha,\beta}(u_2, V) - p_-^{\alpha,\beta}(0, V)}{u_2}.$$

It follows from (3.84) that the inequality $\frac{\partial}{\partial u} p_-^{\alpha,\beta}(u) > \Delta_2^{\alpha,\beta}$ holds for all $u \geq u_2$. Therefore, the graph of the function $p_-^{\alpha,\beta}(u),\ u \geq u_2$ is situated above the secant line joining the points of the graph $(0, p_-^{\alpha,\beta}(0))$ and $(u_2, p_-^{\alpha,\beta}(u_2))$; this implies that the first connected component $\mathcal{O}^{\alpha,\beta}$ is contained in $(0, u_2)$. Further, it follows from (3.85) that $\frac{\partial}{\partial u^2} p_-^{\alpha,\beta}(u) < 0$ at the indicated point $\tan\varphi$; therefore, this point belongs to the set $\mathcal{O}^{\alpha,\beta}$. Hence this set has at least two connected components.

We set

$$P^\beta(\varphi) = \max\{\sup_{0 \leq u \leq \tan\varphi} |p'_-(u, \beta V)|,\ -\inf_{u > \tan\varphi} p'_-(u, \beta V)\}. \tag{3.86}$$

The remainder of this text (till the end of Sect. 3.5.1) is devoted to the proof of the following result: for each $\varepsilon > 0$ there exist $\phi,\ \varphi$, and β such that

$$\pi/2 - \varepsilon \leq \phi \leq \varphi \leq \pi/2 \quad \text{and} \quad 0 < \frac{P^\beta(\phi)}{-g_-(\varphi, \beta V)} < \varepsilon. \tag{3.87}$$

Then taking $\varepsilon < \min\{\omega/g,\ \pi/2 - \arctan u_-^0\}$ and setting $u_2 = \tan\phi$ one can find α such that (3.84) and (3.85) hold and statement (b) of Lemma 3.2 holds for the α and β in question.

We carry out auxiliary calculations. For $\varphi \in (\pi/10,\ \pi/2),\ \beta > 0$ we define the quantity $R = R(\varphi, \beta)$ by the formula

$$\frac{R}{\beta V} = \sin\left(\frac{5\pi}{8} - \frac{5\varphi}{4}\right). \tag{3.88}$$

Then the straight line drawn from the origin and making the angle $5\pi/8 - 5\varphi/4$ with the vertical line touches the circle of radius R with center at $(0, -\beta V)$ (Fig. 3.10). Next we draw two rays from the origin downward at the angles $\pi/2 - \varphi$ and $\frac{5}{4}(\pi/2 - \varphi)$ with respect to the vertical line, as shown in Fig. 3.10. As we already noticed, the second ray touches the circle. The first ray intersects the circle, and after some algebra one concludes that it cuts an arc of angular size at least $\pi/5$.

Let

$$\mathcal{U} := \left\{ v = (v_1, v_2) : v_1 > 0,\ -\tan\left(\frac{5\pi}{8} - \frac{5\varphi}{4}\right) \leq \frac{v_1}{v_2} \leq -\tan\left(\frac{\pi}{2} - \varphi\right) \right\}$$

be the angle formed by the rays indicated previously. We choose constants $\alpha > 1$, $c > 0$ such that there exists a circular sector Σ of angle c, outer radius R, and inner radius R/α with center at $(0, -\beta V)$ lying entirely in $\mathcal{U}$ (Fig. 3.10). One can choose, for instance, arbitrary $c \in (0,\ \pi/5)$ and $\alpha = \frac{\cos(c/2)}{\cos(\pi/10)}$.

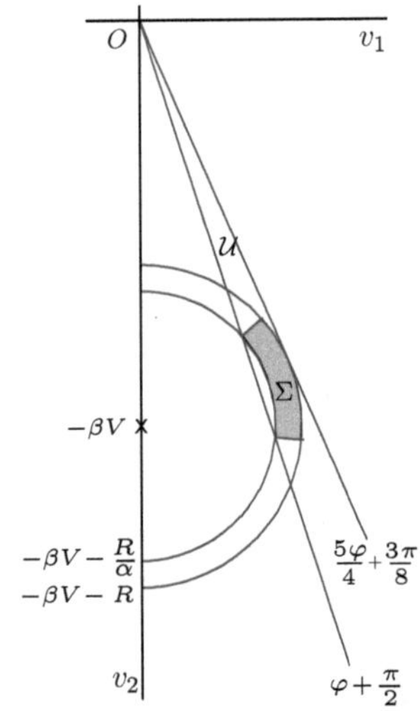

Fig. 3.10 An auxiliary figure. The angles formed by the rays with the vertical axis Ov_2 are indicated near these rays

Next, using the definition of ϱ and proceeding to the variables $v_1 = r\sin\chi$, $v_2 = r\cos\chi$ one obtains

$$\int_{\frac{5\varphi}{4}+\frac{3\pi}{8}}^{\varphi+\frac{\pi}{2}} \varrho(\cos\chi, \beta V)\, d\chi = \iint_{\mathcal{U}} (v_1^2 + v_2^2)\, \sigma(|v + \beta V e_2|)\, dv_1 dv_2. \tag{3.89}$$

Since for $v \in \Sigma$ one has the relation $|v| \geq \beta V - R$, and bearing in mind that $\beta V = R/\sin(\frac{5\pi}{8} - \frac{5\varphi}{4})$ and $\Sigma \subset \mathcal{U}$, we obtain after going over to polar coordinates that

$$\begin{aligned}
&\iint_{\mathcal{U}} (v_1^2 + v_2^2)\, \sigma(|v + \beta V e_2|)\, dv_1 dv_2 \geq \\
&\geq R^2 \left(\frac{1}{\sin^2(\frac{5\pi}{8} - \frac{5\varphi}{4})} - 1 \right) \iint_{\Sigma} \sigma(|v + \beta V e_2|)\, dv_1 dv_2 = \\
&= R^2 \left(\frac{1}{\sin^2(\frac{5\pi}{8} - \frac{5\varphi}{4})} - 1 \right) \cdot c \int_{R/\alpha}^{R} \sigma(r)\, r\, dr \geq \\
&\geq \left(\frac{1}{\sin^2(\frac{5\pi}{8} - \frac{5\varphi}{4})} - 1 \right) \cdot c \int_{R/\alpha}^{R} r^3\, \sigma(r)\, dr.
\end{aligned} \tag{3.90}$$

From (3.89) and (3.90) we conclude that

$$\int_{\frac{5\varphi}{4}+\frac{3\pi}{8}}^{\varphi+\frac{\pi}{2}} \varrho(\cos\chi, \beta V)\, d\chi \geq \left(\frac{1}{\sin^2(\frac{5\pi}{8} - \frac{5\varphi}{4})} - 1 \right) \cdot c \int_{R/\alpha}^{R} r^3\, \sigma(r)\, dr. \tag{3.91}$$

Furthermore,

$$\int_{\varphi-\frac{\pi}{2}}^{\frac{3\varphi}{2}+\frac{\pi}{4}} \varrho(\cos\chi,\beta V)\,d\chi \le \int_{-\frac{5\varphi}{4}-\frac{3\pi}{8}}^{\frac{5\varphi}{4}+\frac{3\pi}{8}} \varrho(\cos\chi,\beta V)\,d\chi \le$$
$$\le \iint_{\mathcal{V}} (v_1^2+v_2^2)\,\sigma(|v+\beta V e_2|)\,dv_1 dv_2, \tag{3.92}$$

where $\mathcal{V} = \{v = (v_1, v_2) : |v+\beta V e_2| \ge R\}$ is the complement of the disc of radius R with center at $(0, -\beta V)$. Bearing in mind that

$$\beta V \le \frac{|v+\beta V e_2|}{\sin(5\pi/8 - 5\varphi/4)}$$

for $v \in \mathcal{V}$ one obtains

$$|v| \le \beta V + |v+\beta V e_2| \le |v+\beta V e_2| \left(1 + 1/\sin\left(\frac{5\pi}{8} - \frac{5\varphi}{4}\right)\right);$$

hence

$$\iint_{\mathcal{V}} (v_1^2+v_2^2)\,\sigma(|v+\beta V e_2|)\,dv_1 dv_2 \le$$
$$\le \left(\frac{1}{\sin(\frac{5\pi}{8}-\frac{5\varphi}{4})}+1\right)^2 \iint_{\mathcal{V}} |v+\beta V e_2|^2\,\sigma(|v+\beta V e_2|)\,dv_1 dv_2 =$$
$$= \left(\frac{1}{\sin(\frac{5\pi}{8}-\frac{5\varphi}{4})}+1\right)^2 \cdot 2\pi \int_R^\infty r^2\sigma(r)\,r dr. \tag{3.93}$$

It follows from (3.92) and (3.93) that

$$\int_{\varphi-\frac{\pi}{2}}^{\frac{3\varphi}{2}+\frac{\pi}{4}} \varrho(\cos\chi,\beta V)\,d\chi \le \left(\frac{1}{\sin(\frac{5\pi}{8}-\frac{5\varphi}{4})}+1\right)^2 \cdot 2\pi \int_R^\infty r^3\sigma(r)\,dr. \tag{3.94}$$

Now, by hypothesis (b) of Lemma 3.2, the function $\gamma(r) := r^{n+3}\sigma(r)$ decreases for $n > 0$ and sufficiently large r; therefore, for sufficiently large R one has

$$\int_R^\infty r^{-n}\gamma(r)\,dr \le \gamma(R)\,\frac{R^{-n+1}}{n-1}, \quad \int_{R/\alpha}^R r^{-n}\gamma(r)\,dr \ge \gamma(R)(\alpha^{n-1}-1)\,\frac{R^{-n+1}}{n-1},$$

so that

$$\int_R^\infty r^3\sigma(r)\,dr \le \frac{1}{\alpha^{n-1}-1}\int_{R/\alpha}^R r^3\sigma(r)\,dr.$$

Here n is arbitrary, and one concludes that

$$\lim_{R\to+\infty} \frac{\int_R^\infty r^3\sigma(r)\,dr}{\int_{R/\alpha}^R r^3\sigma(r)\,dr} = 0. \tag{3.95}$$

It follows from (3.95), (3.91), (3.94), and (3.88) that for each $\varphi \in (\pi/10,\ \pi/2)$,

$$\lim_{\beta\to+\infty} \frac{\int_{\varphi-\frac{\pi}{2}}^{\frac{3\varphi}{2}+\frac{\pi}{4}} \varrho(\cos\chi, \beta V)\,d\chi}{\int_{\frac{5\varphi}{4}+\frac{3\pi}{8}}^{\varphi+\frac{\pi}{2}} \varrho(\cos\chi, \beta V)\,d\chi} = 0 \tag{3.96}$$

and

$$\int_{\frac{5\varphi}{4}+\frac{3\pi}{8}}^{\varphi+\frac{\pi}{2}} \varrho(\cos\chi, \beta V)\,d\chi \ge c_1(\varphi) \int_{R/\alpha}^R r^3\,\sigma(r)\,dr, \tag{3.97}$$

where $c_1(\varphi) = c\,(\sin^{-2}(\frac{5\pi}{8} - \frac{5\varphi}{4}) - 1)$.

Setting "$-$" for $\pm$ and making the change of variable $\chi = \psi + \theta$ for $\psi \in (0,\ \pi/2)$ in (3.74) one obtains

$$g_-(\psi, \beta V) = \int_{\psi-\pi/2}^{\psi+\pi/2} \varrho(\cos\chi, \beta V)\,\cos(2\chi - 3\psi)\,d\chi. \tag{3.98}$$

Let $\mathcal{U}_1 := \{v = (v_1, v_2) : v_2/|v_1| \ge -\tan\psi\}$; let $\mathcal{V}_1 := \{v = (v_1, v_2) : |v + \beta V e_2| \ge \beta V\cos\psi\}$ be the complement of a disc of radius $\beta V\cos\psi$ with center at $(0, -\beta V)$ (Fig. 3.11).

Since $\mathcal{U}_1 \subset \mathcal{V}_1$ and since $\beta V \le |v + \beta V e_2|/\cos\psi$ and $|v| \le |v + \beta V e_2| + \beta V \le |v + \beta V e_2|\,(1 + 1/\cos\psi)$ for $v \in \mathcal{V}_1$, it follows that

$$\begin{aligned}
|g_-(\psi, \beta V)| &\le \int_{-\psi-\pi/2}^{\psi+\pi/2} \varrho(\cos\chi, \beta V)\,d\chi = \iint_{\mathcal{U}_1} (v_1^2 + v_2^2)\,\sigma(|v + \beta V e_2|)\,dv_1 dv_2 \le \\
&\le \left(1 + \frac{1}{\cos\psi}\right)^2 \iint_{\mathcal{V}_1} |v + \beta V e_2|^2\,\sigma(|v + \beta V e_2|)\,dv_1 dv_2 = \\
&= \left(1 + \frac{1}{\cos\psi}\right)^2 \cdot 2\pi \int_{\beta V\cos\psi}^\infty r^2\,\sigma(r)\,r dr.
\end{aligned} \tag{3.99}$$

Taking account of the inequalities

$$\cos(2\chi - 3\psi) \le 0, \qquad \chi \in \left[\frac{3\psi}{2} + \frac{\pi}{4},\ \frac{5\psi}{4} + \frac{3\pi}{8}\right],$$

$$\cos(2\chi - 3\psi) \le \cos\left(\frac{3\pi}{4} - \frac{\psi}{2}\right) = -\cos\left(\frac{\pi}{4} + \frac{\psi}{2}\right) < 0,$$

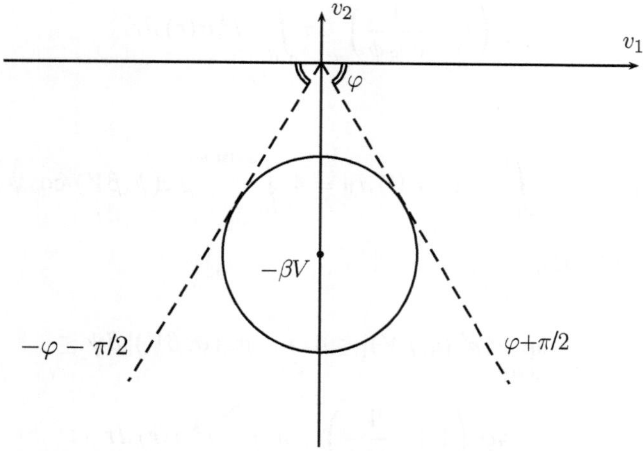

Fig. 3.11 An auxiliary construction. The rays in the figure make angles $\varphi + \pi/2$ and $-\varphi - \pi/2$ with the vertical axis Ov_2. The domain $\mathcal{V}_1$ is the complement of the *circle*, and the domain $\mathcal{U}_1$ is bounded below by two rays

$$\chi \in \left[\frac{5\psi}{4} + \frac{3\pi}{8},\ \psi + \frac{\pi}{2}\right],$$

and using (3.98) one obtains

$$g_-(\psi, \beta V) \le \int_{\psi-\frac{\pi}{2}}^{\frac{3\psi}{2}+\frac{\pi}{4}} \varrho(\cos\chi, \beta V)\, d\chi - \cos\left(\frac{\pi}{4} + \frac{\psi}{2}\right) \cdot \int_{\frac{5\psi}{4}+\frac{3\pi}{8}}^{\psi+\pi/2} \varrho(\cos\chi, \beta V)\, d\chi. \tag{3.100}$$

It follows from (3.96), (3.97), and (3.100) that for each φ and sufficiently large β,

$$g_-(\varphi, \beta V) \le -c_2(\varphi) \int_{R/\alpha}^{R} r^3\, \sigma(r)\, dr, \tag{3.101}$$

where $R = R(\varphi, \beta)$ and $c_2(\varphi) = \frac{1}{2} \cos\left(\frac{\pi}{4} + \frac{\varphi}{2}\right) \cdot c_1(\varphi)$.

Next, consider $\phi = \phi(\varphi) := \frac{5\varphi}{4} - \frac{\pi}{8}$; one has

$$\beta V \cos\phi = \beta V \sin\left(\frac{5\varphi}{4} + \frac{3\pi}{8}\right) = R;$$

hence by (3.99), for each $\psi \in [0, \phi]$,

$$|g_-(\psi, \beta V)| \le \left(1 + \frac{1}{\cos\psi}\right)^2 2\pi \int_{\beta V \cos\psi}^{\infty} r^3\, \sigma(r)\, dr \le$$

$$\leq \left(1 + \frac{1}{\cos\phi}\right)^2 2\pi \int_R^\infty r^3\, \sigma(r)\, dr. \tag{3.102}$$

Further,

$$p'_-(u, \beta V) = \int_0^u p''_-(v, \beta V)\, dv = 4 \int_0^{\arctan u} g_-(\psi, \beta V) \cos\psi\, d\psi;$$

therefore, by (3.102),

$$\sup_{0\leq u\leq \tan\phi} |p'_-(u, \beta V)| \leq 4 \int_0^\phi |g_-(\psi, \beta V)|\, d\psi \leq$$

$$\leq 4\phi \left(1 + \frac{1}{\cos\phi}\right)^2 2\pi \int_R^\infty r^3\, \sigma(r)\, dr \tag{3.103}$$

and

$$-\inf_{u>\tan\phi} p'_-(u, \beta V) = -\int_0^{\tan\phi} p''_-(v, \beta V)\, dv - \inf_{u>\tan\phi} \int_{\tan\phi}^u p''_-(v, \beta V)\, dv \leq$$

$$\leq 4\phi \left(1 + \frac{1}{\cos\phi}\right)^2 2\pi \int_R^\infty r^3\, \sigma(r)\, dr + 4(\pi/2 - \phi) \cdot \sup_{\psi\in[\phi,\pi/2]} (-g_-(\psi, \beta V)). \tag{3.104}$$

From (3.86), (3.103), (3.104), and (3.101) one sees that for sufficiently large β,

$$0 < \frac{P^\beta(\phi)}{\sup_{\psi\in[\phi,\pi/2]}(-g_-(\psi, \beta V))} \leq c_3(\varphi)\, \frac{\int_R^\infty r^3\, \sigma(r)\, dr}{\int_{R/\alpha}^R r^3\, \sigma(r)\, dr} + 4(\pi/2 - \phi), \tag{3.105}$$

where $R = R(\varphi, \beta)$ and $c_3(\varphi) := \frac{8\pi}{c_2(\varphi)}\, \phi \left(1 + \frac{1}{\cos\phi}\right)^2$. For arbitrary $\varepsilon > 0$ one chooses φ such that $0 < 2(\pi/2 - \phi(\varphi)) < \varepsilon/4$, and then taking account of (3.95) one chooses β such that the first term on the right-hand side of (3.105) is less than $\varepsilon/4$. Finally, choose $\psi \in [\phi, \pi/2]$ such that $-g_-(\psi, \beta V) > \frac{1}{2} \sup_{\psi\in[\phi,\pi/2]}(-g_-(\psi, \beta V))$. Then we have

$$0 < \frac{P^\beta(\phi)}{-g_-(\psi, \beta V)} < 2\, \frac{P^\beta(\phi)}{\sup_{\psi\in[\phi,\pi/2]}(-g_-(\psi, \beta V))} \leq \varepsilon.$$

The proof of relation (3.87) is complete.

3.5.3 Proof of Formula (3.41)

The case $d = 2$. Here formula (3.11) takes the following form:

$$p_\pm(u,V) = \pm 2 \iint \frac{(v_1 u \pm v_2)_-^2}{1+u^2}\, \sigma(|v + Ve_2|)\, dv_1 dv_2.$$

Passing to the polar coordinates $v = (-r\sin\varphi, \mp r\cos\varphi)$ and taking account of the relation $\sigma(|v+Ve_2|) = \sigma(r) \mp V\cos\varphi\,\sigma'(r) + o(V)$ as $V \to 0^+$ one obtains

$$\begin{aligned} p_\pm(u,V) &= \pm 2 \int_0^{2\pi}\int_0^\infty \frac{r^2(\sin\varphi\, u + \cos\varphi)_+^2}{1+u^2}\,(\sigma(r) \mp V\cos\varphi\,\sigma'(r) \\ &\quad + o(V))\, r\, dr\, d\varphi = \pm 2 \int_0^\infty \sigma(r)\, r^3\, dr \cdot I^{(2)} - 2V \\ &\quad \times \int_0^\infty \sigma'(r)\, r^3\, dr \cdot J^{(2)} + o(V), \end{aligned} \tag{3.106}$$

where

$$I^{(2)} = \int_0^{2\pi} (\cos(\varphi - \varphi_0))_+^2\, d\varphi, \qquad J^{(2)} = \int_0^{2\pi} (\cos(\varphi-\varphi_0))_+^2 \cos\varphi\, d\varphi,$$

$z_+ := \max\{z, 0\}$, $\varphi_0 := \arccos\frac{1}{\sqrt{1+u^2}}$. Making the change of variable $\psi = \varphi - \varphi_0$ in these integrals one obtains

$$I^{(2)} = \int_{-\pi/2}^{\pi/2} \cos^2\psi\, d\psi = \pi/2, \qquad J^{(2)} = \cos\varphi_0 \int_{-\pi/2}^{\pi/2} \cos^3\psi\, d\psi = \frac{4}{3\sqrt{1+u^2}}.$$

Substituting the values thus obtained in (3.106) and using the equality

$$-\int_0^\infty \sigma'(r)\, r^3\, dr = 3\int_0^\infty \sigma(r)\, r^2\, dr,$$

one arrives at formula (3.41) with coefficients (3.42).

The case $d = 3$. Formula (3.11) now takes the following form:

$$p_\pm(u,V) = \pm 2 \iiint \frac{(v_1 u \pm v_3)_-^2}{1+u^2}\, \rho(v)\, dv_1 dv_2 dv_3.$$

Passing to the spherical coordinates $v = (-r\sin\varphi\cos\theta, -r\sin\varphi\sin\theta, \mp r\cos\varphi)$ one obtains

$$p_{\pm}(u,V) = \pm 2 \int_0^{\pi}\int_0^{2\pi}\int_0^{\infty} \frac{r^2(\sin\varphi\cos\theta\, u + \cos\varphi)_+^2}{1+u^2}\,(\sigma(r) \mp V\cos\varphi\,\sigma'(r) +$$
$$+ o(V))\, r^2\, dr\, d\theta\, \sin\varphi d\varphi = \pm 2\int_0^{\infty} \sigma(r)\, r^4\, dr \cdot I^{(3)} - 2V$$
$$\int_0^{\infty} \sigma'(r)\, r^4\, dr \cdot J^{(3)} + o(V),$$

where

$$I^{(3)} = \int_0^{2\pi}\int_0^{\pi} (\cos(\varphi - \phi_0(\theta)))_+^2\, \frac{1+u^2\cos^2\theta}{1+u^2}\, \sin\varphi d\varphi\, d\theta,$$
$$J^{(3)} = \int_0^{2\pi}\int_0^{\pi} (\cos(\varphi - \phi_0(\theta)))_+^2\, \frac{1+u^2\cos^2\theta}{1+u^2}\, \cos\varphi\,\sin\varphi d\varphi\, d\theta,$$

$\phi_0(\theta) = \arccos\frac{1}{\sqrt{1+u^2\cos^2\theta}}$. Making the change of variable $\psi = \varphi - \phi_0(\theta)$ one obtains

$$I^{(3)} = \int_0^{2\pi} d\theta\, \frac{1+u^2\cos^2\theta}{1+u^2}\, \sin\phi_0(\theta) \int_{-\phi_0(\theta)}^{\pi/2} \cos^3\psi\, d\psi =$$
$$= \int_0^{2\pi} \frac{1+u^2\cos^2\theta}{1+u^2}\, \frac{(1+\sin\phi_0(\theta))^2}{3}\, d\theta =$$
$$= \int_0^{2\pi} \frac{(\sqrt{1+u^2\cos^2\theta} + u\cos\theta)^2}{3(1+u^2)}\, d\theta = \frac{2\pi}{3},$$
$$J^{(3)} = \int_0^{2\pi} d\theta\, \frac{1+u^2\cos^2\theta}{1+u^2} \int_{-\phi_0(\theta)}^{\pi/2} \cos^2\psi\, \frac{\sin(2\psi + 2\phi_0(\theta))}{2}\, d\psi =$$
$$= \int_0^{2\pi} d\theta\, \frac{1+u^2\cos^2\theta}{2(1+u^2)} \left[\cos 2\phi_0(\theta) \int_{-\phi_0(\theta)}^{\pi/2} \cos^2\psi\, \sin 2\psi\, d\psi + \right.$$
$$\left. + \sin 2\phi_0(\theta) \int_{-\phi_0(\theta)}^{\pi/2} \cos^2\psi\, \cos 2\psi\, d\psi \right] = \int_0^{2\pi} \frac{1 + u\,\phi_0(\theta)\cos\theta}{4(1+u^2)}\, d\theta =$$
$$= \frac{1}{4(1+u^2)} \int_0^{2\pi} [1 + (u\cos\theta)\, \arctan(u\cos\theta)]\, d\theta = \frac{\pi}{2\sqrt{1+u^2}}.$$

Substituting these values into (3.5.3) and using the equality

$$-\int_0^{\infty} \sigma'(r)\, r^4\, dr = 4\int_0^{\infty} \sigma(r)\, r^3\, dr$$

one obtains formula (3.41) with coefficients (3.43).

Chapter 4
Scattering in Billiards

In Chap. 2, we constructed bodies with arbitrarily small aerodynamic resistance. We used a simplified physical model by imposing several fairly restrictive assumptions. In particular, the particles have no thermal motion, that is, the medium has temperature zero, and the body is not allowed to oscillate or rotate. Of course, one can pose new questions about our solutions: are they sensitive to the thermal motion of the particles or to rotational motion of the body? Is it possible to construct bodies with arbitrarily small resistance if the temperature is positive or the body rotates? If not, then what is the least resistance the body can display? We can also ask about *maximizing* the resistance. In Newton's original setting, this is a trivial problem: one can produce a body with upper front orthogonal to the direction of motion, but when there is rotation, the problem is not so simple.

We now proceed to study these and similar questions. In this chapter we develop machinery that will be used to solve concrete problems. In Sect. 4.1 we confine ourselves to the two-dimensional case. We define a *hollow*, the *law of billiard scattering* in a hollow and on a two-dimensional body, and characterize scattering laws in hollows (Sect. 4.1.1) and on bodies (Sect. 4.1.5). A scattering law here is a measure describing the joint distribution of a pair (*angle of entry, angle of exit*).

The study of scattering by rough bodies occupies an important place in this chapter (and in the book as a whole). In Sect. 4.2 we define a *rough body* in a space of arbitrary dimension and a scattering law on a rough body. In several dimensions a scattering law describes a joint distribution of a triple of vectors: the initial and final velocities of a particle and the outward normal to a body at the point of reflection. We characterize such laws and believe that these results are also of independent interest for billiard theory.

The natural-science literature on rough surfaces is vast, and obviously this topic is intimately connected with diverse practical problems. Much work has been done on heat transport through a rough interface of two media (see, e.g., [13]), wetting of rough surfaces [15], and scattering of electromagnetic and acoustic waves on rough surfaces [41]. The contact of two rough surfaces is studied in the framework of contact mechanics [42].

A. Plakhov, *Exterior Billiards: Systems with Impacts Outside Bounded Domains*, DOI 10.1007/978-1-4614-4481-7_4,

Various mathematical models have been proposed to describe rough surfaces: from relatively simple ones, where the surface is represented by a periodic function, to quite complicated ones utilizing functions with a fractal structure. Models of a surface given by a random function with a prescribed function of joint distribution are widely used in the physical literature. The nature of papers also varies considerably: from papers on the mathematical level of rigor to those written on a physical or engineering level.

The models presented in the literature aim at a description of the essential properties of *real* surfaces. In contrast, this chapter considers all *geometrically* possible kinds of roughness and ignores the discrete (molecular) nature of bodies.

First, we assume that the asperity on a body surface is quite small; our body is obtained from a convex body by making hollows, grooves, and cracks of small size. Second, we consider the *billiard*, that is, perfectly elastic scattering of particles by the surface. We state that a unique way to find the difference between two rough surfaces consists in observation of the difference in scattering on these surfaces. Thus we identify two surfaces with identical scattering laws on them.

This approach suggests the following definition. We say that a sequence of sets contained in a given convex body C represents a rough body if, first, the volumes of these sets converge to the volume of C and, second, the sequence of scattering laws on these sets weakly converges. We state that two sequences of bodies define the same rough body if the limits of the corresponding scattering laws coincide. This limit is called the *law of billiard scattering by a rough body.*

Recall that $B \subset \mathbb{R}^d$ is a body, that is, a bounded set with a piecewise smooth boundary, and C is a bounded convex body containing B. We introduce some notation used below. The map $T_{B,C} = (\xi^+_{B,C}, v^+_{B,C})$ (Sect. 1.1) induces measures $\nu_{B,C}$ and $\nu'_{B,C}$ on $S^{d-1} \times S^{d-1} \times S^{d-1} = \left(S^{d-1}\right)^3$ as follows. Let A be a Borel subset of $\left(S^{d-1}\right)^3$. Then by definition

$$\nu_{B,C}(A) := \mu\left(\{(\xi, v) \in (\partial C \times S^{d-1})_- : (v, v^+_{B,C}(\xi, v), n(\xi)) \in A\}\right),$$

$$\nu'_{B,C}(A) := \mu\left(\{(\xi, v) \in (\partial C \times S^{d-1})_- : (v, v^+_{B,C}(\xi, v), n(\xi^+_{B,C}(\xi, v))) \in A\}\right).$$

The measures $\nu_{B,C}$ and $\nu'_{B,C}$ contain information about the scattering of billiard particles on the body B, provided that the observer can register data about a particle as it intersects, when entering or leaving, the boundary of the ambient convex body C. The measure $\nu_{B,C}$ ($\nu'_{B,C}$) describes the joint distribution of three quantities: the initial and final velocities of the body and the outward normal to ∂C at the moment when the (incoming or outgoing) particle intersects the boundary.

Let τ_C be the surface measure of C.[1] Let $\pi_{v,n}$ and $\pi_{v^+,n} : \left(S^{d-1}\right)^3 \to \left(S^{d-1}\right)^2$ be the natural projections

[1]This measure is defined as follows. Let $|\cdot|$ denote the $(d-1)$-dimensional Hausdorff measure on ∂C; then by definition for each Borel set $A \subset S^{d-1}$ we have $\tau_C(A) = |\{\xi : n(\xi) \in A\}|$.

$$\pi_{v,n}(v, v^+, n) = (v, n), \quad \pi_{v^+,n}(v, v^+, n) = (v^+, n),$$

and let $\pi_{ad} : (S^{d-1})^3 \to (S^{d-1})^3$ be the map interchanging v and $-v^+$, that is,

$$\pi_{ad}(v, v^+, n) = (-v^+, -v, n).$$

Let $n \in S^{d-1}$. We define a probability measure λ_n on S^{d-1} by

$$d\lambda_n(v) = b_d\,(v \cdot n)_+\,dv, \tag{4.1}$$

where dv is the $(d-1)$-dimensional Lebesgue measure on S^{d-1} and b_d is the normalization coefficient defined in Sect. 1.1. In particular, $b_2 = 1/2$ and $b_3 = 1/\pi$.

We introduce further notation: $\hat{\tau}_C^+$ and $\hat{\tau}_C^-$ are the measures on $(S^{d-1})^2$ such that for each continuous function f on $(S^{d-1})^2$ holds

$$\int_{(S^{d-1})^2} f(v, n)\,d\hat{\tau}_C^{\pm}(v, n) = \int_{S^{d-1}} \int_{S^{d-1}} f(v, n)\,b_d\,(v \cdot n)_{\pm}\,dv\,d\tau_C(n). \tag{4.2}$$

Informally speaking, $\hat{\tau}_C^+$ measures the quantity of particles impinging on C with velocity of incidence v and with outward normal at the point of reflection n. Further, $\hat{\tau}_C^-$ measures the quantity of particles reflected from C with reflection velocity v and outward normal at reflection point n.

In what follows (starting with Sect. 4.1.5), we require the following definition.

Definition 4.1. We denote by Γ_C the set of measures ν on $(S^{d-1})^3$ such that

(Γ1) $\pi_{v,n}^{\#}\nu = \hat{\tau}_C^-$ and $\pi_{v^+,n}^{\#}\nu = \hat{\tau}_C^+$;
(Γ2) $\pi_{ad}^{\#}\nu = \nu$.

In other words, Γ_C is the set of measures with prescribed projections on the subspaces $\{v, n\}$ and $\{v^+, n\}$ and invariant under an interchange of the components v and $-v^+$. It is clear that each measure $\nu \in \Gamma_C$ projects onto $\{n\}$ as τ_C.

We also define natural projections $\pi_n : (v, v^+, n) \mapsto n$, $\pi_v : (v, v^+) \mapsto v$, $\pi_{v^+} : (v, v^+) \mapsto v^+$. Slightly abusing the language, we shall also use the notation π_{ad} for a map of $(S^{d-1})^2$ onto itself given by

$$\pi_{ad}(v, v^+) = (-v^+, -v). \tag{4.3}$$

For each measure ν on $(S^{d-1})^3$ we consider $\tau = \pi_n^{\#}\nu$ and define the family of conditional measures $\nu\rfloor_n$, $n \in S^{d-1}$ by the equation

$$\iiint_{(S^{d-1})^3} f(v, v^+, n)\,d\nu(v, v^+, n) = \int_{S^{d-1}} d\tau \iint_{(S^{d-1})^2} f(v, v^+, n)\,d\nu\rfloor_n(v, v^+) \tag{4.4}$$

for each continuous function f on $(S^{d-1})^3$. Equation (4.4) determines the family $\nu\rfloor_n$ uniquely for almost all (in the sense of the measure τ) values of n.

The following proposition (its proof is easy and left to the reader) can serve as an equivalent definition of the family Γ_C.

Proposition 4.1. *Consider a measure* ν *on* $(S^{d-1})^3$. *Then* $\nu \in \Gamma_C$ *if and only if* $\pi_n^{\#}\nu = \tau_C$, *and for almost all (in the sense of* τ_C*) values of* n

(a) $\pi_v^{\#}\nu\rfloor_n = \lambda_{-n}$, $\pi_{v+}^{\#}\nu\rfloor_n = \lambda_n$,
(b) $\pi_{\text{ad}}^{\#}\nu\rfloor_n = \nu\rfloor_n$.

The results of this chapter were first presented in [45, 50–52].

4.1 Scattering in the Two-Dimensional Case

Consider a body $B \subset \mathbb{R}^2$ bounded by a closed non-self-intersecting piecewise smooth curve ∂B. The set $\partial(\text{Conv}B) \setminus \partial B$ is the union of a finite or countable (perhaps empty) system of disjoint open intervals

$$\partial(\text{Conv}B) \setminus \partial B = I_1 \cup I_2 \cup \ldots.$$

The set $\text{Conv}B \setminus B$ falls into several connected components; let Ω_i be the component with a boundary containing I_i, $i \geq 1$. The sets Ω_i are disjoint. We let $I_0 := \partial(\text{Conv}B) \cap \partial B$ be the "convex part" of the boundary ∂B; thus, $\partial(\text{Conv}B) = I_0 \cup I_1 \cup I_2 \cup \ldots$. In Fig. 4.1 we show a body B, the three corresponding intervals $I_1,\ I_2,\ I_3$, the three sets $\Omega_1,\ \Omega_2,\ \Omega_3$, and the convex part I_0 of the boundary of the body.

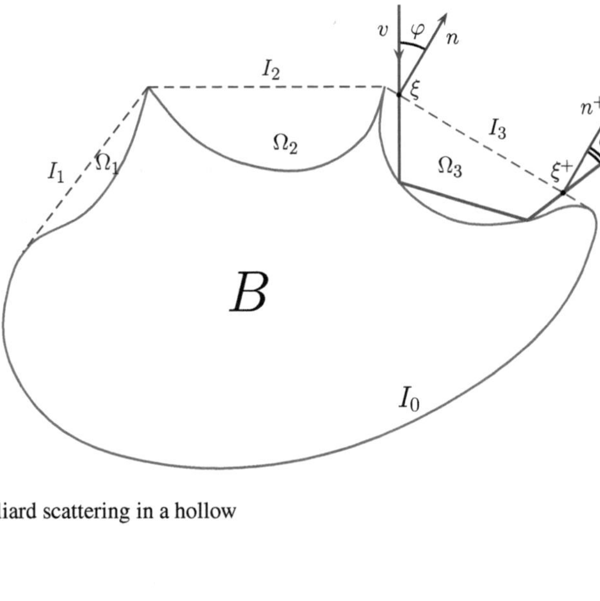

Fig. 4.1 Billiard scattering in a hollow

For brevity we set $\xi^+ := \xi^+_{B,\mathrm{Conv}B}(\xi, v)$, $v^+ := v^+_{B,\mathrm{Conv}B}(\xi, v)$, $n := n(\xi)$, and $n^+ := n(\xi^+_{B,\mathrm{Conv}B}(\xi, v))$. If $\xi \in I_0$, then $\xi^+ = \xi$, and we obtain v^+ from v by reflection from I_0, that is, $v^+ = v - 2(v \cdot n)\, n$. On the other hand, if $\xi \in I_i$, $i \geq 1$, then the corresponding billiard particle goes into the ith hollow Ω_i, reflects several times from $\partial\Omega_i \setminus I_i$, crosses I_i for a second time at $\xi^+ \in I_i$ moving "outwards," and then moves freely. In Fig. 4.1 we plot the trajectory of the particle in the third hollow $i = 3$. In both cases $n(\xi) = n(\xi^+)$, and hence in two dimensions we have $\nu_{B,\mathrm{Conv}B} = \nu'_{B,\mathrm{Conv}B}$ for connected B. (In the case of higher dimensions or when B is disconnected, both of the preceding equalities fail in general.)

Definition 4.2. The measure $\nu_B := \nu_{B,\mathrm{Conv}B}$ is called the *law of billiard scattering* on a body B. It is defined on $(S^1)^3 = \mathbb{T}^3$.

The measure ν_B plays an important role in what follows.

4.1.1 Measures Associated with Hollows

Definition 4.3. A pair of sets of the form (Ω_i, I_i) is called a *hollow*. In other words, a hollow is a pair of sets (Ω, I) such that $I \subset \partial\Omega$ and the sets Ω and I are connected components of $\mathrm{Conv}B \setminus B$ and $\partial(\mathrm{Conv}B) \setminus \partial B$, respectively, for some connected body B. The interval I is called the *inlet of a hollow*.

Remark 4.1. We can provide another definition of a hollow that is more convenient in applications. A hollow is a pair of sets (Ω, I) such that:

(a) Ω is bounded by a closed non-self-intersecting piecewise smooth curve $\partial\Omega$;
(b) I is a straight line interval, and $I \subset \partial\Omega$;
(c) Ω lies to one side of the line containing I; moreover, $\Omega \setminus I$ lies in an open half-plane bounded by this line.

We agree to measure the angle between the normal n (or $-n$) and other vectors v from $\pm n$ to v. The angle is positive if we measure it counterclockwise and negative if measure clockwise. Let $\Omega = \Omega_i$, $I = I_i$ be a hollow, and let $n = n(\xi)$ be the outward normal to $\partial(\mathrm{Conv}B)$ at a point $\xi \in I$. Assume that when entering the hollow a particle intersects I at some point ξ, and its velocity v at the moment of intersection makes an angle φ with $-n$, $-\pi/2 < \varphi < \pi/2$. The particle reflects from $\Omega \setminus I$ at several points and then again intersects I at a point ξ^+ on its way out. The velocity v^+ at the moment of this second intersection makes an angle φ^+ with the normal n. In Fig. 4.1, $\varphi > 0$ and $\varphi^+ < 0$.

This description determines the map $T_{\Omega,I} : (\xi, \varphi) \mapsto (\xi^+, \varphi^+)$ with the components $\xi^+ = \xi^+_{\Omega,I}(\xi, \varphi)$ and $\varphi^+ = \varphi^+_{\Omega,I}(\xi, \varphi)$. It is a one-to-one correspondence of a full-measure subset of $I \times [-\pi/2, \pi/2]$ onto itself; further, it is involutive and preserves the probability measure $\tilde{\mu}_I$ defined by

$$d\tilde{\mu}_I(\xi, \varphi) = \frac{1}{2|I|} \cos\varphi \, d\xi \, d\varphi. \tag{4.5}$$

The map $\mathcal{T}_{\Omega,I} : (\xi,\varphi) \mapsto (\varphi,\ \varphi^+_{\Omega,I}(\xi,\varphi))$ induces the probability measure

$$\eta_{\Omega,I} = \mathcal{T}^{\#}_{\Omega,I}\tilde{\mu}_I \tag{4.6}$$

on $\square := [-\pi/2,\ \pi/2] \times [-\pi/2,\ \pi/2]$. In other words, we have

$$\eta_{\Omega,I}(A) = \tilde{\mu}_I\left(\{(\xi,\varphi) : (\varphi,\varphi^+_{\Omega,I}(\xi,\varphi)) \in A\}\right)$$

for each Borel subset A of $\square$.

Definition 4.4. $\eta_{\Omega,I}$ is called the *measure associated with the hollow* (Ω, I).

The measure associated with a hollow defines the joint distribution of the pair (angle of entry, angle of exit) for a randomly chosen particle reflected in this hollow.

We define the probability measure λ on $[-\pi/2,\ \pi/2]$ by $d\lambda(\varphi) = \frac{1}{2}\cos\varphi\, d\varphi$. Let π_φ and π_{φ^+} be the projections to φ and φ^+, respectively; $\pi_\varphi(\varphi,\varphi^+) = \varphi$ and $\pi_{\varphi^+}(\varphi,\varphi^+) = \varphi^+$. We also consider the map π_d interchanging the variables φ and φ^+; $\pi_d(\varphi,\varphi^+) = (\varphi^+,\varphi)$.

Definition 4.5. Let $\mathcal{M}$ be the set of Borel measures η on $\square$ such that

(M1) $\pi^{\#}_{\varphi}\eta = \lambda = \pi^{\#}_{\varphi^+}\eta$;
(M2) $\pi^{\#}_{d}\eta = \eta$.

In other words, $\eta \in \mathcal{M}$ is a measure such that both its projections on the φ and φ^+ axes coincide with λ, and moreover, η is invariant under an interchange of the variables $(\varphi,\varphi^+) \mapsto (\varphi^+,\varphi)$.

It follows from the measure-preservation property and the involutiveness of the map $T_{\Omega,I} = (\xi^+_{\Omega,I},\ \varphi^+_{\Omega,I})$ that any measure $\eta_{\Omega,I}$ satisfies conditions (M1) and (M2). Thus,

$$\eta_{\Omega,I} \in \mathcal{M}. \tag{4.7}$$

Let us define two important measures from $\mathcal{M}$—η_0 and η_{retr}, which will be systematically used in what follows—by

$$d\eta_0(\varphi,\varphi^+) = \frac{1}{2}\cos\varphi\,\delta(\varphi+\varphi^+), \tag{4.8}$$

$$d\eta_{\mathrm{retr}}(\varphi,\varphi^+) = \frac{1}{2}\cos\varphi\,\delta(\varphi-\varphi^+). \tag{4.9}$$

The measure η_0 describes the law of "elastic reflection": a particle exits at an angle equal in magnitude but opposite to the angle of entry. The measure η_{retr} is called a *retroreflector* measure; it corresponds to the (still hypothetical) scattering reversing the direction of incidence of each particle.

We introduce a functional R on the set of probability measures in $\square$ by

$$R(\eta) = \frac{3}{4}\iint_{\square}(1+\cos(\varphi-\varphi^+))\,d\eta(\varphi,\varphi^+). \tag{4.10}$$

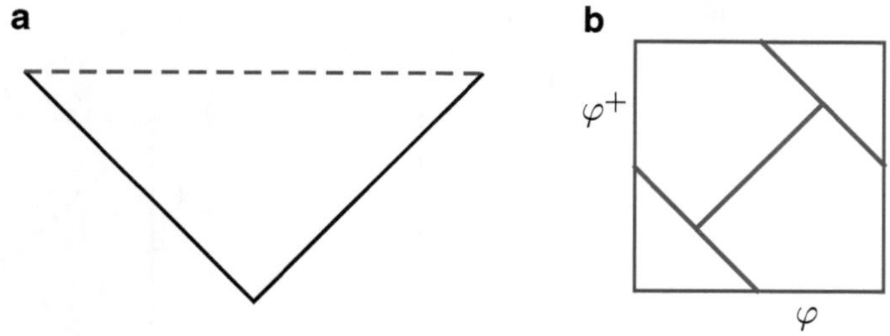

Fig. 4.2 (**a**) The hollow is a right isosceles triangle. (**b**) Support of associated measure

Definition 4.6. The value $R(\eta_{\Omega,I})$ is called the *resistance of the hollow* (Ω, I).

One easily finds that $R(\eta_0) = 1$ and $R(\eta_{\text{retr}}) = 3/2$.

Remark 4.2. According to formulae (4.5), (4.6), and (4.10), the resistance $R(\eta_{\Omega,I})$ of a hollow equals

$$R(\Omega, I) = \frac{3}{4}\int_{-\pi/2}^{\pi/2}\int_I (1 + \cos(\varphi - \varphi^+_{\Omega,I}(\xi,\varphi)))\,\frac{1}{2|I|}\,\cos\varphi\,d\xi\,d\varphi.$$

Sometimes this formula is more convenient for calculation than (4.10).

4.1.2 Examples

Consider several examples of hollows, measures associated with them, and their resistances.

Example 4.1. The hollow is a right isosceles triangle, and the inlet is the hypotenuse of the triangle (Fig. 4.2a). The associated measure η_∇ is supported on the union of three segments: $-\pi/4 \le \varphi = \varphi^+ \le \pi/4$, $0 \le \varphi^+ = \pi/2 - \varphi \le \pi/2$, and $-\pi/2 \le \varphi^+ = -\pi/2 - \varphi \le 0$ (Fig. 4.2b). The measure η_∇ is uniquely defined by its support and the condition $\eta_\nabla \in \mathcal{M}$; the density of η_∇ equals

$$\begin{aligned}
&\frac{1}{2}\cos\varphi\left[\chi_{[-\pi/2,0]}(\varphi)\cdot\delta(\varphi+\varphi^+ + \pi/2) + \chi_{[0,\pi/2]}(\varphi)\cdot\delta(\varphi+\varphi^+ - \pi/2)\right] + \\
&\quad + \frac{1}{2}|\sin\varphi|\left[\chi_{[-\pi/2,\pi/2]}(\varphi)\cdot\delta(\varphi-\varphi^+) - \chi_{[-\pi/4,0]}(\varphi)\cdot\delta(\varphi+\varphi^+ + \pi/2)\right. \\
&\qquad \left. -\chi_{[0,\pi/4]}(\varphi)\cdot\delta(\varphi+\varphi^+ - \pi/2)\right].
\end{aligned} \tag{4.11}$$

One easily calculates that

$$R(\eta_\nabla) = \sqrt{2} \approx 1.414.$$

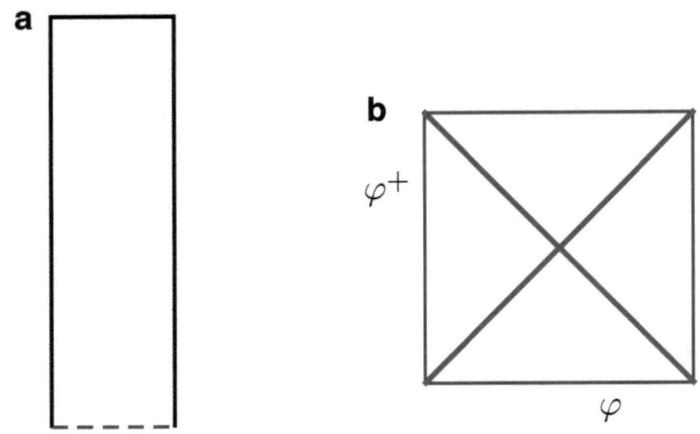

Fig. 4.3 (**a**) Rectangle. (**b**) Support of associated measure

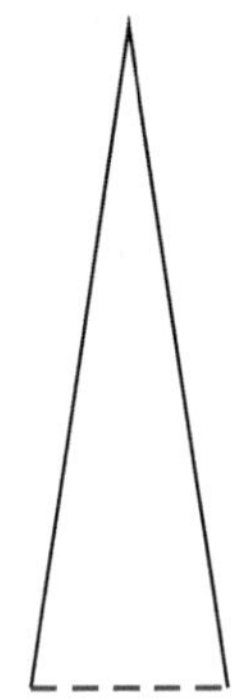

Fig. 4.4 Isosceles triangle

Example 4.2. The hollow is a rectangle with length ε and height 1. The inlet is the lower horizontal side of the rectangle (Fig. 4.3a). The associated measure $\eta_{\sqcup}^{\varepsilon}$ is supported on the union of diagonals of $\square$, $\{\varphi^+ = \varphi\} \cup \{\varphi^+ = -\varphi\}$ (Fig. 4.3b). As $\varepsilon \to 0$, $\eta_{\sqcup}^{\varepsilon}$ converges weakly to $\frac{1}{2}(\eta_0 + \eta_{\text{retr}})$ (Proposition 9.3), and the resistance of the hollow converges to 1.25,

$$\lim_{\varepsilon \to 0} R(\eta_{\sqcup}^{\varepsilon}) = 1.25.$$

Example 4.3. The hollow is an isosceles triangle with an angle α at the apex (Fig. 4.4). The inlet is the base of the triangle. Let $\eta_{\vee}^{\alpha}$ be the associated measure. According to Proposition 9.3, $\eta_{\vee}^{\alpha}$ also converges weakly to $\frac{1}{2}(\eta_0 + \eta_{\text{retr}})$; therefore, the resistance of the triangle also converges to 1.25:

$$\lim_{\alpha \to 0} R(\eta_{\vee}^{\alpha}) = 1.25.$$

Example 4.4. Consider now an isosceles triangle with an obtuse angle $\pi/2 < \alpha < \pi$ at the apex (triangle ABC in Fig. 4.5a). The inlet is the side AC with the coordinate $\xi \in [-2, 1/2]$ varying from $-1/2$ at A to $1/2$ at C. Notice that φ is negative in the figure.

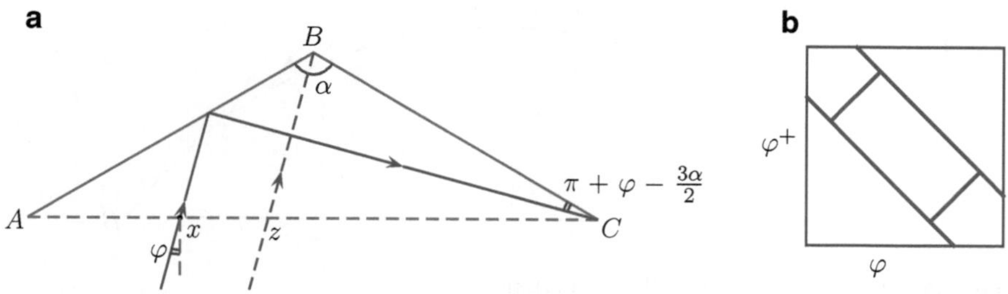

Fig. 4.5 (**a**) The hollow is an obtuse isosceles triangle. (**b**) Support of the associated measure

Introduce $z(\alpha,\varphi) = -\tan\varphi/(2\tan\alpha/2)$ and $x(\alpha,\varphi) = -\frac{1}{2} - \cos\alpha - \sin\alpha\tan\varphi$. Let

$$\begin{aligned}
\mathcal{A} &= \{(\varphi,\xi) : -\pi/2 < \varphi < -\alpha/2,\ -1/2 < \xi < 1/2\} \cup \\
&\quad \cup \{(\varphi,\xi) : -\alpha/2 < \varphi < \alpha/2,\ -1/2 < \xi < z(\alpha,\varphi)\} \\
&\quad \subset [-\pi/2,\ \pi/2] \times [-1/2,\ 1/2], \\
\mathcal{A}_{LR} &= \{(\varphi,\xi) : 3\alpha/2 - \pi < \varphi < \alpha - \pi/2,\ x(\alpha,\varphi) < \xi < z(\alpha,\varphi)\} \cup \\
&\quad \cup \{(\varphi,\xi) : \alpha - \pi/2 < \varphi < \alpha/2,\ -1/2 < \xi < z(\alpha,\varphi)\}, \\
\mathcal{A}_L &= \mathcal{A} \setminus \mathcal{A}_{LR},
\end{aligned}$$

and let $\mathcal{A}_R$ be symmetric to $\mathcal{A}_L$ and $\mathcal{A}_{RL}$ symmetric to $\mathcal{A}_{LR}$ with respect to the origin. The regions $\mathcal{A}_L$, $\mathcal{A}_R$, $\mathcal{A}_{LR}$, and $\mathcal{A}_{RL}$ are disjoint, and their union is a full-measure subset of $[-\pi/2,\ \pi/2] \times [-1/2,\ 1/2]$.

Each incident particle makes one or two reflections in a hollow. If $(\varphi,\xi) \in \mathcal{A}_L$, then the corresponding particle makes exactly one reflection from the left side AB of the triangle, and one has $\varphi^+(\xi,\varphi) = \alpha - \pi - \varphi$. If $(\varphi,\xi) \in \mathcal{A}_R$, then the particle makes a reflection from the right side BC, and $\varphi^+(\xi,\varphi) = \pi - \alpha - \varphi$. If $(\varphi,\xi) \in \mathcal{A}_{LR}$, then the particle makes two successive reflections from the left and right sides, and $\varphi^+(\xi,\varphi) = \varphi + \pi - 2\alpha$. Finally, if $(\varphi,\xi) \in \mathcal{A}_{RL}$, then there are two reflections from the right and left sides, and $\varphi^+(\xi,\varphi) = \varphi - \pi + 2\alpha$. The calculation of resistance is easy but cumbersome. As a result, we obtain

$$\begin{aligned}
R_\alpha = R(\eta_\vee^\alpha) &= \frac{3}{2}\left(1 - \sin\frac{\alpha}{2}\right) + \frac{3}{4}\sin\frac{3\alpha}{2} + \frac{1}{4}\sin\frac{5\alpha}{2} - \frac{1}{2}\cos\alpha + \\
&\quad + \frac{3(1-\cos 2\alpha)\left(1 - \sin\frac{\alpha}{2}\right)}{4\sin\frac{\alpha}{2}} + \frac{\cos 3\alpha - 9\cos\alpha}{8\sin\frac{\alpha}{2}}.
\end{aligned}$$

In particular, $R_\pi = 1$, $R_{2\pi/3} = \frac{5}{8} + \frac{1}{\sqrt{3}} \approx 1.2024$, and $R_{\pi/2} = \sqrt{2} \approx 1.4142$.

The support of the associated measure $\eta_\vee^\alpha$ is shown in Fig. 4.5b; it is the union of four segments lying on the straight lines $\varphi^+ = -\varphi \pm (\pi - \alpha)$ and $\varphi^+ = \varphi \pm (\pi - 2\alpha)$.

Note that the function R_α, $\alpha \in (0,\ \pi)$ is piecewise analytical; the corresponding formulas are different and become more and more complicated on the intervals

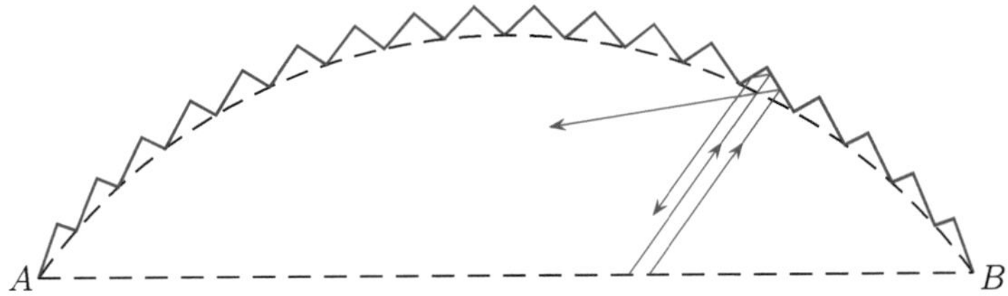

Fig. 4.6 Notched arc with two billiard trajectories

$[\pi/2, \pi)$, $[\pi/3, \pi/2]$, $[\pi/4, \pi/3]$, etc. R_α oscillates and approaches 1.25 as $\alpha \to 0$; it takes the minimum $R_0 = 1$ at $\alpha = 0$ and the maximum $R_{\alpha^*} \approx 1.426$ at $\alpha^* \approx 83.6° \approx 0.464\,\pi$ [56].

Example 4.5. Finally, we present the most complicated shape whose resistance was calculated analytically. The shape was first found numerically at an intermediate step of studying a resistance maximization problem (in collaboration with Gouveia, see [56]) and remained for a while a hollow with the maximum resistance.

Choose a coordinate system x_1Ox_2, with the x_1-axis being horizontal and the x_2-axis vertical. A right triangle ABC is called *canonical* if vertex B is situated above the hypotenuse AC and the median drawn from B to AC is vertical. Fix $\psi \in (0, \pi/2]$ and consider the arc of angular size 2ψ contained in the upper half-plane $x_2 \geq 0$, with the endpoints $(0,0)$ and $(1,0)$. Select a positive integer m and mark points $x^0 = (x_1^0, x_2^0) = (0,0)$, $x^2 = (x_1^2, x_2^2), \ldots, x^{2i} = (x_1^{2i}, x_2^{2i}), \ldots, x^{2m} = (x_1^{2m}, x_2^{2m}) = (1,0)$ on the arc, with $0 = x_1^0 < x_1^2 < \ldots < x_1^{2m} = 1$. For $i = 1, \ldots, m$, draw the canonical triangle $\Delta x^{2i-2}x^{2i-1}x^{2i}$ with the hypotenuse $[x^{2i-2}, x^{2i}]$. Thus, one has $x_1^{2i-1} = \frac{1}{2}(x_1^{2i-2} + x_1^{2i})$, $x_2^{2i-1} = \frac{1}{2}(x_2^{2i-2} + x_2^{2i}) + \frac{1}{2}\sqrt{(x_1^{2i} - x_1^{2i-2})^2 + (x_2^{2i} - x_2^{2i-2})^2}$. Consider the broken line $x^0x^1 \ldots x^{2m-1}x^{2m}$ composed of legs of all triangles obtained in this way. Denote $X = (x^0, x^1, \ldots, x^{m-1}, x^m)$, and let $\delta = \delta(X)$ be the maximum of values $x_1^{2i} - x_1^{2i-2}$.

Now, let I be the line segment joining $(0,0)$ and $(1,0)$, and let Ω_X be the region bounded by the broken line $x^0x^1 \ldots x^{2m-1}x^{2m}$ from above and by I from below. The hollow (Ω_X, I) will be called a *notched arc* and is shown in Fig. 4.6 along with two typical trajectories reflected in a canonical triangle.

As $\delta(X) \to 0$, the resistance of the hollow $R(\Omega_X, I)$ converges to

$$R_\psi = 1 + \frac{1}{6}\sin^2\psi + \frac{2\sqrt{2}\sin\frac{\psi}{2} - 2\sin^4\frac{\psi}{2} - \psi}{\sin\psi}. \tag{4.12}$$

The proof of this convergence is given in the last Sect. 4.2.5 of this chapter. The maximum value of R_ψ is attained at $\psi_0 \approx 0.6835 \approx 39.16°$ and is equal to

$R_{\psi_0} \approx 1.445$. In the limiting case $\psi = 0$, the resistance $R_{\psi=0} = \sqrt{2}$ coincides with that of the right isosceles triangle.

Informally speaking, the boundary of the arc-shaped hollow is formed by small triangular hollows of second order. We believe that it may be of interest to consider the general case of rough boundaries with a hierarchical structure (that is, the boundary is formed by small hollows, the boundary of the hollows is formed by hollows of second order, these hollows are formed by hollows of third order, etc.). This example provides the simplest case of a two-level hierarchy.

4.1.3 Basic Theorem

First we state the following preparatory lemma.

Lemma 4.1. *There exists a family of hollows* $(\Omega_\varepsilon, I_\varepsilon)$, $\varepsilon > 0$ *such that the family of the corresponding measures* $\eta_{\Omega_\varepsilon, I_\varepsilon}$ *weakly converges to the retroreflector measure* η_{retr}.

Remark 4.3. For a continuous function f on $\square$ we have $\iint_\square f(\varphi, \varphi^+)\, d\eta_{\text{retr}} (\varphi, \varphi^+) = \int_{-\pi/2}^{\pi/2} f(\varphi, \varphi)\, \frac{1}{2} \cos\varphi\, d\varphi$. This, the statement of the lemma, means that for each continuous function f,

$$\lim_{\varepsilon\to 0} \iint_\square f(\varphi, \varphi^+)\, d\eta_{\Omega_\varepsilon, I_\varepsilon} = \int_{-\pi/2}^{\pi/2} f(\varphi, \varphi)\, \frac{1}{2} \cos\varphi\, d\varphi.$$

Remark 4.4. The idea of the proof of this lemma, in modified form, will be used in the proof of the following Theorem 4.1. The construction used in the lemma also plays an important role in the construction of an asymptotically perfect retroreflector (Chap. 9).

Proof. We shall use a property of a billiard in ellipse: a particle crossing the line segment joining the foci again crosses the same segment in the opposite direction after reflection from the boundary of the ellipse. If the eccentricity of the ellipse is small, then the velocity of the reflected particle is almost opposite to the initial velocity.

Consider a *mushroom*, a shape Ω_ε formed by the upper half of the ellipse $x_1^2/(1+\varepsilon^2) + x_2^2 = 1$, $x_2 \geq 0$, and the rectangle $-\varepsilon \leq x_1 \leq \varepsilon$, $-\varepsilon^2 \leq x_2 \leq 0$ (Fig. 4.7). Let I_ε be the "base" of the mushroom: the segment $-\varepsilon \leq x_1 \leq \varepsilon$, $x_2 = -\varepsilon^2$. Note that we cannot take just the half of the ellipse with an inlet between the foci: this is not a hollow according to the definition.

Only some of the trajectories, of total measure $O(\varepsilon)$, hit the lateral sides of the "mushroom stem": the segments $x_1 = \pm\varepsilon$, $-\varepsilon^2 \leq x_2 \leq 0$. For the other trajectories the reflected particle has the velocity $\varphi^+_{\Omega_\varepsilon, I_\varepsilon}(\xi, \varphi) = \varphi + O(\varepsilon)$, where the estimate for $O(\varepsilon)$ is uniformly small with respect to ξ and φ as $\varepsilon \to 0$. Hence for some function $\alpha(\varepsilon) = O(\varepsilon)$, $\varepsilon \to 0$ we have the equality

Fig. 4.7 A mushroom

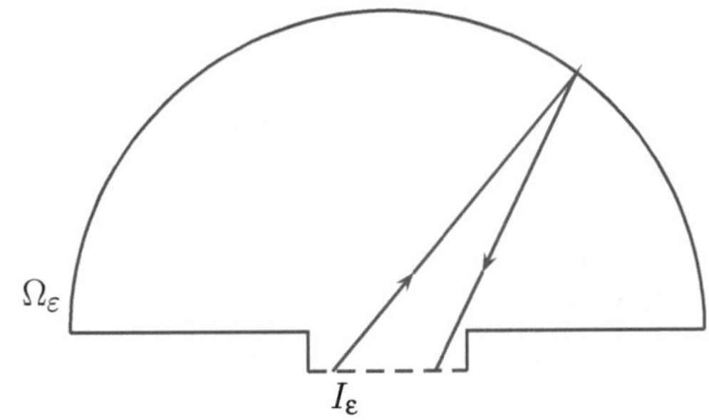

$$\eta_{\Omega_\varepsilon, I_\varepsilon}\left(\{(\varphi,\varphi^+) : |\varphi-\varphi^+| > \alpha(\varepsilon)\}\right) = O(\varepsilon) \quad \text{as } \varepsilon \to 0,$$

so that, in view of property (M1), we easily see that $\eta_{\Omega_\varepsilon, I_\varepsilon}$ weakly converges to η_{retr}. □

Remark 4.5. Besides an elliptic mushroom we can also consider a circular mushroom, the union of the upper semicircle $x_1^2 + x_2^2 \le 1,\ x_2 \ge 0$ and the rectangle $-\varepsilon \le x_1 \le \varepsilon,\ -\varepsilon^2 \le x_2 \le 0$. The measures associated with the circular mushroom also approximate η_{retr}. Moreover, the construction involving a circular mushroom can immediately be generalized to higher dimensions $d \ge 3$.

For a more detailed discussion of the mushroom see [56], and a multidimensional generalization is carried out in [49].

Remark 4.6. A billiard in mushrooms was first considered by Bunimovich [10] in his construction of a dynamical system with divided phase space.

Now we turn again to the system of hollows (Ω_i, I_i) on the boundary of B. Recall (Sect. 1.1) that the measure $\mu = \mu_{\partial C}$ on $\partial C \times S^1$ is defined by $d\mu(\xi, v) = \frac{1}{2}\,|n(\xi)\cdot v|\,d\xi\,dv$. Let

$$\left(I_i \times S^1\right)_\pm := \{(\xi, v) \in I_i \times S^1 : \pm n(\xi)\cdot v \ge 0\}$$

be measurable spaces with induced measure μ. The measure ν_B can be represented as

$$\nu_B = |\partial(\mathrm{Conv} B)| \cdot \sum c_i \nu_i, \tag{4.13}$$

where $c_i = |I_i|/|\partial(\mathrm{Conv}B)|$ is the relative length of I_i, $\sum c_i = 1$, and the probability measures ν_i on $\mathbb{T}^3$ are defined by the following relation: for each $A \subset \mathbb{T}^3$,

$$\nu_i(A) := \frac{1}{|I_i|}\,\mu\left(\{(\xi, v) \in \left(I_i \times S^1\right)_- : (v, v_B^+(\xi, v), n(\xi)) \in A\}\right).$$

If, in particular, $c_0 = 0$, that is, the length of the convex part of ∂B equals zero, then ν_0 can still be defined.

The measures ν_i are closely related with the measures associated with hollows. That is, using the complex representation $n = e^{i\theta}$ for vectors $n \in S^1$, we define the maps $\sigma : \square \times S^1 \to \mathbb{T}^3$ and $\sigma_n : \square \to \mathbb{T}^3$ by

$$\sigma(\varphi, \varphi^+, n) := (-e^{i\varphi} n, e^{i\varphi^+} n, n) \quad \text{and} \quad \sigma_n(\varphi, \varphi^+) := \sigma(\varphi, \varphi^+, n). \tag{4.14}$$

For $i \geq 1$ let n_i be the outward normal to $\partial(\text{Conv} B)$ at a point in I_i. Then

$$\nu_i = \sigma_{n_i}^{\#} \eta_{\Omega_i, I_i}. \tag{4.15}$$

Next, we set

$$\tilde{\tau}_{\text{Conv} B} = \tau_{\text{Conv} B} - \sum_{i \geq 1} |I_i| \delta_{n_i}$$

(recall that δ_{n_i} is the atomic measure concentrated at n_i) and

$$\eta_{I_0} = \eta_0 \otimes \tau_{\text{Conv} B}. \tag{4.16}$$

This means that for each $A_1 \subset \square$ and $A_2 \subset S^1$ we have $\eta_{I_0}(A_1 \times A_2) = \eta_0(A_1) \cdot \tau_{\text{Conv} B}(A_2)$. The measure η_{I_0} is the *law of elastic scattering from the convex part of the boundary of* B. Then we can write ν_0 as follows:

$$|I_0| \nu_0 = \sigma^{\#} \eta_{I_0}. \tag{4.17}$$

We see that information about the measures associated with hollows is important for our further analysis of billiard scattering. The next theorem is a key result.

Theorem 4.1. $\overline{\{\eta_{\Omega,I}\}} = \mathcal{M}$, *where the bar denotes weak closure.*

Remark 4.7. The theorem implies that for each measure $\eta \in \mathcal{M}$ there exists a family of hollows $(\Omega_\varepsilon, I_\varepsilon)$ such that

$$\lim_{\varepsilon \to 0} \int_\square f \, d\eta_{\Omega_\varepsilon, I_\varepsilon} = \int_\square f \, d\eta$$

for any continuous function f on $\square$.

Proof. By (4.7), we have the inclusion $\{\eta_{\Omega,I}\} \subset \mathcal{M}$. It remains to show that each measure $\eta \in \mathcal{M}$ is a weak limit of measures $\eta_{\Omega,I}$ associated with hollows. The proof takes several steps.

Step 1. Let $\eta \in \mathcal{M}$. First we discretize the measure η, that is, partition $[-\pi/2, \pi/2]$ into m intervals of equal measure λ. Correspondingly, $\square$ is partitioned into m^2 rectangles $\square_{ij}$. A discretization of η is the matrix $\left(\eta(\square_{ij})\right)_{i,j=1}^m$, which is symmetric and has the property that $\sum_j \eta(\square_{ij}) = 1/m$ for each i. In turn, we approximate it by a symmetric matrix

$$D = \left(\frac{1}{M} \delta_{j-\sigma(i)}\right)_{i,j=1}^M$$

of larger size $M = ml$, where

$$\delta_i = \begin{cases} 1 & \text{for } i = 0, \\ 0 & \text{for } i \neq 0, \end{cases}$$

and where σ is a permutation of the set $\{1, \ldots, M\}$. Since D is symmetric, σ is involutive: $\sigma^2 = \text{id}$. "Approximation" means that if we represent D as an $m \times m$ block matrix with blocks of size $l \times l$, then the sum of the entries in the (i, j)th block approximates the quantity $\eta(\square_{ij})$. The matrix D contains a single 1 in each row and each column. The construction is not difficult, but it is cumbersome, and therefore is omitted here; it is described in detail in [50].

Finally, we turn to the construction of a family of hollows $(\Omega[\varepsilon], I[\varepsilon])$ such that the discretization of $\eta_{\Omega[\varepsilon],I[\varepsilon]}$ converges to D as $\varepsilon \to 0^+$.

We shall construct a family of hollows depending on two parameters ε and δ and looking like mushrooms with corroded "caps." On the cap of each mushroom are "hollows of second order." Their purpose is to send a particle arriving from the inlet of the mushroom in the required direction. The parameters ε and δ are the size of the inlet of the hollow and the maximum relative size of the inlets of the "second-order hollows," respectively. As a result, the hollows $(\Omega_{\varepsilon,\delta(\varepsilon)}, I_\varepsilon)$, where $\delta(\varepsilon)/\varepsilon \to 0$ as $\varepsilon \to 0$, are associated with measures approximating D.

We start our construction with a mushroom, the union of the upper semicircle $x_1^2 + x_2^2 \leq 1$, $x_2 \geq 0$ and the rectangle $-\varepsilon \leq x_1 \leq \varepsilon$, $-\varepsilon^2 \leq x_2 \leq 0$, with inlet I_ε: $-\varepsilon \leq x_1 \leq \varepsilon$, $x_2 = -\varepsilon^2$. Consider the polar coordinates $x_1 = -r \sin\varphi$, $x_2 = r\cos\varphi$, in which $\varphi = 0$ corresponds to the vertical ray Ox_2. We partition $[-\pi/2, \pi/2]$ into M intervals of the same λ-measure, $[-\pi/2, \pi/2] = \cup_{i=1}^M J_i$, $\lambda(J_i) = 1/M$, and we partition the semicircle into M sectors corresponding to the partition into intervals, so that the angular measure of the ith sector in polar variables is J_i. If $\sigma(i) = i$, then we leave the corresponding sector as it is. The remaining values $i = 1, \ldots, M$, $\sigma(i) \neq i$ group into pairs (i, j), where $\sigma(i) = j$ (and therefore also $\sigma(j) = i$). For each pair (i, j) we modify the corresponding pair of sectors.

Next we prove that for all $(\xi, \varphi) \in I_\varepsilon \times [-\pi/2, \pi/2]$ outside a subset of measure $o(1)$, $\varepsilon \to 0$ the corresponding particle displays the following dynamics. (a) If $\varphi \in J_i$, $\sigma(i) = i$, then after a single reflection from an arc of the circle the particle intersects I_ε again, at an angle $\varphi^+ = \varphi^+_{\Omega_{\varepsilon,\delta(\varepsilon)},I_\varepsilon}(\xi, \varphi) \in J_i$. (b) If $\varphi \in J_i$, $\sigma(i) = j \neq i$, then the corresponding particle is reflected twice from the boundary of the modified ith sector, is then reflected twice from the boundary of the modified jth sector, and finally intersects the inlet I_ε at an angle $\varphi^+ \in J_j$. This will complete the proof of the theorem.

The proof of (a) proceeds without difficulty; see [50]. The main problem is in modifying the ith and jth sectors in a pair i, $j = \sigma(i) \neq i$, and the proof of (b).

The modified ith sector is the union of the ith sector and several reflectors, sets that we define below. These will be our *second-order hollows*; their bases are chords with endpoints on the ith arc; they do not intersect and look outward relative to the chord. A particle moving from the inlet of the hollow to the ith arc goes into a reflector, reflects in it twice, and moves on to the jth arc. There it goes into a reflector of the jth sector, reflects in it twice, and finally returns to the inlet of the hollow. For particles displaying this sequence of reflections we have $\varphi \in J_i$, $\varphi^+ \in J_j$. The next step is the definition of a reflector and analysis of its properties.

Step 2. The definition of a $(\varphi_1, \varphi_2, \delta)$-reflector is as follows. We set $e_\varphi = (-\sin\varphi, \cos\varphi)$, suppose that $-\pi/2 < \varphi_1 \neq \varphi_2 < \pi/2$, $\delta > 0$, and consider two rays te_{φ_1} and te_{φ_2}, $t \geq 0$. They form angles φ_1 and φ_2, respectively, with the vector $e_0 = (0, 1)$. Let $x^{(1)}$ and $x^{(2)}$ be the points of intersection of these rays with the unit circumference $|x - e_0| = 1$, and let $O = (0, 0)$ be the origin. We consider two parabolas p_1 and p_2 with the same focus O and a common axis parallel to $x^{(2)} - x^{(1)}$. We also assume that p_1 contains $x^{(1)}$ and p_2 contains $x^{(2)}$, and that the intersection of the convex sets bounded by these parabolas contains the segment $[x^{(1)}, x^{(2)}]$. Let $R(\varphi_1, \varphi_2, \delta)$ be the convex set bounded by p_1, p_2, and the three lines $x_2 = 0$, $x \cdot e_{-\delta} + \delta \sin\delta = 0$, and $x \cdot e_\delta + \delta \sin\delta = 0$. We also set $I(\delta) := [-\delta, \delta] \times \{0\}$. It is easy to see that the pair $(R(\varphi_1, \varphi_2, \delta),\ I(\delta))$ is a hollow.

Definition 4.7. A copy of $R(\varphi_1, \varphi_2, \delta)$ obtained by means of an isometry and a dilation is called a $(\varphi_1, \varphi_2, \delta)$-*reflector*. The copy of the segment $I(\delta)$ obtained by the same transformations is called an *inlet of this reflector.* The image of O under these transformations is called the *center of the reflector.* The dilation coefficient k is called the *contraction coefficient*; throughout, we assume that $0 < k \leq 1$ (Fig. 4.8).

Also, let $R(\varphi_1, \varphi_2) := R(\varphi_1, \varphi_2, 0)$ be the set bounded by the parabolas p_1 and p_2 and by the horizontal line $x_2 = 0$ from below. The inlet $I := I(0)$ to it is the intersection of $R(\varphi_1, \varphi_2)$ with $x_2 = 0$. For simplicity we set

$$(\xi^+_{R(\varphi_1,\varphi_2,\delta),I(\delta)},\ \varphi^+_{R(\varphi_1,\varphi_2,\delta),I(\delta)}) =: (\xi^+,\ \varphi^+),$$

$$(\xi^+_{R(\varphi_1,\varphi_2),I},\ \varphi^+_{R(\varphi_1,\varphi_2),I}) =: (\xi^+_0,\ \varphi^+_0),$$

$$\eta_{R(\varphi_1,\varphi_2,\delta),I(\delta)} =: \eta_\delta.$$

Clearly, the map ξ^+, φ^+ coincides with ξ^+_0, φ^+_0 in a neighborhood of $(0, \varphi_1)$ formed by the points (ξ, φ) such that the corresponding billiard particle, upon several reflections from p_1 and p_2, intersects $I(\delta)$. We shall see below that the η_δ-measure of this neighborhood approaches 1 as $\delta \to 0$.

Note that

$$R(\varphi_1, \varphi_2, \delta) = R(\varphi_2, \varphi_1, \delta) \quad \text{and} \quad R(\varphi_1, \varphi_2) = R(\varphi_2, \varphi_1).$$

The reflector $R(\varphi_1, \varphi_2, \delta)$ has the following property: a particle moving from O in the direction φ_1, after reflections at the points $x^{(1)}$ and $x^{(2)}$, returns to O moving in the direction φ_2. Moreover, if φ belongs to some neighborhood $\mathcal{O}(\varphi_1)$ of φ_1 independent of δ, then a particle starting from O, after two reflections from the parabolas, returns to O. In our notation, by identifying points $\xi \in I(\delta)$ with real numbers $\xi \in [-\delta, \delta]$ we can express these properties as follows:

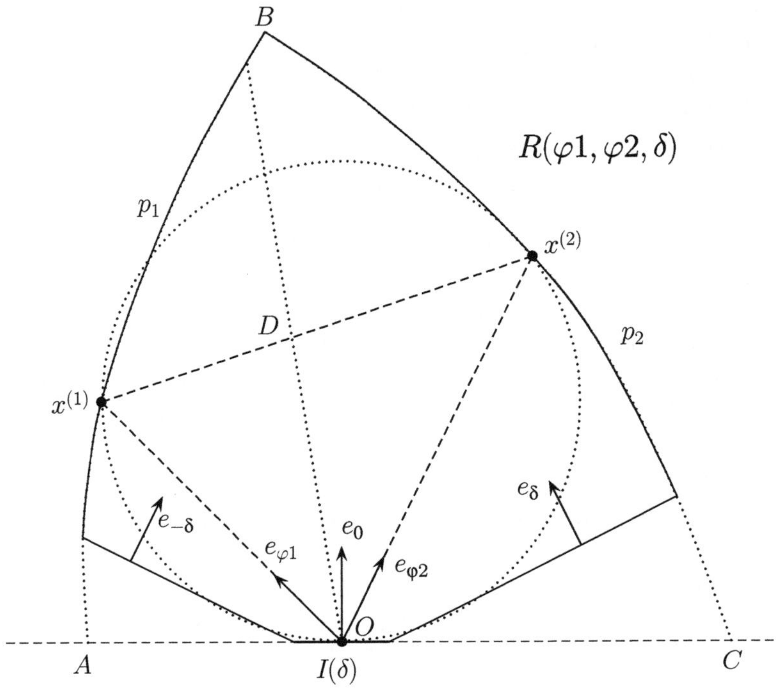

Fig. 4.8 $(\varphi_1, \varphi_2, \delta)$-reflector

$$\varphi^+(0, \varphi_1) = \varphi_2, \tag{4.18}$$

$$\xi^+(0, \varphi) = 0 \quad \text{for } \varphi \in \mathcal{O}(\varphi_1). \tag{4.19}$$

Note, in particular, that $\varphi_0^+(0, \varphi_1) = \varphi_2$ and $\xi_0^+(0, \varphi) = 0$ for $\varphi \in \mathcal{O}(\varphi_1)$.

The map $\xi_0^+,\ \varphi_0^+$ has two further important properties:

$$\frac{\cos\varphi_2}{\cos\varphi_1}\frac{\partial\varphi_0^+}{\partial\varphi}(0, \varphi_1) = -1, \tag{4.20}$$

$$\left|\frac{\partial\xi_0^+}{\partial\xi}(0, \varphi_1)\right| = 1. \tag{4.21}$$

We defer the proof of the property (4.20) to Sect. 4.1.4. It essentially uses the fact that the circle through $O,\ x^{(1)}$, and $x^{(1)}$ is tangent to $x_2 = 0$ and thus justifies the use of a circle in the definition of a reflector. Property (4.21) is a consequence of (4.18)–(4.20) and the fact that the map $\xi_0^+,\ \varphi_0^+$ preserves the measure $\cos\varphi\, d\xi\, d\varphi$. A more detailed discussion allows us to specify the sign in (4.21) and to conclude that

$$\frac{\partial \xi_0^+}{\partial \xi}(0, \varphi_1) = 1.$$

Property (4.21) ensures that most particles moving from the reflector inlet $R(\varphi_1, \varphi_2, \delta)$ in directions close to φ_1 leave the reflector after reflecting from p_1 and p_2. Indeed,

$$\xi_0^+(\xi, \varphi) = \xi_0^+(0, \varphi) + \xi \cdot \frac{\partial \xi_0^+}{\partial \xi}(0, \varphi) + O(\xi^2).$$

By (4.19) and the observation after it, $\xi_0^+(0, \varphi) = 0$. By (4.21),

$$\left|\frac{\partial \xi_0^+}{\partial \xi}(0, \varphi)\right| = \left|\frac{\partial \xi_0^+}{\partial \xi}(0, \varphi_1) + O(\varphi - \varphi_1)\right| = 1 + O(\varphi - \varphi_1).$$

Hence

$$|\xi_0^+(\xi, \varphi)| = |\xi| \cdot (1 + O(\varphi - \varphi_1) + O(|\xi|)).$$

This means that all the particles intersecting the inlet $I(\delta)$ at a distance larger than $\delta \cdot (O(\varphi - \varphi_1) + O(\delta))$ from its endpoints, after reflecting from p_1 and p_2, intersect $I(\delta)$ again in the opposite direction. Thus, for all values $(\xi, \varphi) \in I(\delta) \times [\varphi_1 - \varepsilon, \varphi_1 + \varepsilon]$, except for a fraction of them of order $O(\varepsilon) + O(\delta)$, the value of $\xi^+(\xi, \varphi)$, $\varphi^+(\xi, \varphi)$ coincides with that of $\xi_0^+(\xi, \varphi)$, $\varphi_0^+(\xi, \varphi)$.

Next we obtain

$$\varphi_0^+ := \varphi_0^+(\xi, \varphi) = \varphi_0^+(0, \varphi_1) + \xi \cdot \frac{\partial \varphi_0^+}{\partial \xi}(0, \varphi_1) +$$
$$+(\varphi - \varphi_1) \cdot \frac{\partial \varphi_0^+}{\partial \varphi}(0, \varphi_1) + O(\xi^2) + O((\varphi - \varphi_1)^2).$$

By (4.18) and the observation after it, $\varphi_0^+(0, \varphi_1) = \varphi_2$, and by (4.20),

$$\frac{\partial \varphi_0^+}{\partial \varphi}(0, \varphi_1) = -\frac{\cos \varphi_1}{\cos \varphi_2}.$$

Hence the particle flies away in the direction

$$\varphi_0^+ = \varphi_2 - (\varphi - \varphi_1) \cos \varphi_1 / \cos \varphi_2 + O(|\xi|) + O((\varphi - \varphi_1)^2).$$

It follows from this and our conclusions in the previous paragraph that for particles with initial data $(\xi, \varphi) \in I(\delta) \times [\varphi_1 - \varepsilon, \varphi_1 + \varepsilon]$ entering the reflector $R(\varphi_1, \varphi_2, \delta)$, except for a fraction of initial data of order $O(\varepsilon) + O(\delta)$, we have

$$\varphi^+ = \varphi_2 - (\varphi - \varphi_1) \frac{\cos \varphi_1}{\cos \varphi_2} + O(\delta) + O(\varepsilon^2). \tag{4.22}$$

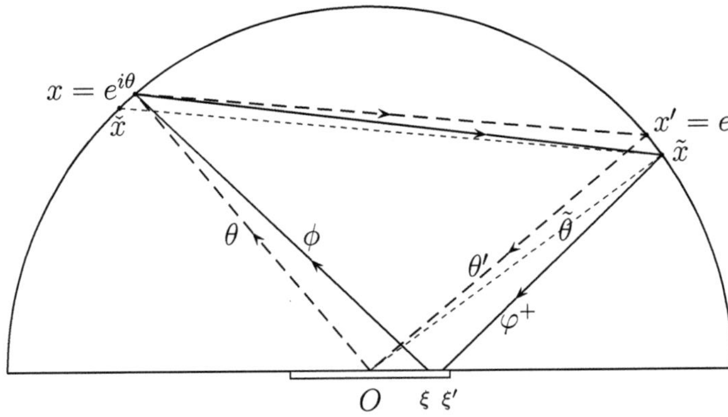

Fig. 4.9 Trajectory of a particle in a hollow (mushroom with corroded cap)

Thus, if the direction of the entry into $R(\varphi_1, \varphi_2, \delta)$ is close to φ_1, then the direction of the exit from the reflector is close to φ_2.

Step 3: Dynamics related to a pair of sectors i, $j = \sigma(i) \neq i$. We shall show that for all initial data (ξ, φ), except for a fraction of data of order $o(1)$, it follows from $\varphi \in J_i$ that

$$\varphi^+ = \varphi^+_{\Omega_{\varepsilon,\delta(\varepsilon)}, I_\varepsilon}(\xi, \varphi) \in J_j .$$

This will complete the proof of the theorem.

Without loss of generality assume that $i < j$. We consider a map $\theta' : J_i \to J_j$ that is monotonically decreasing and preserves the measure λ; it is determined by

$$\sin\theta' - \sin\theta_{j-1} = \sin\theta_i - \sin\theta,$$

where $\theta_i = \arcsin(-1 + 2i/M)$ is the right endpoint of J_i. To each point $x = e^{i\theta}$, $\theta \in J_i$ in the ith arc we assign the point $x' = e^{i\theta'}$ in the jth arc, where $\theta' = \theta'(\theta) \in J_j$ (Fig. 4.9). Consider the triangle Oxx'. Let $\varphi_1(\theta)$ and $\varphi_2(\theta)$ be the angles between the inward normal to the circle at x (that is, the vector $-x$) and the sides Ox and $x'x$. It is easy to see that $\varphi_1(\theta) = 0$ and $\varphi_2(\theta) = \pi/2 - (\theta - \theta')/2$. We also denote by $\varphi_1(\theta')$ and $\varphi_2(\theta')$ the angles between the inward normal vector $-x'$ and the sides Ox' and xx'. Here $\varphi_1(\theta') = -\pi/2 + (\theta - \theta')/2$ and $\varphi_2(\theta') = 0$.

On the ith and jth arcs we place a finite system of reflectors whose inlets are chords with endpoints on these arcs. These reflectors are disjoint and point outward, that is, the lines containing the chords separate them from O. Moreover, the reflectors lie entirely in the ith and jth sectors defined in the polar coordinates as $\varphi \in J_i$ and $\varphi \in J_j$, respectively. A reflector on the ith (jth) arc with center at $re^{i\theta}$, $0 < r < 1$, $\theta \in J_i$ (J_j), is a $(\varphi_1(\theta), \varphi_2(\theta), \delta)$-reflector, which means that θ corresponds to the midpoint of the arc based on the inlet of the reflector. The inlets of reflectors cover almost the whole of the ith arc and jth arc, and the free part of each arc has a total length of at most ε. (For brevity we call reflectors of the ith and jth arcs i- and j-reflectors, respectively.)

We now show that this construction is indeed possible. The corresponding procedure is inductive. We confine ourselves to the ith arc. Let θ_0 be its midpoint. In the first step, we place a $(\varphi_1(\theta_0), \varphi_2(\theta_0), \delta)$-reflector whose inlet is the chord subtending an arc with midpoint at θ_0. This is a reflector of original size (that is, $k_0 = 1$). If there is insufficient space for it inside the ith sector, then we take a sufficiently small copy of it, with $k_0 < 1$. The inlet of the reflector splits off two smaller arcs from the ith arc; let θ_1^1 and θ_1^2 be their midpoints. In the second step, we place two $(\varphi_1(\theta_1^i), \varphi_2(\theta_1^i), \delta)$-reflectors with centers at θ_1^i, $i = 1, 2$ and with the same contraction coefficient $k_1 \le k_0$, sufficiently small that they lie in the ith sector and do not intersect each other or the reflector of the first step. The inlets of the reflectors of the first and second steps split off four smaller arcs from the ith arc; we denote by θ_2^i, $i = 1, 2, 3, 4$ their midpoints and construct four reflectors of the fourth generation. Continuing in this way, we obtain a hierarchy of reflectors. In each step the contraction coefficient is not larger than in the previous step. The total length of the part of the arc not occupied by the inlets of reflectors decreases at least in powerlike fashion. We terminate the process when the length of the free part is less than ε.

A particle moving from a point $\xi \in I_\varepsilon$ in the inlet of a hollow in a direction $\varphi \in J_i$ enters an i-reflector, reflects in it twice, and leaves it in the direction of the jth arc. Then it enters some j-reflector, reflects in it twice, and returns to the inlet I_ε of the mushroom. The fraction of particles striking a point of the ith or jth arc outside the reflectors has order $O(\varepsilon)$ and can be ignored. It is easy to see from our estimates that the direction of the reverse motion of a particle is $\varphi' = \theta'(\varphi) + O(\varepsilon)$. Let ξ' be the point at which the particle intersects the line $x_2 = 0$. We must show that $|\xi'| < \varepsilon$ for most of the particles, so that after four reflections they leave through I_ε. A priori, it is clear that $\xi' = O(\varepsilon)$, but this is not enough. We show below that $\xi' = \xi + O(\delta) + O(\varepsilon^2)$. Thus, taking $\delta = \delta(\varepsilon)$ such that $\delta/\varepsilon \to 0$ as $\varepsilon \to 0$, we will be able to conclude that for most $(\xi, \varphi) \in I_\varepsilon \times J_i$ we have $\xi' \in [-\varepsilon, \varepsilon]$, and therefore $\xi' = \xi^+_{\Omega_{\varepsilon,\delta(\varepsilon)}, I_\varepsilon}(\xi, \varphi)$ and $\varphi^+_{\Omega_{\varepsilon,\delta(\varepsilon)}, I_\varepsilon}(\xi, \varphi) \in J_j$.

Consider the motion of a particle starting from $\xi \in I_\varepsilon$ in a direction $\phi \in J_i$, entering an i-reflector at a point $x = e^{i\theta}$, then a j-reflector at $\tilde{x} = e^{i\tilde{\theta}}$, and finally intersecting the line $x_2 = 0$ at ξ' (Fig. 4.9). We identify points $\xi \in [-\varepsilon, \varepsilon] \times \{-\varepsilon^2\}$ and $\xi' \in \mathbb{R} \times \{0\}$ with the corresponding numbers $\xi \in [-\varepsilon, \varepsilon]$ and $\xi' \in \mathbb{R}$. The "mushroom stem" has height ε^2, so the x_1-coordinate of the first intersection of the trajectory with $x_2 = 0$ is $\xi + O(\varepsilon^2)$. For brevity, in what follows we carry out intermediate calculations up to terms of order ε, dropping terms of higher orders $O(\varepsilon^2)$ and $O(\delta)$. Looking at the triangle $\xi O x$, we conclude that $\phi = \theta + \xi \cos\theta$.

Let us consider an auxiliary "trajectory" beginning and ending at O. (In contrast to the previous *true* trajectory, we say that this trajectory is *imaginary*.) It has one reflection point at $x = e^{i\theta}$ and the other at $x' = e^{i\theta'}$, where $\theta' = \theta'(\theta)$. The imaginary particle enters an i-reflector at an angle $\varphi_1(\theta) = 0$ and leaves it at an angle $\varphi_2(\theta) = \pi/2 - \Delta\theta$, where $\Delta\theta = (\theta - \theta')/2$. Substituting into (4.22) the values $\varphi_1 = \varphi_1(\theta) = 0$, $\varphi_2 = \varphi_2(\theta)$, and $\varphi = \phi - \theta = \xi\cos\theta$, we see that the true particle is reflected from the reflector at an angle $\pi/2 - \Delta\theta - \xi\cos\theta/\sin\Delta\theta$. Thus, the angle between the directions of motion of the imaginary and true particles after reflection from i-reflectors is $\xi\cos\theta/\sin\Delta\theta$. Hence

$$\theta' - \tilde{\theta} = 2\xi\cos\theta/\sin\Delta\theta.$$

Let $\check{\theta}$ be the inverse image of $\tilde{\theta}$, that is, $\theta'(\check{\theta}) = \tilde{\theta}$. It follows from the relations $\sin\check{\theta} + \sin\tilde{\theta} = \sin\theta + \sin\theta'$ and $\tilde{\theta} = \theta' - 2\xi\cos\theta/\sin\Delta\theta$ that

$$\check{\theta} = \theta + 2\xi\cos\theta'/\sin\Delta\theta,$$

so that

$$\Delta\tilde{\theta} = (\check{\theta} - \tilde{\theta})/2 = \Delta\theta + \xi(\cos\theta + \cos\theta')/\sin\Delta\theta.$$

At $\tilde{x}$ the true particle encounters a j-reflector, with angle of entry $\varphi_1(\tilde{\theta}) = -\pi/2 + \Delta\tilde{\theta}$ and angle of exit $\varphi_2(\tilde{\theta}) = 0$. The incidence angle of its collision with the reflector is $-\pi/2 + \Delta\theta + \xi\cos\theta/\sin\Delta\theta$. We use formula (4.22) again, this time with $\varphi_1 = \varphi_1(\tilde{\theta})$, $\varphi_2 = \varphi_2(\tilde{\theta}) = 0$, and $\varphi = -\pi/2 + \Delta\theta + \xi\cos\theta/\sin\Delta\theta$. We then see that the reflection angle of the true particle after the collision with the j-reflector is $\xi\cos\theta'$. This is the angle between the direction of its reverse motion and the radius $O\tilde{x}$, that is, $\measuredangle O\tilde{x}\xi' = \xi\cos\theta'$. Considering the triangle $O\tilde{x}\xi'$, we see that $\xi' = \xi$. In effect, with lower-order corrections taken into account, $\xi' = \xi + O(\delta) + O(\varepsilon^2)$. This proves the theorem. □

4.1.4 *Proof of Formula (4.20)*

We fix φ_1 and φ_2 with $-\pi/2 < \varphi_2 < \varphi_1 < \pi/2$. The case $\varphi_1 < \varphi_2$ can be studied in a similar way. Let $A = x^{(1)}$ and $B = x^{(2)}$. In Fig. 4.10, we depict angles $\varphi_1 > 0$ and $\varphi_2 < 0$.

A particle going away from O in the direction $\varphi_1 + \Delta\varphi$ is first reflected from p_1 at some point A' and then from p_2 at some point B', after which it returns to O moving in the direction $\varphi_2 + \Delta\varphi^+$. Thus, $\varphi_2 + \Delta\varphi^+ = \varphi_0^+(0, \varphi_1 + \Delta\varphi)$. For definiteness we take $\Delta\varphi > 0$, so that $\Delta\varphi^+ < 0$. Note that points A and B lie on the circle $|x - e_0| = 1$, but A' and B' do not necessarily lie on it. Moreover, the line $A'B'$ is parallel to AB.

For convenience we introduce additional points L_1 and L_2 lying on the line $x_2 = 0$, to the left and to the right of O. The angles $\measuredangle OAB$ and $\measuredangle L_2OB$ are inscribed in the same arc of the circle and therefore are equal: $\measuredangle OAB = \measuredangle L_2OB$. Similarly, $\measuredangle OBA = \measuredangle L_1OA$. It is known that $\measuredangle L_2OB = \pi/2 + \varphi_2$ and $\measuredangle L_1OA = \pi/2 - \varphi_1$, hence

$$\measuredangle OAB = \pi/2 + \varphi_2 \quad \text{and} \quad \measuredangle OBA = \pi/2 - \varphi_1.$$

By the law of sines,

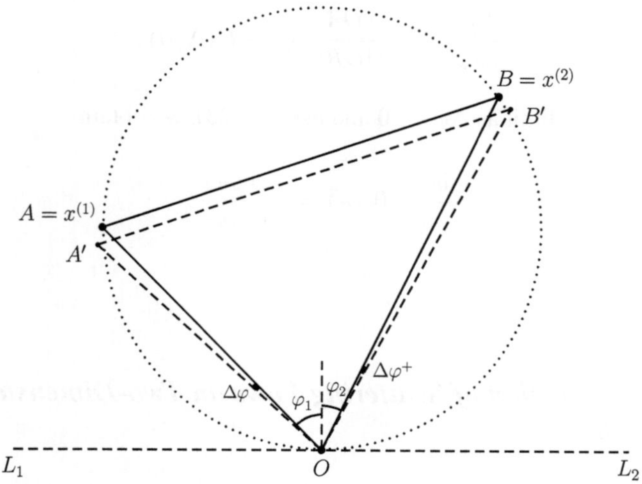

Fig. 4.10 Dynamics in a reflector

$$\frac{|OA|}{|OB|} = \frac{\sin \measuredangle OBA}{\sin \measuredangle OAB} = \frac{\cos\varphi_1}{\cos\varphi_2}. \tag{4.23}$$

The line AA' makes an angle $O(\Delta\varphi)$ with the tangent to the parabola p_1 at A. Hence $\measuredangle OAA' = \frac{1}{2}(\pi - \measuredangle OAB) + O(\Delta\varphi) = \pi/4 - \varphi_2/2 + O(\Delta\varphi)$. Consequently,

$$\measuredangle OA'A = \pi - \measuredangle OAA' - \measuredangle AOA' = \frac{3\pi}{4} + \frac{\varphi_2}{2} + O(\Delta\varphi), \tag{4.24}$$

$$\measuredangle A'AB = \measuredangle OAA' + \measuredangle OAB = \frac{3\pi}{4} + \frac{\varphi_2}{2} + O(\Delta\varphi). \tag{4.25}$$

Applying the law of sines to the triangle OAA' and taking (4.24) into account, we obtain

$$\frac{|AA'|}{|OA|} = \frac{\sin \measuredangle AOA'}{\sin \measuredangle OA'A} = \frac{\sin\Delta\varphi}{\sin(\frac{3\pi}{4} + \frac{\varphi_2}{2} + O(\Delta\varphi))}. \tag{4.26}$$

Let Δx be the distance between the lines AB and $A'B'$. We have $\Delta x = |AA'| \cdot \sin \measuredangle A'AB$. Using (4.25) and (4.26), we obtain

$$\Delta x = |OA| \cdot \Delta\varphi \cdot (1 + O(\Delta\varphi)) \quad \text{as } \Delta\varphi \to 0. \tag{4.27}$$

Considering similarly the triangle OBB' and taking into account that $\Delta\varphi^+ < 0$, we see that

$$\Delta x = -|OB| \cdot \Delta\varphi^+ \cdot (1 + O(\Delta\varphi)) \quad \text{as } \Delta\varphi \to 0. \tag{4.28}$$

By (4.27) and (4.28),

$$\frac{\Delta\varphi^+}{\Delta\varphi} = -\frac{|OA|}{|OB|} \cdot (1 + O(\Delta\varphi)).$$

Passing here to the limit as $\Delta\varphi \to 0$ and using (4.23), we obtain

$$\frac{\partial \varphi_0^+}{\partial \varphi}(0, \varphi_1) = -\frac{\cos\varphi_1}{\cos\varphi_2},$$

as required.

4.1.5 *Classification of Scattering Laws on Two-Dimensional Bodies*

Recall that τ_C is the surface measure of the convex body $C \subset \mathbb{R}^2$, λ_n is the probability measure on S^1 with density $d\lambda_n(v) = \frac{1}{2}(v \cdot n)_+ dv$, the maps $\pi_{v,n}$, $\pi_{v^+,n}$, π_{ad} and the set Γ_C of measures are defined in the beginning of this chapter, and the set $\mathcal{M}$ of measures is defined in Sect. 4.1.1.

Let C' and C'' be bounded convex bodies such that

$$C' \subset C'' \quad \text{and} \quad \text{dist}(\partial C', \partial C'') > 0.$$

Let $\mathcal{B}(C', C'')$ be the class of connected bodies B such that $C' \subset B \subset C''$.

Theorem 4.2. *(a) If $B \in \mathcal{B}(C', C'')$, then $\nu_B \in \Gamma_{\text{Conv}B}$.*
(b) If $C' \subset C \subset C''$, then $\Gamma_C \subset \overline{\{\nu_B : B \in \mathcal{B}(C', C'')\}}$.

The following corollary follows immediately from the theorem.

Corollary 4.1. $\overline{\{\nu_B : B \in \mathcal{B}(C', C'')\}} = \cup\{\Gamma_C : C' \subset C \subset C''\}$.

In some problems it is convenient to deal with the "reduced" scattering law: the measure

$$\eta_B = c_0\eta_0 + \sum_{i\geq 1} c_i \eta_{\Omega_i, I_i}$$

in $\square$. This measure describes the joint distribution of the pair (*incidence angle, reflection angle*) for a random particle incident on B. The following useful relation links the "complete" scattering law ν_B and the reduced scattering law.

We define a map $\varpi : \mathbb{T}^3 \to \Box$ by

$$\varpi(v, v^+, n) = (\varphi, \varphi^+), \tag{4.29}$$

where φ is the angle between $-n$ and v and φ^+ is the angle between n and v^+. (Note that ϖ is the right inverse of each map σ_n.) Then

$$\varpi^{\#}\nu_B = |\partial(\mathrm{Conv}B)|\, \eta_B. \tag{4.30}$$

For the measure η_B we have the following result.

Theorem 4.3. $\mathcal{M} = \overline{\{\eta_B : B \in \mathcal{B}(C', C'')\}}$.

We shall prove only Theorem 4.2. Theorem 4.3 can be obtained from it by a slight modification of the proof.

Proof. (a) First we verify the property (Γ1) for ν_B. For this we must show that (i) $\pi^{\#}_{v,n}\nu_B = \hat{\tau}^-_{\mathrm{Conv}B}$ and (ii) $\pi^{\#}_{v^+,n}\nu_B = \hat{\tau}^+_{\mathrm{Conv}B}$. It is sufficient that we verify this for functions $f(v,n) = f_i(v) \cdot \chi_{n_i}(n)$ vanishing for $n \neq n_i$ and for functions $f(v,n)$ vanishing for $n = n_i$, $i = 1, 2, \ldots$.

Recall that $\sigma_n(\varphi, \varphi^+) = (-e^{i\varphi}n, e^{i\varphi^+}n, n)$. We consider the auxiliary function $\tilde{\sigma}_n(\varphi) = (-e^{i\varphi}n, n)$. It is easy to see that

$$\tilde{\sigma}_n\, \pi_\varphi = \pi_{v,n}\, \sigma_n, \tag{4.31}$$

$$\pi_{\mathrm{ad}}\, \sigma_n = \sigma_n\, \pi_d \quad \text{and} \quad \pi_{\mathrm{ad}}\, \sigma = \sigma\, (\pi_d \otimes \mathrm{id}), \tag{4.32}$$

$$\tilde{\sigma}^{\#}_n \lambda = \lambda_{-n} \otimes \delta_n, \tag{4.33}$$

$$\pi^{\#}_{v,n}\, \sigma^{\#}\, \eta_0 \otimes \tau_C = \hat{\tau}^-_C. \tag{4.34}$$

For $f(v,n) = f_i(v) \cdot \chi_{n_i}(n)$ we take (4.15), (4.7), (4.31), and (4.33) into account, let $\nu(f) := \int_\Box f d\nu$, and obtain the chain of equalities

$$\pi^{\#}_{v,n}\nu_B(f) = |I_i|\pi^{\#}_{v,n}\nu_i(f) = |I_i|\pi^{\#}_{v,n}\sigma^{\#}_{n_i}\eta_{\Omega_i,I_i}(f) = |I_i|\tilde{\sigma}^{\#}_{n_i}\pi^{\#}_\varphi\eta_{\Omega_i,I_i}(f) = |I_i|\tilde{\sigma}^{\#}_{n_i}\lambda(f) =$$

$$= |I_i|\lambda_{-n_i} \otimes \delta_{n_i}(f) = |I_i|\lambda_{-n_i}(f_i) \cdot \delta_{n_i}(\chi_{n_i}) = |I_i| \int_{S^1} f_i(v)\, \frac{1}{2}\, (v \cdot n_i)_-\, dv = \hat{\tau}^-_{\mathrm{Conv}B}(f).$$

In the last equality, we used the equality $\delta_{n_i}(\chi_{n_i}) = 1$, the relations (4.1) and (4.2) defining the measures λ_n and $\hat{\tau}^{\pm}_{\mathrm{Conv}B}$, and the fact that $\tau_{\mathrm{Conv}B}$ has an atom $|I_i|\delta_{n_i}$ at point n_i.

Assume now that $f(v,n)$ vanishes for $n = n_i$, $i = 1, 2, \ldots$ and additionally $|I_0| \neq 0$. Then

$$\tau_{\mathrm{Conv}B}(f) = \tilde{\tau}_{\mathrm{Conv}B}(f). \tag{4.35}$$

According to (4.13)–(4.17),

$$\pi^{\#}_{v,n}\nu_B(f) = \pi^{\#}_{v,n}|\partial(\mathrm{Conv}B)|c_0\nu_0(f) = \pi^{\#}_{v,n}\sigma^{\#}\eta_0 \otimes \tilde{\tau}_{\mathrm{Conv}B}(f).$$

Taking (4.34) and (4.35) into account, we now get that

$$\pi^{\#}_{v,n}\nu_B(f) = \hat{\tau}^{-}_{\mathrm{Conv}B}(f).$$

Thus, we have proved (i), and the proof of (ii) is similar.

To prove ($\Gamma 2$), it is sufficient to show that (iii) $\pi^{\#}_{\mathrm{ad}}\nu_i = \nu_i$, $i \geq 1$ and (iv) $\pi^{\#}_{\mathrm{ad}}\nu_0 = \nu_0$. By (4.32), (4.15), and (4.7), we have the chain of equalities

$$\pi^{\#}_{\mathrm{ad}}\nu_i = \pi^{\#}_{\mathrm{ad}}\sigma^{\#}_{n_i}\eta_{\Omega_i,I_i} = \sigma^{\#}_{n_i}\pi^{\#}_{d}\eta_{\Omega_i,I_i} = \sigma^{\#}_{n_i}\eta_{\Omega_i,I_i} = \nu_i,$$

which yields (iii). Similarly, taking (4.16) and (4.17) into account, we verify (iv):

$$\begin{aligned}\pi^{\#}_{\mathrm{ad}}\nu_0 &= \frac{1}{|I_0|}\pi^{\#}_{\mathrm{ad}}\sigma^{\#}\eta_0 \otimes \tilde{\tau}_{\mathrm{Conv}B} = \frac{1}{|I_0|}\sigma^{\#}\left(\pi^{\#}_{d} \otimes \mathrm{id}\right)\left(\eta_0 \otimes \tilde{\tau}_{\mathrm{Conv}B}\right)\\ &= \frac{1}{|I_0|}\sigma^{\#}\left(\eta_0 \otimes \tilde{\tau}_{\mathrm{Conv}B}\right) = \nu_0.\end{aligned}$$

This proves ($\Gamma 2$).

(b) We assert that each $\nu \in \Gamma_C$ can be approximated by measures ν_B, $B \in \mathcal{B}(C', C)$. In the general case, it is sufficient to see that ν is the limit as $\varepsilon \to 0$ of the measures $(1+\varepsilon)\nu \in \Gamma_{(1+\varepsilon)C}$, where $(1+\varepsilon)C$ is the dilation of the body C with coefficient $1+\varepsilon$ relative to a point $O \in C$, and we have $\mathrm{dist}(\partial C', \partial((1+\varepsilon)C)) > 0$ and $C' \subset (1+\varepsilon)C \subset C''$ for sufficiently small $\varepsilon > 0$.

The proof proceeds in two steps: (v) each measure $\nu \in \Gamma_C$ can be approximated by measures of the form $\nu_k \in \Gamma_{C_k}$, where C_k is a convex polygon, $C' \subset C_k \subset C$, and $\mathrm{dist}(\partial C', \partial C_k) > 0$; (vi) in turn, any $\nu_k \in \Gamma_{C_k}$ can be approximated by the measures ν_B, where $\mathrm{Conv}B = C_k$ and $C' \subset B$. The combination of (v) and (vi) immediately yields the converse statement of the theorem.

We prove (v). A partitioning of S^1 into finitely many arcs, $S^1 = \cup_i S^i$, induces a partitioning of the boundary ∂C into arcs: $\partial C^i = \{\xi \in \partial C : n(\xi) \in S^i\}$. Consider a polygon $\check{C}$ inscribed in ∂C, with sides inscribed in the arcs ∂C^i, and let n_i be the outward normal to its ith side. Let s_{v_1,v_2} be the rotation of S^1 taking v_1 to v_2, and consider the map $\Upsilon_i : (S^1)^2 \times S^i \to (S^1)^2$ defined by $\Upsilon_i(v, v^+, n) = (s_{n,n_i}v, s_{n,n_i}v^+)$. Let l^i be the ith side of the polygon $\check{C}$ and define the measure

$$\check{\nu} = \sum_i \frac{|l^i|}{|\partial C^i|}\,\Upsilon^{\#}_i\nu \otimes \delta_{n_i}.$$

It is easy to verify that $\check{\nu} \in \Gamma_{\check{C}}$.

We consider a sequence $\{\mathcal{S}_k^i\}_i,\ k = 1, 2, \ldots$ of partitions of the circle, where the size of the largest arc approaches zero as $k \to \infty$. Let $\{\partial C_k^i\}_i,\ k = 1, 2, \ldots$ be the sequence of induced partitions ∂C, let C_k be a polygon with sides inscribed in the arcs of the kth partition, let l_k^i be the ith side of this polygon, and let n_{ik} be the outward normal to the ith side. It is clear that $C_k \subset C$, and starting from some k we also have $C' \subset C_k$ and $\operatorname{dist}(\partial C', \partial C) > 0$. Moreover,

$$\max_i \frac{|l_k^i|}{|\partial C_k^i|} \to 1 \quad \text{as} \quad k \to \infty. \tag{4.36}$$

As above, we define the maps $\Upsilon_{ik} : (S^1)^2 \times \mathcal{S}_k^i \to (S^1)^2$ and the measures $\nu_k = \sum_i \frac{|l_k^i|}{|\partial C_k^i|} \Upsilon_{ik}^{\#} \nu \otimes \delta_{n_{ik}} \in \Gamma_{C_k}$.

We assert that the ν_k converge weakly to ν. Indeed, for a continuous function f on $\mathbb{T}^3$

$$\iiint_{\mathbb{T}^3} f(v, v^+, n)\, d\nu_k(v, v^+, n) = \sum_i \frac{|l_k^i|}{|\partial C_k^i|} \iint_{\mathbb{T}^2} f(v, v^+, n_{ik})\, d\Upsilon_{ik}^{\#}\nu(v, v^+) =$$

$$= \sum_i \frac{|l_k^i|}{|\partial C_k^i|} \iiint_{(S^1)^2 \times \mathcal{S}_k^i} f(\Upsilon_{ik}(v, v^+, n), n_{ik})\, d\nu(v, v^+, n). \tag{4.37}$$

The map of $\mathbb{T}^3$ onto itself, which for $n \in \mathcal{S}_k^i$ is defined by

$$(v, v^+, n) \mapsto (\Upsilon_{ik}(v, v^+, n), n_{ik}),$$

converges uniformly to the identity map as $k \to \infty$. Consequently, the integrand in (4.37) converges uniformly to $f(v, v^+, n)$ as $k \to \infty$. Hence it follows from (4.36) that the expression in (4.37) approaches $\iiint_{\mathbb{T}^3} f(v, v^+, n)\, d\nu(v, v^+, n)$ as $k \to \infty$. Thus, we have shown that $\int f d\nu_k \to \int f d\nu$, as required.

We now prove (vi). Let C be a polygon. We must show that each measure $\nu \in \Gamma_C$ can be approximated by measures ν_B, where $\operatorname{Conv} B = C$ and $C' \subset B$.

The surface measure of C is $\tau_C = \sum_i |l_i| \delta_{n_i}$, where l_i is the ith side of C and n_i is the outward normal to l_i. Hence each $\nu \in \Gamma_C$ has the form $\nu = \sum_i |l_i| \sigma_{n_i}^{\#} \eta_i$, where $\eta_i \in \mathcal{M}$. Recall that $\sigma_{n_0}(\varphi, \varphi^+) = (-e^{i\varphi} n_0, e^{i\varphi^+} n_0, n_0)$. By Theorem 4.1, each measure η_i can be approximated by measures associated with hollows. Assume that the measure η_{Ω_i, I_i} associated with a hollow (Ω_i, I_i) approximates η_i. For each i we construct a system $\{(\Omega_i^k, I_i^k)\}_k$ of copies (Ω_i, I_i) (obtained by isometries and dilations) of the hollow Ω_i, I_i such that these copies are mutually disjoint and do not intersect C' but lie in C; their inlets I_i^k lie on l_i, and the proportion $\delta_i = 1 - |\cup_k I_i^k| / |l_i|$ of this side not covered by such inlets is small. This construction is similar to the construction of a system of hollows on a mushroom cap in Theorem 4.1; for details see [50]. See also Fig. 4.11, where we depict a system of hollows on one side of a polygon.

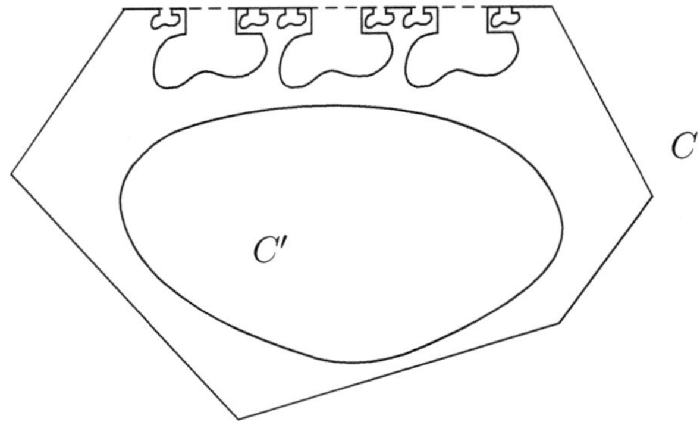

Fig. 4.11 System of similar hollows on a side of a polygon

Let $B = C \setminus \left(\cup_{i,k} \Omega_i^k\right)$. The measure

$$\nu_B = \sum_i |l_i| \sigma^{\#}_{n_i} \left((1-\delta_i)\eta_{\Omega_i, I_i} + \delta_i \eta_0\right)$$

approximates ν. Thus, the proof of assertion (vi) is also complete, which finishes the proof of the theorem. □

4.2 Scattering by the Surface of Rough Bodies

It is much more difficult to characterize scattering in the exterior of a body in dimensions $d \geq 3$ than in the two-dimensional case. The trick using hollows does not work in higher dimensions. We explain this by two examples of three-dimensional bodies: (a) a cylindrical shell B_1 and (b) a "spool" B_2 (Fig. 4.12). In both cases, $\Omega = \mathrm{Conv} B \setminus B$ is a connected set and $I = \partial(\mathrm{Conv} B) \setminus \partial B$ lies on its boundary. Thus, Ω is an analog of a hollow and I is an analog of its inlet. In case (a), Ω_1 is a cylinder and I_1 is the union of its two ends, whereas in case (b), Ω_2 is a cylindrical shell and I_2 is its outer cylindrical surface. In both cases, we do not have the main property of two-dimensional hollows, which is the existence of a vector n such that $v \cdot n < 0$ and $v^+ \cdot n > 0$ for each particle that enters the hollow, where v and v^+ are the initial and final velocities of the particle.

To cope with this complication, we can reasonably simplify the problem so that there is still an analogy with hollows in the new setting. We shall define a new object, a *convex rough body*. At first sight, a rough body looks like a convex body C, but its surface has microscopic flaws that influence the scattering of impinging particles. It is as if a convex part of some device has gotten small pock marks and cracks on its surface after long usage. The "microscopic structure" of the surface of a rough body reveals itself only in observations of the scattering of particles striking it. From this point of view two rough bodies are viewed as equivalent if they scatter flows of particles in the same way. Bearing in mind these heuristics, we now present the definition of a rough body.

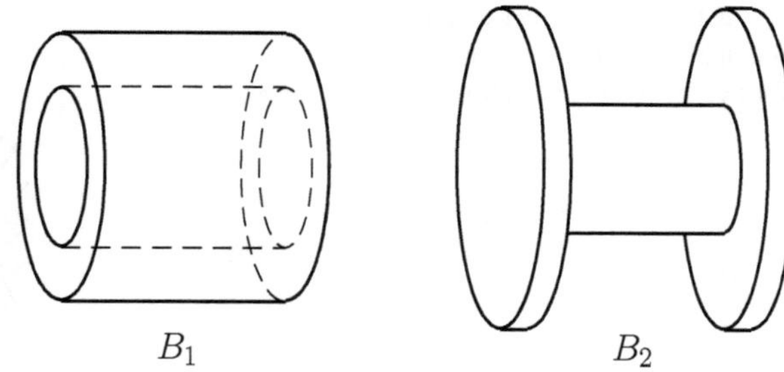

Fig. 4.12 A cylindrical shell B_1 and a "spool" B_2

Definition 4.8. We say that a sequence $\{B_m,\ m = 1, 2, \ldots\}$ of bodies represents a *rough body* obtained by *grooving* a convex body C if:

(R1) $B_m \subset C$ and $\text{Vol}(C \setminus B_m) \to 0$ as $m \to \infty$.
(R2) The sequence of measures $\nu_{B_m,C}$ is weakly convergent.

Two such sequences are said to be equivalent if the corresponding limit measures are equal and the ambient body C is the same. A rough body is an equivalence class of such sequences. We denote a rough body by $\mathcal{B}$ and the corresponding limit measure by $\nu_{\mathcal{B}}$. The measure $\nu_{\mathcal{B}}$ is called the law of billiard scattering by the rough body $\mathcal{B}$.

For brevity we also say that $\mathcal{B}$ is obtained by *grooving* the convex body C.

Remark 4.8. Since $(S^{d-1})^3$ is compact and the total measure $\nu_{B,C}\left((S^{d-1})^3\right)$ is at most $|\partial C|$, for fixed C the sequence $\nu_{B_m,C}$ is a weakly precompact set and therefore contains a weakly convergent subsequence. In this sense, we can think of a sequence of bodies satisfying condition (R1) but not (R2) as representing more than one body.

Remark 4.9. Note that the bodies B_m in Definition 4.8 are not necessarily connected. In informal terms, we allow rough bodies to "split" by microscopic cracks. However, if we add the condition that B_m is connected to the definition, then in fact the class of rough bodies will not get smaller; "microscopic cracks" have no effect on scattering.

Recall that the conditional measures $\nu\rfloor_n$ were defined at the beginning of this chapter by (4.4).

Definition 4.9. Assume that $\xi \in \partial C$ and $n = n(\xi)$ is not an atom of τ_C; then the conditional measure $\nu_{\mathcal{B}}\rfloor_n$ is called the *billiard scattering law on $\mathcal{B}$ at point ξ*.

We now consider two examples.

Example 4.6 (Nonrough Case). A rough body represented by the constant sequence $B_m = C$ is identified with the convex body C itself. The corresponding measure ν_C^{e} can be expressed as

$$\int_{(S^{d-1})^3} f(v, v^+, n)\, d\nu_C^{\text{e}}(v, v^+, n) = \int_{S^{d-1}} \int_{S^{d-1}} f(v,\ v - 2(v \cdot n)n,\ n)\, (v \cdot n)_-\, b_d\, dv\, d\tau_C(n). \tag{4.38}$$

Fig. 4.13 A two-dimensional rough body with the surface formed by triangular hollows

It describes the elastic reflection of particles by C. The scattering law at a point on a body surface $\nu_C^{\mathrm{e}}\rfloor_n$ is the well-known law of elastic reflection; it is written as

$$d\nu_C^{\mathrm{e}}\rfloor_n(v, v^+) = b_d\,\delta(v^+ - v + 2(v\cdot n)n)\,d\lambda_{-n}(v).$$

Example 4.7 (Rough Surface Formed by Triangular Hollows). Consider a two-dimensional convex body $C \subset \mathbb{R}^2$ and a sequence of convex m-gons inscribed in C, with sides of maximal length tending to zero as $m \to \infty$. A set B_m is obtained from the corresponding m-gon by removing m isosceles right triangles (hollows) based on the sides of the m-gon (as hypotenuses) and pointing inward. See Fig. 4.13 in the case where C is a disc.

The measure of the rough set represented by the sequence of B_m has the form $\nu_{\mathcal{B}} = \sigma^{\#}\eta_\nabla \otimes \tau_C$, where $\eta_\nabla \in \mathcal{M}$ is a measure with support shown in Fig. 4.13b (and is uniquely determined by this support). The density of η_∇ is defined by (4.11), and the support of η_∇ is depicted in Fig. 4.2b.

The scattering law at each point on the surface of $\mathcal{B}$ is generated by the measure η_∇; more precisely, $\nu_{\mathcal{B}}\rfloor_n = \pi^{\#}_{v,v^+}\sigma^{\#}_n\eta_\nabla$.

We now define a d-dimensional hollow. Let $n \in S^{d-1}$ and $S^{d-1}_n := \{v \in S^{d-1} : v\cdot e \geq 0\}$.

Definition 4.10. A pair (Ω, I) is called a d-dimensional hollow (or an n-hollow) if

(i) Ω is a subset of $\mathbb{R}^d$, is homeomorphic to a ball, and has a piecewise smooth boundary;
(ii) $I \subset \partial\Omega$, and, moreover,
(iii) I lies in a hyperplane $x \cdot n = c$;
(iv) $\Omega \setminus I$ lies in the open half-space $x \cdot n < c$.

The set I is called the *inlet of the hollow*.

See Fig. 4.14 for an example of a three-dimensional hollow.

Although a multidimensional hollow does not arise as naturally as in two dimensions, an analysis of scattering in a hollow is nevertheless useful.

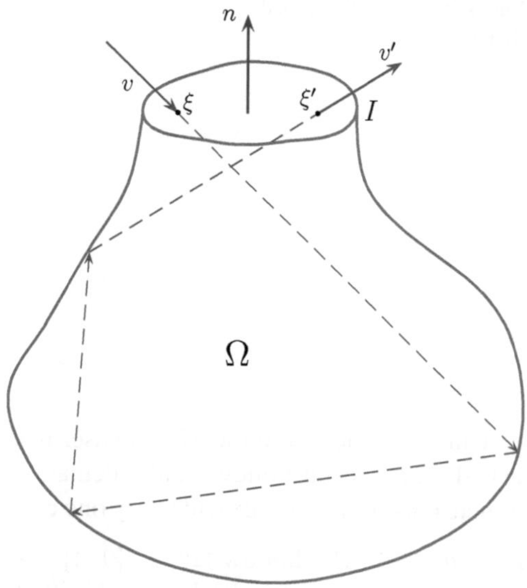

Fig. 4.14 A three-dimensional hollow

This is because the whole variety of scattering laws reduces to bodies with hollows on their surface, that is, to bodies B such that the difference $\mathrm{Conv}\, B \setminus B$ is the union of finitely many disjoint hollows. Indeed, any rough body can be represented by a sequence of bodies of this type, as follows from the proof of Theorem 4.5.

The measure associated with a hollow is defined as in the two-dimensional case. A billiard particle in Ω starts its motion from a point $\xi \in I$ and with velocity $v \in S^{d-1}_{-n}$. After reflecting from $\partial\Omega \setminus I$ several times it intersects I again, at some point ξ', and its velocity just before this is $v' \in S^{d-1}_n$. We consider the map $\xi' = \xi'_{\Omega,I}(\xi, v)$, $v' = v'_{\Omega,I}(\xi, v)$. It is one-to-one outside a zero-measure subset and preserves the probability measure $\tilde{\mu}_I$ defined by

$$d\,\tilde{\mu}_I(\xi, v) = b_d \cdot \frac{1}{|I|}\, |v \cdot n|\, d\xi\, dv.$$

This map induces a probability measure $\nu_{\Omega,I}$ on $(S^{d-1})^2$ supported on $S^{d-1}_{-n} \times S^{d-1}_n$ and given by

$$\nu_{\Omega,I}(A) := \tilde{\mu}_I\left(\{(\xi, v) : (v, v'_{\Omega,I}(\xi, v)) \in A\}\right)$$

for any Borel subset $A \subset S^{d-1}_{-n} \times S^{d-1}_n$.

Definition 4.11. The measure $\nu_{\Omega,I}$ is called the *measure associated with the hollow* (Ω, I).

Definition 4.12. The resistance of the hollow (Ω, I) equals $R(\nu_{\Omega,I})$, where

$$R(\nu) = \frac{d+1}{4} \iint_{(S^{d-1})^2} (1 - v \cdot v')\, d\nu(v, v').$$

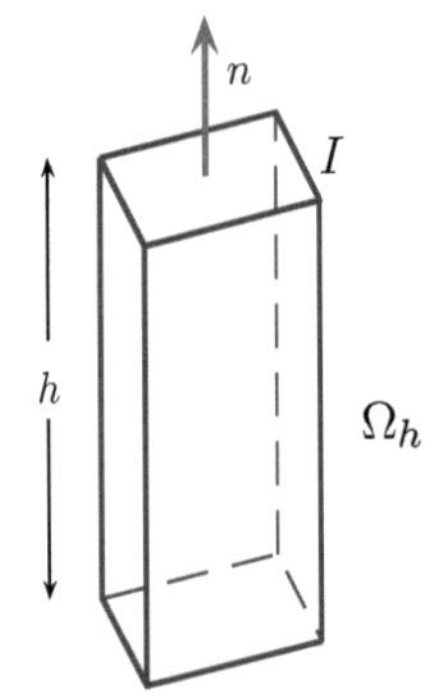

Fig. 4.15 A parallelepiped-shaped hollow

Unlike in the two-dimensional case, it is not easy to describe measures associated with special hollows and calculate the corresponding resistances in many dimensions. Below we provide only one example in three dimensions.

Example 4.8. The hollow $\Omega^h = [0,\, 1] \times [0,\, 1] \times [-h,\, 0]$ is a rectangular parallelepiped in $\mathbb{R}^3$, and its inlet $I = [0,\, 1]{\times}[0,\, 1]{\times}\{0\}$ is a unit square (Fig. 4.15). Thus, we have $n = (0, 0, 1)$. We adopt the notation $v = (v_x, v_y, v_z)$, $v' = (v'_x, v'_y, v'_z)$; the associated measure $\nu_{\Omega^h, I}$ is supported in $\{(v, v') : \ |v'_x| = |v_x|,\ |v'_y| = |v_y|,\ v'_z = -v_z\}$. A calculation similar to that in the two-dimensional case yields

$$\lim_{h\to\infty} R(\nu_{\Omega^h, I}) = 1.5.$$

Recall that the measure λ_n is defined at the beginning of this chapter by (4.1) and the map π_{ad} is given by (4.3).

Definition 4.13. $\mathcal{M}_n$ denotes the set of measures ν in $(S^{d-1})^2$ such that

(M1$_\text{n}$) $\pi_v^\# \nu = \lambda_{-n},\ \pi_{v+}^\# \nu = \lambda_n$;
(M2$_\text{n}$) $\pi_{\text{ad}}^\# \nu = \nu$.

If (Ω, I) is an n-hollow, then the measure $\nu_{\Omega, I}$ satisfies conditions (M1$_\text{n}$) and (M2$_\text{n}$), so that $\nu_{\Omega, I} \in \mathcal{M}_n$. A stronger result also holds.

Theorem 4.4. $\overline{\{\nu_{\Omega, I} : (\Omega, I) \textit{ is an } n\textit{-hollow}\}} = \mathcal{M}_n$.

That is, each measure in $\mathcal{M}_n$ is a weak limit of measures associated with hollows. The next basic result characterizes the laws of scattering by rough bodies.

Theorem 4.5. $\{\nu_{\mathcal{B}},\ \mathcal{B} \textit{ is obtained by grooving } C\} = \Gamma_C$.

A two-dimensional version of Theorem 4.5 was proved in [51], and its general form was published in [52].

The next corollary is a direct consequence of Theorem 4.5 and Proposition 4.1.

Corollary 4.2. *For almost all (in the sense of measure τ_C) values n the conditional measure $\nu = \nu_{\mathcal{B}}\rfloor_n$ describing the scattering at a point of a rough body (or in a flat face of a body orthogonal to n) satisfies the following relations:*

(a) $\pi_v^{\#}\nu = \lambda_{-n}$, $\pi_{v^+}^{\#}\nu = \lambda_n$;
(b) $\pi_{\mathrm{ad}}^{\#}\nu = \nu$.

4.2.1 Proof of Theorem 4.4

The general scheme of the proof is the same as for Theorem 4.1.

1. Approximating a retroreflecting measure. First we consider a retroreflector measure $\nu_{retr}^n \in \mathcal{M}_n$ supported on the subspace $v^+ = -v$. Hence for any continuous function f on $(S^{d-1})^2$

$$\nu_{\mathrm{retr}}^n(f) = \int_{S^{d-1}} f(v, -v)\, d\lambda_{-n}(v).$$

The measure ν_{retr}^n is a weak limit of measures $\nu_{\Omega_\varepsilon, I_\varepsilon}$ associated with mushrooms. Here a d-dimensional mushroom Ω_ε is the union of the half-ball $x_1 \le 0$, $|x| \le 1$ and the parallelepiped $[0, \varepsilon^2] \times [-\varepsilon, \varepsilon]^{d-1}$ in a reference system chosen so that $n = (1, 0, \dots, 0)$. The base of the mushroom is $I_\varepsilon = \{\varepsilon^2\} \times [-\varepsilon, \varepsilon]^{d-1}$. The proof of the indicated convergence is easy and can be found in [49].

2. Discretizing the measure ν. First, for arbitrary measure $\nu \in \mathcal{M}_n$ we discretize it as follows. We choose a reference system so that $n = e_1 := (1, 0, \dots, 0)$ and use the shorthand notation $\lambda := \lambda_{-e_1}$ and $S_\pm^{d-1} := S_{\pm e_1}^{d-1}$. We partition the hemisphere S_-^{d-1} into a large number m of subsets of small diameter having the same λ-measure $1/m$. For instance, we can take $m = \tilde{m}^{d-1}$ and carry out the partitioning by hyperplanes parallel to coordinate hyperplanes. Hyperplanes orthogonal to e_2 partition the hemisphere into $\tilde{m}$ subsets of the same measure, after which we partition each of these subsets into $\tilde{m}$ equal parts by hyperplanes orthogonal to e_3, and so on. We use central symmetry to obtain the corresponding partition of S_+^{d-1}. Accordingly, $S_-^{d-1} \times S_+^{d-1}$ is partitioned into m^2 subsets $\Box_{ij}$ of the same $(\lambda \otimes \lambda)$-measure. A discretization of the measure ν is the matrix $\big(\nu(\Box_{ij})\big)_{i,j=1}^m$. It is symmetric and $\sum_j \nu(\Box_{ij}) = 1/m$.

In turn, we approximate this matrix by a symmetric matrix

$$D = \left(\frac{1}{M}\,\delta_{j-\sigma(i)}\right)_{i,j=1}^{M}$$

of larger size $M = ml$. Recall that

$$\delta_i = \begin{cases} 1 & \text{if } i = 0, \\ 0 & \text{if } i \ne 0, \end{cases}$$

and σ is an involutive permutation of the set $\{1, \dots, M\}$. As in the proof of Theorem 4.1, approximation here means that if we represent D by an $m \times m$ block matrix with blocks of size $l \times l$, then the sum of the entries in the (i, j)th block approximates $\nu(\Box_{ij})$. The matrix D contains a single 1 in each column and each row.

Accordingly, we further partition each element in the partition of the hemisphere S_-^{d-1} into l subsets of the same λ-measure (for instance, by parallel hyperplanes $x_2 = \text{const}$). Numbering the elements of this finer partition $J_1^-, \dots, J_M^-$, we denote by J_i^+ the set centrally symmetric to J_i^-. Thus, we have the partitions

$$S_-^{d-1} = \cup_{i=1}^M J_i^- \quad \text{and} \quad S_+^{d-1} = \cup_{i=1}^M J_i^+,$$

which are centrally symmetric to each other.

3. The scheme of the end of the proof. Our aim is to construct a one-parameter family of e_1-hollows $(\Omega[\varepsilon], I[\varepsilon])$ approximating D as $\varepsilon \to 0$. In other words, for all $(\xi, v) \in I[\varepsilon] \times S_-^{d-1}$ outside a subset of measure tending to zero we must have the following: if $v \in J_i^-$, then $v^+ = v'_{\Omega[\varepsilon], I[\varepsilon]}(\xi, v) \in J_{\sigma(i)}^+$. Hence the discretization of the measure $\nu_{\Omega[\varepsilon], I[\varepsilon]}$ associated with this hollow approaches D, which will prove Theorem 4.4.

We construct a two-parameter family of hollows $(\Omega_{\varepsilon,\delta}, I_\varepsilon)$, "mushrooms with corroded caps": the cap of each mushroom will carry "hollows of second order." The parameter ε is the size of the hollow's inlet, and δ is the maximum of the relative sizes of inlets of "hollows of second order." Next we select a diagonal family of hollows $\Omega_{\varepsilon,\delta(\varepsilon)}$, $\lim_{\varepsilon \to 0}(\delta(\varepsilon)/\varepsilon) = 0$ and show that it approximates the matrix D in the sense described above.

For $i = 1, \dots, M$ we define a system of i-reflectors, second-order hollows with inlets in the spherical domain J_i^-. Two cases are possible: (a) $\sigma(i) = i$ and (b) $\sigma(i) \neq i$.

(a) If $\sigma(i) = i$, then the corresponding system of i-reflectors is empty. For all $(\xi, v) \in I_\varepsilon \times J_i^-$ outside a subset of measure $O(\varepsilon)$ the corresponding particle reflects once from J_i^- and flies out of I_ε. The velocity v^+ of the reflected particle belongs to J_i^+.

The main difficulties are related to case (b) when $j = \sigma(i) \neq i$. We fix i and j and focus on constructing systems of i- and j-reflectors.

4. Constructing a reflector and its properties. Let v_1, v_2, e_0 be a triple of unit coplanar vectors such that $v_1 \cdot e_0 > 0$ and $v_2 \cdot e_0 < 0$. We shall define a (v_1, v_2, e_0, δ)-reflector using Fig. 4.8 from Sect. 4.1.1 for illustration. The plane of the figure is identified with the two-dimensional subspace spanned by v_1 and v_2, and point O is at the origin. Furthermore, the vector e_{φ_1} is identified with v_1, and e_{φ_2} is identified with $-v_2$. Thus, according to the figure vectors v_1 and $-v_2$ form angles φ_1 and φ_2 with e_0.

Recall that $I(\delta)$ is a horizontal segment of length 2δ with midpoint O, the circle in the figure has radius 1 and is tangent to $I(\delta)$ at O, and p_1 and p_2 are arcs of parabolas with focus at O and axis parallel to $x^{(1)}x^{(2)}$. We consider the truncated cone K defined by

$$x \cdot e_0 \geq 0, \qquad \frac{x + \delta \tan \delta \cdot e_0}{|x + \delta \tan \delta \cdot e_0|} \cdot e_0 \geq \sin \delta.$$

Let OD be the bisector of the angle $x^{(1)}Ox^{(2)}$, and consider the hyperplane through it orthogonal to the plane of the figure (that is, the hyperplane containing OD and the $(d-2)$-dimensional subspace $\{v_1, v_2\}^{\perp}$). This hyperplane partitions the cone K into two subsets: let Π_L be the left-hand and Π_R the right-hand one. We consider the curvilinear triangle ABC bounded by the segment AC of the horizontal line and arcs of the parabolas p_1 and p_2. The line OD partitions it into two figures, the left one in Π_L and the right one in Π_R. Let B_L and B_R be the bodies obtained by revolution of the left and right figures, respectively, about the axis OD, and let

$$R(v_1, v_2, e_0, \delta) = (B_L \cap \Pi_L) \cup (B_R \cap \Pi_R).$$

The inlet of the reflector $R(v_1, v_2, e_0, \delta)$ is its intersection with the hyperplane $x \cdot e_0 = 0$. This is a $(d-1)$-dimensional disc $I(e_0, \delta)$ obtained by rotating the segment $I(\delta)$ about e_0.

A set obtained from $R(v_1, v_2, e_0, \delta)$ by isometry and dilation will also be called a (v_1, v_2, e_0, δ)-reflector.

In the three-dimensional case $d = 3$, a reflector $R(v_1, v_2, e_0, \delta)$ is easy to imagine: take a body obtained by revolution of the curvilinear triangle ABC about the line OD through 90° counterclockwise and through 90° clockwise and consider its intersection with the truncated cone K. The inlet of the reflector is a disc of radius δ with center at O lying in the plane orthogonal to e_0. The truncated cone is centrally symmetric relative to the axis Oe_0, and its intersection with ABC is the union of three segments indicated by bold lines in Fig. 4.8 [$I(\delta)$ is the central segment].

Consider the map

$$\xi' = \xi'_{R(v_1,v_2,e_0,\delta),I(e_0,\delta)}(\xi, v), \quad v' = v'_{R(v_1,v_2,e_0,\delta),I(e_0,\delta)}(\xi, v)$$

generated by the reflector. As in the two-dimensional case, $v'(0, v_1) = v_2$ and $\xi'(0, v) = 0$ for all v lying in some neighborhood of $v_1 : v \in \mathcal{O}(v_1) \subset S^{d-1}$. Hence

$$\frac{D\xi'}{Dv}(0, v_1) = 0.$$

The differential $\frac{Dv'}{Dv}(0, v_1)$ is a linear map between the tangent spaces $T_{v_1}S^{d-1}$ and $T_{v_2}S^{d-1}$ of S^{d-1} at points v_1 and v_2. These tangent spaces contain the subspace $\{v_1, v_2\}^{\perp}$, so they can be written as direct sums:

$$T_{v_1}S^{d-1} = \{w_1\} \otimes \{v_1, v_2\}^{\perp}, \quad T_{v_2}S^{d-1} = \{w_2\} \otimes \{v_1, v_2\}^{\perp},$$

where vectors w_1 and w_2 lie in the plane $\{v_1, v_2\}$ of Fig. 4.8 and are orthogonal to v_1 and v_2, respectively (the pairs v_1, w_1 and v_2, w_2 define the same orientation).

Let $\tilde{K} \subset S^{d-1}$ be the set obtained by rotating v_1 about OD. This set is a $(d-2)$-dimensional sphere containing $-v_2$. The tangent spaces of $\tilde{K}$ at v_1 and $-v_2$ coincide with $\{v_1, v_2\}^\perp$. Since our construction has rotational symmetry, the triangle obtained by rotating $x^{(1)}Ox^{(2)}$ about OD by an angle less than $\pi/2$ is also the trajectory of a billiard particle. Hence if $v \in \tilde{K}$, then also $-v' \in \tilde{K}$ and $v - v_1 = v' - v_2$. This means that the restriction of the differential $\frac{Dv'}{Dv}(0, v_1)\rfloor_{\{v_1,v_2\}^\perp}$ is the identity map

$$\frac{Dv'}{Dv}(0, v_1)\bigg\rfloor_{\{v_1,v_2\}^\perp} = \mathrm{id}. \tag{4.39}$$

Finally, considering the two-dimensional reflector, we see that on the additional vector w_1 the differential acts as follows:

$$\frac{Dv'}{Dv}(0, v_1)\, w_1 = -\frac{\cos\varphi_1}{\cos\varphi_2}\, w_2. \tag{4.40}$$

We look at the trajectory of a perturbed particle reflecting from points $x^{(1)} + z$ and $x^{(2)} + z$, where $z \in \{v_1, v_2\}^\perp$ is an infinitesimal perturbation. (Henceforth in this paragraph we neglect quantities of order z^2 or higher.) The middle part of this trajectory is parallel to the line $x^{(1)}x^{(2)}$ and passes through $D + z$. The initial and final points, ξ and ξ', of the trajectory are infinitesimally close to O. Considering the initial part of the trajectory marked by the points ξ, $x^{(1)} + z$, $D + z$, we conclude that the vector $\frac{D-x^{(1)}}{|D-x^{(1)}|}$ is obtained from $\frac{x^{(1)}+z-\xi}{|x^{(1)}+z-\xi|}$ by reflection at the point $x^{(1)}+z$. On the other hand, the points O, $x^{(1)} + z$, D define the trajectory of another billiard particle [the triangle $x^{(1)}Ox^{(2)}$ rotated through an angle z around OD]; therefore, $\frac{D-x^{(1)}-z}{|D-x^{(1)}-z|}$ is obtained from the vector $\frac{x^{(1)}+z}{|x^{(1)}+z|}$ by reflection at the same point $x^{(1)}+z$. Comparing these two reflections and bearing in mind that z is approximately orthogonal to the normal at $x^{(1)} + z$, we see that $-z/|D - x^{(1)}| = \xi/|x^{(1)}|$, and therefore $\xi = -\frac{|x^{(1)}|}{|D-x^{(1)}|}z$. Analyzing similarly the final part of the trajectory, which is marked by the points $D + z$, $x^{(2)} + z$, ξ', we finally get that $\xi' = -\frac{|x^{(2)}|}{|D-x^{(2)}|}z$. Since OD is a bisector in $x^{(1)}Ox^{(2)}$, it follows that $\frac{|x^{(1)}|}{|D-x^{(1)}|} = \frac{|x^{(2)}|}{|D-x^{(2)}|}$, and therefore $\xi = \xi'$.

Hence we obtain

$$\frac{D\xi'}{D\xi}(0, v_1)\xi + \frac{D\xi'}{Dv}(0, v_1)\delta v = \xi,$$

where $\delta v = (\frac{1}{|x^{(1)}|} + \frac{1}{|D-x^{(1)}|})z$ is the difference between the initial velocity of the particle and v_1. Taking into account that $\frac{D\xi'}{Dv}(0, v_1) = 0$ and that ξ is an arbitrary infinitesimal vector in $\{v_1, v_2\}^\perp$, we see that the restriction of $\frac{D\xi'}{D\xi}(0, v_1)$ to $\{v_1, v_2\}^\perp$ is the identity map

$$\frac{D\xi'}{D\xi}(0, v_1)\Big|_{\{v_1,v_2\}^\perp} = \text{id}.$$

Let e_2 be a vector parallel to $I(\delta)$. We can see from analyzing a two-dimensional reflector [formula (4.21) and the specification of the sign after it] that $\frac{D\xi'}{D\xi}(0, v_1)$ also acts as the identity on this vector—$\frac{D\xi'}{D\xi}(0, v_1)e_2 = e_2$—so that the map $\frac{D\xi'}{D\xi}(0, v_1)$ is the identity $\frac{D\xi'}{D\xi}(0, v_1) = \text{id}$. Now an important conclusion: if we consider incident particles with velocities lying in an interval of size $O(\varepsilon)$ about v_1, $v = v_1 + \delta v$, $\delta v = O(\varepsilon)$, then after two reflections near $x^{(1)}$ and $x^{(2)}$, all but a fraction of them of order $O(\varepsilon) + O(\delta)$ go out across the inlet, and the velocity v' of an outgoing particle is close to v_2 and can be calculated by the formula

$$v' = v_2 + \frac{Dv'}{Dv}(0, v_1)\delta v + O(\delta) + O(\varepsilon^2).$$

5. Completing the construction. Below we establish a one-to-one correspondence $\theta : J_i^- \to J_j^-$ preserving the measure λ, and we place finitely many reflectors in such a way that each reflector with center at rv, $v \in J_i^-$ ($v \in J_j^-$) is a $(v, \frac{\theta(v)-v}{|\theta(v)-v|}, v, \delta)$-reflector [resp. a $(v, \frac{\theta^{-1}(v)-v}{|\theta^{-1}(v)-v|}, v, \delta)$-reflector]. Here $\delta = \delta(\varepsilon) = o(\varepsilon)$ as $\varepsilon \to 0$. Each reflector lies in one of the cones with vertex at O and generatrices J_i^- and J_j^- (correspondingly, we distinguish between the i- and j-reflectors). The reflectors are disjoint, their inlets are bases of spherical segments, and the reflectors are directed outside with respect to the hemisphere S_-^{d-1}. The subdomains of J_i^- and J_j^- not covered by bases of reflectors have a proportion of at most ε.

We carry out further calculations to within $O(\varepsilon^2) + O(\delta)$. Let us identify a point $\xi = (\xi_1, \xi_2, \dots, \xi_d) \in I_\varepsilon = \{\varepsilon^2\} \times [-\varepsilon, \varepsilon]^{d-1}$ with a point $(\xi_2, \dots, \xi_d) \in \mathbb{R}^{d-1}$. Assume that a particle goes from the point $\xi = 0$ in the direction $v \in J_i^-$. It will be reflected from an i-reflector at v and from a j-reflector at the point $\tilde{v} = \theta(v)$, after which it returns to 0 with velocity $v^+ \in J_j^+$.

Consider now a particle flying from an arbitrary point $\xi \in [-\varepsilon, \varepsilon]^{d-1}$. Our purpose is to ensure that after being reflected from an i-reflector and a j-reflector it passes through $-\xi$ and, therefore, goes out through the inlet of the reflector with velocity $v^+ \in J_j^+$. [Here and in what follows, our assertions hold for all values outside a subset of measure $o(1)$ as $\varepsilon \to 0^+$]. This will complete the proof of the theorem.

To ensure this property, we construct a map θ of a special form. Assume that a particle flying from a point re_k, $k = 2, \dots, d$, $|r| < \varepsilon$, and reflected from an i-reflector at $v \in J_i^-$ then intersects the hemisphere S_-^{d-1} at a point $\tilde{v} + r\check{p}_k$, where $\tilde{v} = \theta(v)$. The projection of $\check{p}_k = (p_k^1, p_k^2, \dots, p_k^d)$ on the plane $\mathbb{R}^{d-1}_{\{e_2,\dots,e_d\}}$, that is, $p_k = (p_k^2, \dots, p_k^d)$, can be expressed as a function of v and $\tilde{v}$: $p_k = p_k(v, \tilde{v})$.

In a similar way, if a particle leaves re_k in the direction of J_j^-, then after reflection from a j-reflector at a point $\tilde{v}$ it intersects S_-^{d-1} at $v + r\check{p}_k$, where

$v = \theta^{-1}(\tilde{v})$. The projection of $\check{p}_k$ on $\mathbb{R}^{d-1}_{\{e_2,\dots,e_d\}}$ is also a function of $\tilde{v}$ and v: $p_k = p_k(\tilde{v}, v)$. Generally speaking, it is not symmetric: $p_k(v, \tilde{v},) \neq p_k(\tilde{v}, v)$. The volume of a parallelepiped generated by the system of vectors $\{p_k(v, \tilde{v}),\ k = 2, \dots, d\}$ is equal to the volume of the parallelepiped generated by $\{p_k(\tilde{v}, v),\ k = 2, \dots, d\}$. We prove this equality of volumes for the three-dimensional case in the last part, item 6.

Let $\tilde{J}_i^-$ and $\tilde{J}_j^-$ be the orthogonal projections of J_i^- and J_j^- on $\mathbb{R}^{d-1}_{\{e_2,\dots,e_d\}}$. The map θ induces a bijective map $\tilde{\theta} : \tilde{J}_i^- \to \tilde{J}_j^-$ that preserves the Lebesgue measure. We define the map $\tilde{\theta}$ so that its derivative takes the vectors $p_k(\tilde{v}, v)$ to the vectors $p_k(v, \tilde{v})$, $k = 2, \dots, d$. That is, first we partition $\tilde{J}_i^-$ and $\tilde{J}_j^-$ into equal numbers of parallelepipeds with edge length of order $\varepsilon^{1/3}$, and with each parallelepiped in $\tilde{J}_i^-$ we associate a parallelepiped in $\tilde{J}_j^-$. In each parallelepiped in $\tilde{J}_i^-$ and in the corresponding parallelepiped in $\tilde{J}_j^-$ we select a point and denote by v and $\tilde{v}$ the lifts of these points to S^{d-1}_-. We consider the partitioning of the first parallelepiped by the lattice formed by periodic translations of the small parallelepiped $\{\varepsilon^{2/3} p_k(\tilde{v}, v),\ k = 2, \dots, d\}$, and the partitioning of the second parallelepiped by the lattice formed by $\{\varepsilon^{2/3} p_k(v, \tilde{v}),\ k = 2, \dots, d\}$ (recall that these parallelepipeds have the same volume). The piecewise linear map $\tilde{\theta}$ takes the small parallelepipeds lying fully in the first parallelepiped to small parallelepipeds lying fully in the second parallelepiped. The part of the first parallelepiped (of negligibly small volume) that we were not able to map in this procedure will be mapped in an arbitrary fashion.

We now return to the consideration of a particle flying from a point re_k and reflecting from an i-reflector at a point v. Thus, the scheme of its reflections on the initial part of its trajectory has the form $re_k \mapsto v \mapsto \tilde{v} + r\check{p}_k(v, \tilde{v})$. Assume that the ε-neighborhoods of points v and $\tilde{v}$ contain no discontinuities of the function θ [this condition holds outside a subset of measure $O(\varepsilon)$]. Then θ maps $v' := v + r\check{p}_k(\tilde{v}, v)$ to $\tilde{v}' := \tilde{v} + r\check{p}_k(v, \tilde{v})$, so that the scheme of reflections on the final part of the trajectory of some other particle has the form $v' \mapsto \tilde{v}' \mapsto 0$. Perturbing this trajectory, we obtain the scheme $v' - r\check{p}_k(\tilde{v}, v) \mapsto \tilde{v}' \mapsto -re_k$, which enables us to recover the full scheme of reflections of the original particle: $re_k \mapsto v \mapsto \tilde{v}' \mapsto -re_k$. This is what we require.

6. Equality of areas in the three-dimensional case. Let $d = 3$. We consider two points v and $\tilde{v}$ in S^2_- and select a system of coordinates such that these points become

$$v = (\sin\varphi,\ \cos\varphi \sin\vartheta,\ \cos\varphi\cos\vartheta) \quad \text{and} \quad \tilde{v} = (\sin\tilde{\varphi},\ \cos\tilde{\varphi}\sin\vartheta,\ \cos\tilde{\varphi}\cos\vartheta).$$

Using properties (4.39) and (4.40), we consider two partial schemes of reflection. One of them has the form $\xi_1 = re_1 \mapsto v \mapsto \tilde{v} + r\check{p}_1(v, \tilde{v})$. Taking (4.40) into account, we can calculate $\check{p}_1(v, \tilde{v})$ and see that its projection on the horizontal plane is

$$p_1(v, \tilde{v}) = \frac{2\cos\varphi}{\sin\frac{\varphi - \tilde{\varphi}}{2}} (\cos\tilde{\varphi},\ -\sin\tilde{\varphi}\sin\vartheta).$$

Choosing

$$\xi' = r\,\frac{(\sin\varphi\sin\vartheta,\ \cos\varphi)}{\sqrt{\sin^2\varphi\sin^2\vartheta + \cos^2\varphi}}$$

and taking (4.39) into account, we obtain the scheme of reflections $\xi' \mapsto v \mapsto \tilde{v} + r\check{p}'(v, \tilde{v})$, where $\check{p}'$ has the projection

$$p'(v, \tilde{v}) = -\frac{2\cos\varphi\cos^2\vartheta\sin\frac{\varphi-\tilde{\varphi}}{2}}{\sqrt{\sin^2\varphi\sin^2\vartheta + \cos^2\varphi}}\,(0,\ 1).$$

Since the map $\xi \mapsto p$ is linear, we find the quantity $p_2(v, \tilde{v})$ determining a scheme $re_2 \mapsto v \mapsto \tilde{v} + r\check{p}_2(v, \tilde{v})$. The area of the parallelogram generated by the vectors $p_1(v, \tilde{v})$ and $p_2(v, \tilde{v})$ is

$$|p_1(v, \tilde{v}) \times p_2(v, \tilde{v})| = 4\cos\varphi\cos\tilde{\varphi}\cos^2\vartheta.$$

Similar arguments demonstrate that the area of the parallelogram generated by the vectors $p_1(\tilde{v}, v)$ and $p_2(\tilde{v}, v)$ is the same: $|p_1(\tilde{v}, v) \times p_2(\tilde{v}, v)| = 4\cos\varphi\cos\tilde{\varphi}\cos^2\vartheta$.

4.2.2 *Proof of Theorem 4.5*

(a) We assert that for each body $\mathcal{B}$ obtained by grooving C we have $\nu_{\mathcal{B}} \in \Gamma_C$. For the proof we use the following Lemma 4.2, which will be proved below.

For a one-to-one map $(\xi, v) \mapsto (\xi', v')$ between full-measure subsets of $(\partial C \times S^{d-1})_-$ and $(\partial C \times S^{d-1})_+$ that preserves the measure μ we set

$$\overline{|\xi - \xi'|} := \int_{(\partial C\times S^{d-1})_-} |\xi - \xi'(\xi, v)|\, d\mu(\xi, v),$$

$$\overline{|n - n'|} := \int_{(\partial C\times S^{d-1})_-} |n(\xi) - n(\xi'(\xi, v))|\, d\mu(\xi, v).$$

Lemma 4.2. *The inequalities*

$$(a) \qquad \overline{|\xi - \xi^+_{B,C}|} \le b_d|S^{d-1}| \cdot \mathrm{Vol}(C \setminus B)$$

$$(b) \qquad \overline{|n - n'|} \le f(\overline{|\xi - \xi'|})$$

hold, where f is a positive function of a real variable such that $\lim_{x\to 0} f(x) = 0$.

Note that the measures $\nu_{B,C}$ and $\nu'_{B,C}$ have the following properties:

$$\pi^{\#}_{v,n}\nu_{B,C} = \lambda_{-n} \otimes \tau_C, \tag{4.41}$$

$$\pi^{\#}_{v^+,n^+}\nu'_{B,C} = \lambda_n \otimes \tau_C, \tag{4.42}$$

$$\pi^{\#}_{\text{ad}}\nu_{B,C} = \nu'_{B,C}. \tag{4.43}$$

Let us consider a sequence $\{B_m\}$ representing $\mathcal{B}$. We assert that $\nu_{B_m,C} - \nu'_{B_m,C}$ converges weakly to zero. For the proof it suffices to show that for each continuous function f on $(S^{d-1})^3$,

$$\lim_{m\to\infty}\left(\int_{(S^{d-1})^3} f(v,v^+,n)\, d\nu_{B_m,C}(v,v^+,n) - \right.$$
$$\left. - \int_{(S^{d-1})^3} f(v,v^+,n^+)\, d\nu'_{B_m,C}(v,v^+,n^+)\right) = 0. \tag{4.44}$$

Since we have the change-of-variables formulae

$$\int_{(S^{d-1})^3} f(v,v^+,n)\, d\nu_{B,C}(v,v^+,n) = \int_{(\partial C\times S^{d-1})_-} f(v, v^+_{B,C}(\xi,v), n(\xi))\, d\mu(\xi,v),$$

$$\int_{(S^{d-1})^3} f(v,v^+,n^+)\, d\nu'_{B,C}(v,v^+,n^+)$$
$$= \int_{(\partial C\times S^{d-1})_-} f(v, v^+_{B,C}(\xi,v), n(\xi^+_{B,C}(\xi,v)))\, d\mu(\xi,v),$$

equality (4.44) takes the form

$$\lim_{m\to\infty}\int_{(\partial C\times S^{d-1})_-}\left[f(v,v^+_{B_m,C}(\xi,v), n(\xi^+_{B_m,C}(\xi,v))) - f(v,v^+_{B_m,C}(\xi,v),n(\xi))\right] d\mu(\xi,v) = 0. \tag{4.45}$$

It follows from assertion (a) in Lemma 4.2 that the function $\xi^+_{B_m,C}(\xi,v)$ converges to ξ in the mean. Hence it follows from assertion (b) of Lemma 4.2 that the function $n(\xi^+_{B_m,C}(\xi,v))$ converges to $n(\xi)$ in the mean and therefore also in the measure μ. Consequently, the function $f(v,v^+_{B_m,C}(\xi,v), n(\xi^+_{B_m,C}(\xi,v)))$ converges to $f(v,v^+_{B_m,C}(\xi,v),n(\xi))$ in the measure, which proves (4.45).

Thus, both sequences $\nu_{B_m,C}$ and $\nu'_{B_m,C}$ converge to $\nu_{\mathcal{B}}$. Substituting $B = B_m$ into (4.41)–(4.43), we obtain

$$\pi^{\#}_{v,n}\nu_{\mathcal{B}} = \lambda_{-n} \otimes \tau_C,$$

$$\pi^{\#}_{v^+,n}\nu_{\mathcal{B}} = \lambda_n \otimes \tau_C,$$

$$\pi^{\#}_{\text{ad}}\nu_{\mathcal{B}} = \nu_{\mathcal{B}}$$

in the limit $m \to \infty$. This proves that $\nu_{\mathcal{B}} \in \Gamma_C$.

(b) We prove that for each measure $\nu \in \Gamma_C$ there exists a rough body $\mathcal{B}$ obtained by grooving C such that $\nu_{\mathcal{B}} = \nu$.

First we state another lemma, which will be proved below.

Lemma 4.3. *For each measure $\nu \in \Gamma_C$ there exists a sequence of convex polyhedra $C_k \subset C$ and a sequence of measures $\nu_k \in \Gamma_{C_k}$ such that $\nu_k \to \nu$ and $Vol(C \setminus C_k) \to 0$ as $k \to \infty$.*

First we prove assertion (b) in the case where C is a convex polyhedron. We number its faces and denote the $(d-1)$-dimensional volume of the ith face by c_i and the unit outward normal to this face by n_i. Recall that δ_n is the probability measure on S^{d-1} concentrated at n, that is, $\delta_n(n) = 1$. The surface measure on C is $\tau_C = \sum c_i \delta_{n_i}$. Hence each measure $\nu \in \Gamma_C$ has the form $\nu = \sum_i c_i \nu_i \otimes \delta_{n_i}$, where $\nu_i \in \mathcal{M}_{n_i}$. By Theorem 4.4 each ν_i can be approximated by measures associated with n_i-hollows (Ω_{im}, I_{im}) as $m \to \infty$. As in the two-dimensional case we prepare a finite system of hollows produced from (Ω_{im}, I_{im}) by translations and dilations in such a way that (a) the images of the inlet I_{im} lie in the ith face of C and cover all of it but a subset of small $(d-1)$-dimensional volume $o(1)$ as $m \to \infty$; (b) all the images of Ω_{im} lie in C and are disjoint, and their total d-dimensional volume is $o(1)$ as $m \to \infty$.

We define the set B_m as a set-theoretic difference: C minus all the hollows that are the images of Ω_{im} for all i. The sequence B_m represents a rough body $\mathcal{B}$, and $\nu_{\mathcal{B}} = \lim_{m\to\infty} \nu_{B_m,C} = \nu$, as required.

Now let C be an arbitrary convex body and $\nu \in \Gamma_C$. Using Lemma 4.3, we approximate ν by measures $\nu_k \in \Gamma_{C_k}$. Next we find a body $\mathcal{B}_k$ obtained by grooving the convex polyhedron C_k such that $\nu_k = \nu_{\mathcal{B}_k}$. Each $\mathcal{B}_k$ is represented by some sequence of bodies $B_{k,m}$. We pick a diagonal subsequence $B_{k,m(k)}$ such that $m(k) \to \infty$, $\mathrm{Vol}(C \setminus B_{k,m(k)}) \to 0$, and $\nu_{B_{k,m(k)},C} \to \nu$ as $k \to \infty$. It represents a rough body with scattering law ν. The proof of Theorem 4.5 is complete.

4.2.3 Proof of Lemma 4.2

(a) Consider the billiard in $\mathbb{R}^d \setminus B$. For $(\xi, v) \in (\partial C \times S^{d-1})_-$ we denote by $\tau(\xi, v)$ the time during which a billiard particle with initial data (ξ, v) stays in $C \setminus B$. In particular, if $\xi \in \partial C \cap \partial B$ and ξ is a regular point of ∂B, then $\tau(\xi, v) = 0$.

Let D be the set of $(x, w) \in (C \setminus B) \times S^{d-1}$ attainable from $(\partial C \times S^{d-1})_-$. This means that a billiard particle with initial data $(\xi, v) \in (\partial C \times S^{d-1})_-$ corresponding to time zero will be at point x and have velocity w at some time t, $0 \le t \le \tau(\xi, v)$. This definition also determines a change $(\xi, v, t) \mapsto (x, w)$ of the variables in D, where $(\xi, v) \in (\partial C \times S^{d-1})_-$, $t \in [0, \tau(\xi, v)]$, and the phase volume element $dx\, dw$ has the form $\frac{1}{b_d} d\mu(\xi, v)\, dt$ in the new variables. Hence the phase volume of D is equal to

$$\int_D dx\,dw = \frac{1}{b_d}\int_{(\partial C\times S^{d-1})_-} \tau(\xi,v)\,d\mu(\xi,v).$$

Taking into account that $D\subset (C\setminus B)\times S^{d-1}$ and the phase volume of $(C\setminus B)\times S^{d-1}$ is $|S^{d-1}|\cdot \mathrm{Vol}(C\setminus B)$, we obtain

$$\int_{(\partial C\times S^{d-1})_-} \tau(\xi,v)\,d\mu(\xi,v)\;\le\; b_d|S^{d-1}|\cdot \mathrm{Vol}(C\setminus B). \tag{4.46}$$

This is a simple modification of the formula for the average length of a billiard path; see, for example, [14].

We have $\tau(\xi,v)\ge |\xi-\xi^+_{B,C}(\xi,v)|$: the a particle spends in $C\setminus B$ is larger than the distance between the endpoints of its trajectory. This inequality and relation (4.46) yield (a).

(b) Let $N_\xi\subset S^{d-1}$ be the cone of outward normals to C at a point $\xi\in\partial C$. In particular, if ξ is a regular point of ∂C, then $N_\xi=\{n(\xi)\}$ is a singleton. Otherwise, it contains more than one point. Let

$$D_\xi := \mathrm{diam}(N_\xi) = \max\{|n_1-n_2| : n_1,\, n_2\in N_\xi\}$$

be the diameter of N_ξ. The point ξ is singular if and only if $D_\xi>0$. Let S_ε be the set of points $\xi\in\partial C$ such that $D_\xi\ge\varepsilon$, and let S be the set of all singular points. It is clear that the family $\{S_\varepsilon,\ \varepsilon>0\}$ of sets is decreasing and $\cup_{\varepsilon>0}S_\varepsilon = S$.

We introduce some further notation. Let $r_\varepsilon(\xi)$ denote the infimum of the set of $r>0$ such that the oscillation of the normal in the set $\{\xi' : |\xi'-\xi|<r\}$ is less than ε, that is, $|n(\xi')-n(\xi'')|<\varepsilon$ for any ξ' and ξ'' such that $|\xi'-\xi|<r$ and $|\xi''-\xi|<r$. The function r_ε is Lipschitz: $|r_\varepsilon(\xi_1)-r_\varepsilon(\xi_2)|\le|\xi_1-\xi_2|$ and S_ε is the set of ξ such that $r_\varepsilon(\xi)=0$. Hence S_ε is a closed set, and since the set S of singular points has zero Lebesgue measure, so does S_ε. Consider an open neighborhood $\mathcal{U}(S_\varepsilon)\supset S_\varepsilon$ of measure less than ε. The function r_ε is positive on the closed set $\partial B\setminus\mathcal{U}(S_\varepsilon)$; let $\sigma=\sigma(\varepsilon)>0$ be its minimum there. Thus, for a pair of regular points $\xi\in\partial C\setminus\mathcal{U}(S_\varepsilon)$ and $\xi'\in\partial C$ we have $|n(\xi)-n(\xi')|<\varepsilon$ if $|\xi-\xi'|<\sigma$.

Consider the map $(\xi,v)\mapsto(\xi',v')$. The set $\{(\xi,v):$ either ξ or $\xi'(\xi,v)$ is singular$\}$ has measure zero. Let $\mathcal{C}_1=\big(\mathcal{U}(S_\varepsilon)\times S^{d-1}\big)_-$, let $\mathcal{C}_2$ be the set of $(\xi,v)\in\big((\partial C\setminus\mathcal{U}(S_\varepsilon))\times S^{d-1}\big)_-$ such that $|\xi-\xi'(\xi,v)|\ge\sigma$, and let $\mathcal{C}_3$ be the complement of $\mathcal{C}_1\cup\mathcal{C}_2$, that is, $\mathcal{C}_3=(\partial C\times S^{d-1})_-\setminus(\mathcal{C}_1\cup\mathcal{C}_2)$. Then

$$\mu(\mathcal{C}_1)\le\varepsilon\cdot|S^{d-1}|,\quad \mu(\mathcal{C}_2)\le\frac{1}{\sigma}\,\overline{|\xi-\xi'|}\ \text{ and }\ \mu((\partial C\times S^{d-1})_-)=|\partial C|,$$

and therefore

$$\overline{|n-n'|}=\left(\int_{\mathcal{C}_1}+\int_{\mathcal{C}_2}+\int_{\mathcal{C}_3}\right)|n(\xi)-n(\xi'(\xi,v))|\,d\mu(\xi,v)\le 2\varepsilon$$

$$+\frac{2}{\sigma(\varepsilon)}\overline{|\xi - \xi'|} + \varepsilon|\partial C|.$$

Hence if $\overline{|\xi - \xi'|} \le \varepsilon\sigma(\varepsilon)$, then $\overline{|n - n'|} \le f(\varepsilon\sigma(\varepsilon)) := \varepsilon\,(4 + |\partial C|)$. Here we define the function f as the largest increasing function on $\mathbb{R}_+$ satisfying the last equality. This completes the proof of (b).

4.2.4 *Proof of Lemma 4.3*

Consider a sequence of polyhedra $C_k \subset C$ such that $\mathrm{Vol}(C \setminus C_k) \to 0$ as $k \to \infty$. For example, we can take a lattice of cubes of size 2^{-k} and denote by C_k the convex hull of the union of all the cubes lying in C. We number the faces of the polyhedron C_k and denote its ith face by l_{ik} and the outward normal to it by n_{ik}. Let $\partial C'_{ik}$ be the intersection of ∂C with the set obtained by a parallel translation of the face l_{ik} along the vector n_{ik} in the positive direction. We consider a partition of ∂C into sets containing the $\partial C'_{ik}$; $\partial C = \cup_i \partial C_{ik}$, $\partial C_{ik} \supset \partial C'_{ik}$. Clearly, $|\partial C_{ik}| \ge |l_{ik}|$, $\sum_i |\partial C_{ik}| = |\partial C|$, and $\lim_{k\to\infty} \sum_i |l_{ik}| = |\partial C|$. Hence $|l_{ik}|/|\partial C_{ik}|$, as a function on ∂C, converges to 1 in the mean as $k \to \infty$.

Let $n(\partial C_{ik})$ be the set of vectors $n(\xi)$ with $\xi \in \partial C_{ik}$. For each i and k we pick a continuous family of rotations V^n_{ik}, $n \in n(\partial C_{ik})$, taking n to n_{ik}: $V^n_{ik} n = n_{ik}$. Consider the maps $\Upsilon_{ik} : (S^{d-1})^2 \times n(\partial C_{ik}) \to (S^{d-1})^2$ defined by $\Upsilon_{ik}(v, v^+, n) = (V^n_{ik} v, V^n_{ik} v^+)$. Finally, let

$$\nu_k = \sum_i \frac{|l_{ik}|}{|\partial C_{ik}|} \Upsilon^\#_{ik} \nu \otimes \delta_{n_{ik}}.$$

The proof that $\nu_k \in \Gamma_{C_k}$ and that the ν_k converge weakly to ν is similar to the proof of part (v) in the proof of Theorem 4.2.

4.2.5 *Resistance of Notched Arc*

Here we prove that there exists the limit $R_\psi = \lim_{\delta(X)\to 0} R(\Omega_X, I)$ (Example 4.5 in Sect. 4.1.2) and that the limit satisfies (4.12).

Recall that a right triangle is called canonical if (a) it is situated above its hypotenuse and (b) the median dropped on the hypotenuse is vertical. The angle α between the hypotenuse and the horizontal line, $\alpha \in (-\pi/2, \pi/2)$, is called the inclination of the triangle. Consider a particle that intersects the hypotenuse, enters the triangle, makes one or two reflections from its legs, and then intersects the hypotenuse again and leaves the triangle (Fig. 4.16). Let the initial v and final v^+ velocities of the particle be $v = (\sin\varphi, \cos\varphi)$ and $v^+ = -(\sin\varphi^+, \cos\varphi^+)$, with φ and φ^+ ranging from $-\pi/2 + \alpha$ to $\pi/2 + \alpha$.

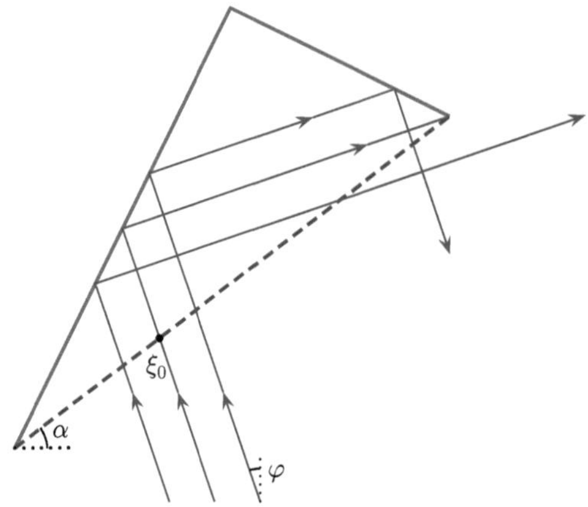

Fig. 4.16 A canonical triangle

Parameterize the hypotenuse by $\xi \in [0, 1]$; the value $\xi = 0$ corresponds to the left endpoint of the hypotenuse, and $\xi = 1$ to the right one. Denote by A_1 the set of values $(\varphi, \xi) \in [-\pi/2+\alpha, \pi/2+\alpha] \times [0, 1]$ corresponding to particles that have a single reflection from the left leg, by A_2 the set of values corresponding to a single reflection from the right leg, and by A_{12} the set corresponding to particles that have double reflections. One easily finds that A_1 is given by the inequality $\xi < -\frac{\sin\varphi}{\cos(\varphi-\alpha)}$, A_2 by the inequality $\xi > 1 - \frac{\sin\varphi}{\cos(\varphi-\alpha)}$, and A_{12} by the double inequality $-\frac{\sin\varphi}{\cos(\varphi-\alpha)} < \xi < 1 - \frac{\sin\varphi}{\cos(\varphi-\alpha)}$. Moreover, for $(\varphi, \xi) \in A_1$ we have $\varphi^+ = -\pi/2 + \alpha - \varphi$, for $(\varphi, \xi) \in A_2$ we have $\varphi^+ = \pi/2 + \alpha - \varphi$, and for $(\varphi, \xi) \in A_{12}$ we have $\varphi^+ = \varphi$. In Fig. 4.16, $\varphi < 0$, $\alpha > 0$ and $\xi_0 = -\frac{\sin\varphi}{\cos(\varphi-\alpha)}$.

Consider the parallel beam of particles falling on the hypotenuse in the direction φ. If $|\sin\varphi| < \cos(\varphi - \alpha)$, then the portion of particles that make a single reflection equals $|\sin\varphi|/\cos(\varphi - \alpha)$ and the direction of reflected particles is $\pm\pi/2 + \alpha - \varphi$; one must choose the "+" sign if $\sin\varphi > 0$ and the "−" sign if $\sin\varphi < 0$. The remaining particles make double reflections; the portion of these particles is $1 - |\sin\varphi|/\cos(\varphi - \alpha)$ and the direction of reflected particles is φ. If $|\sin\varphi| > \cos(\varphi-\alpha)$, then all the particles make a single reflection and the direction of reflected particles is $\pm\pi/2 + \alpha - \varphi$.

Now, consider the arc of circumference of angular size 2ψ contained in the half-plane $x_2 \geq 0$, with endpoints $A = (0,0)$ and $B = (1,0)$. Parameterize this arc with the angular parameter $\alpha \in [-\psi, \psi]$; the value $\alpha = -\psi$ corresponds to point A and $\alpha = \psi$ to point B. Divide it into a large number of small arcs and substitute each of them with two legs of the corresponding canonical triangle. The resulting broken line (Fig. 4.6) defines a notched arc. Recall that δ denotes the maximum distance between the abscissas of successive endpoints of the small arcs.

For small δ, the scheme of billiard reflection can be approximately described as follows. A particle of some mass moving in a direction $\varphi \in (-\pi/2, \pi/2)$ is reflected from the arc $\breve{AB}$. If $|\sin\varphi| < \cos(\varphi-\alpha)$, then it is split into two "splinters" of relative masses $|\sin\varphi|/\cos(\varphi - \alpha)$ and $1 - |\sin\varphi|/\cos(\varphi - \alpha)$. The first splinter

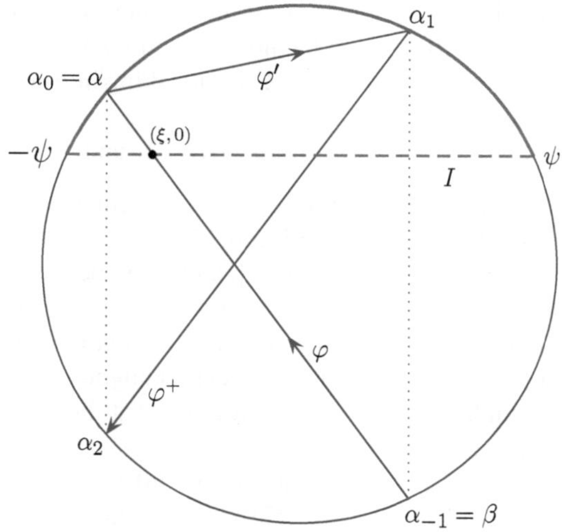

Fig. 4.17 Pseudo-billiard dynamics

is reflected in the direction $\pm\pi/2 + \alpha - \varphi$ and the second in the direction φ. If $|\sin\varphi| \geq \cos(\varphi - \alpha)$, then there is no splitting, and the whole particle is reflected in the direction $\pm\pi/2 + \alpha - \varphi$. The described dynamics will be called the *pseudo-billiard* one. A particle of unit mass starts moving at a point of the inlet $I = AB$ in a direction $\varphi \in (-\pi/2, \pi/2)$, and following several pseudo-billiard reflections, the resulting splinters return to I.

The limiting value of resistance coincides with the resistance resulting from the pseudo-billiard dynamics,

$$R_\psi = \frac{3}{4} \sum_i \int_{-\pi/2}^{\pi/2} \int_0^1 m_i(\xi, \varphi) \left(1 + \cos(\varphi_i^+(\xi, \varphi) - \varphi)\right) \frac{1}{2} \cos\varphi \, \xi \, d\varphi,$$

where $m_i = m_i(\varphi, \xi)$ are masses and $\varphi_i^+ = \varphi_i^+(\xi, \varphi)$ the final directions of the splinters resulting from the particle with the initial data (ξ, φ). (As we will see below, splitting can actually occur only once, after the first reflection; therefore, there are at most two splinters.) The difference between R_ψ and the true value of the resistance $R(\Omega_X, I)$ is $O(\delta)$ as $\delta \to 0$. Below we calculate R_ψ.

To describe the pseudo-billiard motion, it is helpful to change the variables. Consider the circumference containing the arc under consideration and parameterize it with the same angular variable α, which now ranges in $[-\pi, \pi]$. Consider a particle that starts its motion at a point β of the circumference, intersects I at a point $(\xi, 0)$, and then reflects from the arc, according to the pseudo-billiard rule, at a point α (Fig. 4.17). Thus, one has $\psi < |\beta| \leq \pi$ and $|\alpha| < \psi$. If $|\beta| \leq \pi/2$, then there is no splitting, and if $|\beta| > \pi/2$, there is.

Let us describe the dynamics of the first splinter. For a while, change the notation; let $\beta =: \alpha_{-1}$, $\alpha =: \alpha_0$, and let α_1 be the point of intersection of the splinter trajectory with the circumference. Denote by φ the initial direction of the particle and by φ' the direction of the splinter after the first reflection. (We do not call it φ^+ since there may be more reflections.) One has $\varphi = \pm\pi/2 + (\alpha_0 + \alpha_{-1})/2$ and $\varphi' = \pi/2 + (\alpha_0 + \alpha_1)/2$. Then, taking into account that $\varphi' = \pm\pi/2 + \alpha_0 - \varphi$, one obtains

$$\varphi = (\alpha_0 - \alpha_1)/2 \quad \text{and} \quad (\alpha_{-1} + \alpha_1)/2 = \pi/2,$$

the equalities being true mod π. In other words, the points α_{-1} and α_1 lie on the same vertical line; see Fig. 4.17.

If α_1 belongs to $[-\psi,\ \psi]$, then there occurs one more reflection, this time without splitting, since the splinter arrived from the point $\alpha_0 \in [-\pi/2,\ \pi/2]$. Extend the trajectory after the second reflection until the intersection with the circumference at a point α_2. Using an argument analogous to the one stated above, one derives the formula $\alpha_0 + \alpha_2 = \pi$; it follows that the point α_2 does not lie on the arc, that is, there are no more reflections.

Summarizing, the pseudo-billiard dynamics is as follows. After the first reflection from the arc the particle may or may not split into two "splinters." If $\alpha_{-1} \in [-\pi/2,\ -\psi] \cup [\psi,\ \pi/2]$, then there is no splitting, and the reflection is unique. If $\alpha_{-1} \in [-\pi,\ -\pi/2) \cup (\pi/2,\ \pi]$, then there is splitting into two splinters. If $\alpha_{-1} \in [-\pi + \psi,\ -\pi/2) \cup (\pi/2,\ \pi - \psi]$, then the first splinter makes no more reflections. If $\alpha_{-1} \in [-\pi,\ -\pi + \psi] \cup [\pi - \psi,\ \pi]$, then it makes one more reflection (without splitting) from the arc, and the final direction is $\varphi^+ = \pi/2 + (\alpha_1 + \alpha_2)/2$. Taking into account the preceding equalities, one obtains $\varphi - \varphi^+ = \alpha_{-1} + \alpha_0 + \pi$.

Note that the factor $1 + \cos(\varphi - \varphi^+)$, meaning the impact force per unit mass, equals 2 for the second splinter. For the first splinter that makes no reflections, as well as for the reflection without splitting, this factor equals $1 + \cos(\varphi - \varphi') = 1 + |\sin\alpha_{-1}|$. Finally, for the first splinter that makes one more reflection, this factor equals $1 + \cos(\varphi - \varphi^+) = 1 - \cos(\alpha_0 + \alpha_{-1})$.

Let us pass from the variables φ and ξ to $\alpha = \alpha_0$ and $\beta = \alpha_{-1}$ and calculate the integral R_ψ in terms of the new variables. Points α and β on the circumference have the cartesian coordinates $\frac{1}{2\sin\psi}(\sin\psi + \sin\alpha, -\cos\psi + \cos\alpha)$ and $\frac{1}{2\sin\psi}(\sin\psi + \sin\beta, -\cos\psi + \cos\beta)$, respectively. The interval with the endpoints α and β intersects I at the point $(\xi, 0)$, where

$$\xi = \frac{\sin(\psi + \alpha) - \sin(\psi + \beta) + \sin(\beta - \alpha)}{2\sin\psi\,(\cos\alpha - \cos\beta)}. \tag{4.47}$$

Further, one has

$$\varphi = \frac{\alpha + \beta \pm \pi}{2}; \tag{4.48}$$

one must take the sign "−" or "+ if $\beta > 0$ or $\beta < 0$, respectively. Therefore, $\cos\varphi = |\sin\frac{\alpha+\beta}{2}|$.

Point (α, β) ranges in the set $[-\psi,\ \psi] \times ([-\pi,\ -\psi] \cup [\psi,\ \pi])$, and the mapping $(\alpha, \beta) \mapsto (\xi, \varphi)$ given by (4.47) and (4.48) is a one-to-one mapping from this set to $[0,\ 1] \times [-\pi/2,\ \pi/2]$ with the Jacobian

$$\frac{D(\xi,\varphi)}{D(\alpha,\beta)} = \frac{\cos(\psi+\alpha)+\cos(\psi+\beta)-2\cos(\beta-\alpha)}{4\sin\psi\,(\cos\beta-\cos\alpha)} +$$
$$+\frac{\sin(\psi+\beta)-\sin(\psi+\alpha)+\sin(\alpha-\beta)}{4\sin\psi\,(\cos\beta-\cos\alpha)^2}\,(\sin\alpha+\sin\beta) =$$
$$= \frac{1}{4\sin\psi}\,\frac{\sin\frac{\alpha-\beta}{2}}{\sin\frac{\alpha+\beta}{2}}; \tag{4.49}$$

therefore, the factor of integration equals

$$\cos\varphi\, d\varphi\, d\xi = \frac{1}{4\sin\psi}\left|\sin\frac{\beta-\alpha}{2}\right| d\alpha\, d\beta.$$

Further, the mass of the first splinter is $|\cos\frac{\alpha+\beta}{2}|\Big/|\sin\frac{\beta-\alpha}{2}|$ and that of the second one is $1-|\cos\frac{\alpha+\beta}{2}|\Big/|\sin\frac{\beta-\alpha}{2}|$. Note also that integrating over $\beta\in[-\pi,\ -\psi]\cup[\psi,\ \pi]$ can be substituted by integrating over $\beta\in[\psi,\ \pi]$ with subsequent duplication of the result. With this substitution, one always has $\sin\frac{\beta-\alpha}{2}>0$.

The integral R_ψ can be written as the sum $R_\psi = I + II + III + IV$, where

$$I = \frac{3}{16\sin\psi}\int_{-\psi}^{\psi} d\alpha \int_{\psi}^{\pi/2} (1+\sin\beta)\,\sin\frac{\beta-\alpha}{2}\, d\beta,$$
$$II = \frac{3}{16\sin\psi}\int_{-\psi}^{\psi} d\alpha \int_{\pi/2}^{\pi} 2\left(\sin\frac{\beta-\alpha}{2} - \left|\cos\frac{\alpha+\beta}{2}\right|\right) d\beta,$$
$$III = \frac{3}{16\sin\psi}\int_{-\psi}^{\psi} d\alpha \int_{\pi/2}^{\pi-\psi} (1+\sin\beta)\,\cos\frac{\alpha+\beta}{2}\, d\beta,$$
$$IV = \frac{3}{16\sin\psi}\int_{-\psi}^{\psi} d\alpha \int_{\pi-\psi}^{\pi} (1-\cos(\alpha+\beta))\left|\cos\frac{\alpha+\beta}{2}\right| d\beta.$$

As a result of a simple calculation, one obtains

$$I = III = \frac{3}{16\sin\psi}\left[4\sin\psi - \frac{8\sqrt{2}}{3}\sin\frac{\psi}{2} - \frac{16}{3}\sin^4\frac{\psi}{2}\right],$$
$$II = \frac{3}{16\sin\psi}\left[16\sqrt{2}\sin\frac{\psi}{2} - 8\psi\right],$$
$$IV = \frac{3}{16\sin\psi}\left[-\frac{8}{3}\sin\psi + \frac{8}{9}\sin^3\psi + \frac{8}{3}\psi\right].$$

Summing these expressions, one finally comes to formula (4.12).

Chapter 5
Problems of Optimal Mass Transportation

In this chapter, several special problems of optimal mass transportation (or Monge–Kantorovich problems) are considered. Their solution will be used in the next chapter when we solve problems of minimum and maximum resistance for non-convex and rough bodies.

The problem in general consists in minimization of the functional

$$\mathcal{F}(\eta) = \iint_{X \times Y} c(x, y)\, d\eta(x, y), \quad \eta \in \Gamma(\lambda_1, \lambda_2),$$

where $X \subset \mathbb{R}^n$ and $Y \subset \mathbb{R}^m$ are closed subsets of Euclidean spaces, $c : X \times Y \to \mathbb{R}$ is a fixed (usually continuous) function of two variables determining the transportation cost, λ_1 and λ_2 are Borel measures in X and Y, respectively (the initial and final distributions of mass) with $\lambda_1(X) = \lambda_2(Y) < \infty$, and $\Gamma(\lambda_1, \lambda_2)$ is the set of Borel measures η on $X \times Y$ with fixed marginals λ_1 and λ_2. The last conditions means that for any two Borel sets $J \subset X$, $J' \subset Y$ we have $\eta(J \times Y) = \lambda_1(J)$ and $\eta(X \times J') = \lambda_2(J')$. The measure η is called the *plan of mass transportation* with the initial distribution λ_1 and final distribution λ_2: the amount of mass moved from J to J' is $\eta(J \times J')$.

It is well known (e.g., [2]) that if c is continuous, then there exists a measure $\eta_* \in \Gamma(\lambda_1, \lambda_2)$ minimizing $\mathcal{F}$. This measure is called the *plan of optimal mass transportation*.

Note that, although the theory of optimal transportation has grown rapidly in the last three decades, only a few exactly solvable problems are known to date, even in the one-dimensional case.

In Sects. 5.1 and 5.2, we consider a special problem concerning mass transportation from $\mathbb{R}$ to $\mathbb{R}$. We assume that the initial and final mass distributions coincide, $\lambda_1 = \lambda_2$, and the transportation cost is $c(x, y) = f(x + y)$, where the function f is odd, continuous, and strictly concave on $\mathbb{R}_+$. Under some assumptions on the mass distribution we show that an optimal measure belongs to a certain family of measures depending on countably many parameters. Usually the number of

A. Plakhov, *Exterior Billiards: Systems with Impacts Outside Bounded Domains*, DOI 10.1007/978-1-4614-4481-7_5,

parameters is finite (often there is only one parameter), and in these cases the problem reduces to minimizing a function of several (or just one) variables. In Sect. 5.3, we consider several examples related to concrete mass distributions in this problem.

Finally, in Sect. 5.4, we consider a problem of mass transport on a sphere with cost function equal to the squared distance, $c(x, y) = \frac{1}{2}|x - y|^2$. We fix a unit vector n and consider the transfer from the "lower" hemisphere $X = S_{-n}^{d-1} := \{x \in S^{d-1} : x \cdot n \le 0\}$ to the complementary "upper" hemisphere $Y = S_n^{d-1} = \{x \in S^{d-1} : x \cdot n \ge 0\}$. We consider a special case of mass distributions in X and Y, which are symmetric relative to rotations about n. This problem is reduced to the one-dimensional problem of Sects. 5.1 and 5.2 and then explicitly solved.

The results of this chapter were first obtained in [45, 46].

5.1 Statement of the One-Dimensional Problem and the Results

We consider the minimization problem

$$\inf_{\eta \in \Gamma(\lambda,\lambda)} \mathcal{F}(\eta), \quad \text{where } \mathcal{F}(\eta) = \iint_{\mathbb{R}^2} f(x+y)\, d\eta(x,y), \tag{5.1}$$

where f and λ satisfy the following conditions.

(A1) The function f is odd on $\mathbb{R}$ and strictly concave on $x \ge 0$.
(A2) The support of the measure λ is the union of a finite number of compact intervals. Additionally, the measure of each singleton is equal to 0.

In this section, we define a family of measures in $\mathbb{R}^2$ depending on countably many parameters, and in the next section we prove that the optimal measure η_* belongs to that family. The family is determined by the measure λ and does not depend on f.

The following definition is standard in optimal transportation.

Definition 5.1. A set $A \subset X \times Y$ is said to be *c-monotone* if for each pair of points (x_1, y_1) and (x_2, y_2) in A we have

$$c(x_1, y_1) + c(x_2, y_2) \le c(x_1, y_2) + c(x_2, y_1). \tag{5.2}$$

As is well known, if c is continuous, then the support of each optimal measure is c-monotone (e.g., [63]). In particular, this holds in our case of $c(x, y) = f(x + y)$. We will for brevity use the term "f-monotone set."

Definition 5.2. The signed measure $\tilde{\lambda}$ on $\mathbb{R}_+ := [0, +\infty)$ is given by the formula

$$\tilde{\lambda}(B) = \lambda(B) - \lambda(-3B),$$

where $B \subset \mathbb{R}_+ = [0, +\infty)$ is a Borel set and $-3B := \{-3x : x \in B\}$.

Let λ_+ and λ_- be the upper and lower variations of $\tilde{\lambda}$.

Consider a countable nonempty system of intervals $\mathcal{I} = \{I_i\}$. We assume that the indices i are nonnegative integers. For $i \neq 0$ the intervals have the form $I_i = (a_i, b_i)$, where

$$0 < a_i < b_i \leq +\infty \quad (i \neq 0).$$

If $i = 0$ belongs to the index set $\{i\}$, then the corresponding interval has the form $I_0 = [0, b_0)$; thus, we take $a_0 = 0$.

We also assume that the closures of the intervals are disjoint, that is,

$$\bar{I}_i \cap \bar{I}_j = \emptyset \text{ for } i \neq j \tag{5.3}$$

(thus, for each pair of indices $i \neq j$ holds either $a_i < b_i < a_j < b_j$ or $a_j < b_j < a_i < b_i$).

Finally, we assume that

$$\mathbb{R}_+ \setminus (\cup_i I_i) \subset \operatorname{spt} \lambda_+. \tag{5.4}$$

It follows from (5.3) and (5.4), in view of the boundedness of $\operatorname{spt} \lambda_+$, that one of the intervals in $\mathcal{I}$ is semi-infinite. The corresponding index will be denoted by $i = r$. If $r \neq 0$, the interval has the form $I_r = (a_r, +\infty)$, that is, $b_r = +\infty$. It may happen that $r = 0$; in that case system $\mathcal{I}$ contains exactly one interval, $\mathcal{I} = \{[0, +\infty)\}$.

Definition 5.3. A system of intervals $\mathcal{I}$ is said to be *admissible* under the following conditions (in addition to the aforementioned properties):

(a) If $i \neq 0$, then $\tilde{\lambda}(I_i) = 0$;
(b) For each i and for

$$x \in \begin{cases} (a_i, b_i) & \text{if } a_i > 0, \\ (-3b_i, b_i) & \text{if } a_i = 0, \end{cases} \tag{5.5}$$

one has

$$\lambda((-3b_i, -2b_i - x)) \leq \lambda((x, b_i)) \leq \lambda((-3b_i, -2a_i - x)). \tag{5.6}$$

If the index set $\{i\}$ contains 0, then, setting $x = 0$ and $i = 0$ in (5.6), we obtain $\lambda((0, b_0)) \leq \lambda((-3b_0, 0))$, and therefore $\tilde{\lambda}(I_0) \leq 0$.

Note that for $i = r$ the first inequality in (5.6) becomes the trivial relation $\lambda((x, +\infty)) \geq 0$.

Below we define the set $G_{\mathcal{I}}$ on the plane induced by an admissible system $\mathcal{I}$. This definition is rather cumbersome; first we define auxiliary sets G^+, $G^D_{(0)}$, $G^L_{(0)}$, G^D_i, G^L_i, G_0, and then define $G_{\mathcal{I}}$ as their union.

First we define $G^+ = G^+(\mathcal{I})$ by

$$G^+ := \{(x,x) : x \in \mathbb{R}_+ \setminus (\cup_i I_i)\}.$$

Then let

$$G^D_{(0)} := \{(x,y) : y = -3x,\ x \in \mathbb{R}_+ \setminus (\cup_i I_i),\ y \in \operatorname{spt}\lambda\}$$

and let $G^L_{(0)}$ be the set symmetric to $G^D_{(0)}$ relative to the diagonal $\{x = y\}$. Define $G_{(0)} = G_{(0)}(\mathcal{I})$ by

$$G_{(0)} := G^D_{(0)} \cup G^L_{(0)}.$$

For $i \neq 0$ let

$$G^D_i := \{(x,y) : x \in \operatorname{spt}\lambda \cap I_i,\ y \in \operatorname{spt}\lambda \cap (-3I_i),\ \lambda((x,\ b_i)) = \lambda((-3b_i,\ y))\} \tag{5.7}$$

and let G^L_i be the set symmetric to G^D_i relative to the diagonal $\{x = y\}$. Define $G_i = G_i(\mathcal{I})$ by

$$G_i := G^D_i \cup G^L_i.$$

On the other hand, if the index set $\{i\}$ contains 0 (in other words, $\mathcal{I}$ contains an interval $[0,\ b_0)$), then the set

$$G_0 = \{(x,y) : x,\ y \in \operatorname{spt}\lambda \cap (-3b_0,\ b_0),\ \lambda((x,\ b_0)) = \lambda((-3b_0,\ y))\}. \tag{5.8}$$

Define the set $G^- = G^-(\mathcal{I})$ by

$$G^- := G_{(0)} \cup (\cup_i G_i)$$

and set

$$G_{\mathcal{I}} = G^+ \cup G^-.$$

Remark 5.1. It follows from the second inequality in (5.6) and the definition of G^- [in particular, (5.7) and (5.8)] that G^- lies in the half-plane $x + y \leq 0$.

We illustrate these definitions by Fig. 5.1, where $\mathcal{I} = \{I_0,\ I_1,\ I_2\}$. The interval I_2 is semi-infinite, that is, we have $r = 2$. The set G^+ is the union of the segments C_1C_2 and C_3C_4, $G^D_{(0)}$ is the union of the segments D_1D_2 and D_3D_4, and $G^L_{(0)}$ is the union of the segments L_1L_2 and L_3L_4. The sets G_0, G^D_1, G^D_2, G^L_1, and G^L_2 are indicated in the figure. The set G^- is the curve D_5L_5. The intervals $I_0,\ I_1,\ I_2$ and the set $G_{\mathcal{I}}$ are plotted by bold lines. The measure λ has the following properties: $\operatorname{spt}\lambda$ is the orthogonal projection of the curve D_5L_5 onto the x-axis, relations (5.8) and (5.7) hold for $i = 0$ and $i = 1,\ 2$, respectively, and

$$\text{for each interval } J \subset \mathbb{R} \setminus (I_0 \cup I_1 \cup I_2) \text{ we have } \lambda(J) > \lambda(-3J). \tag{5.9}$$

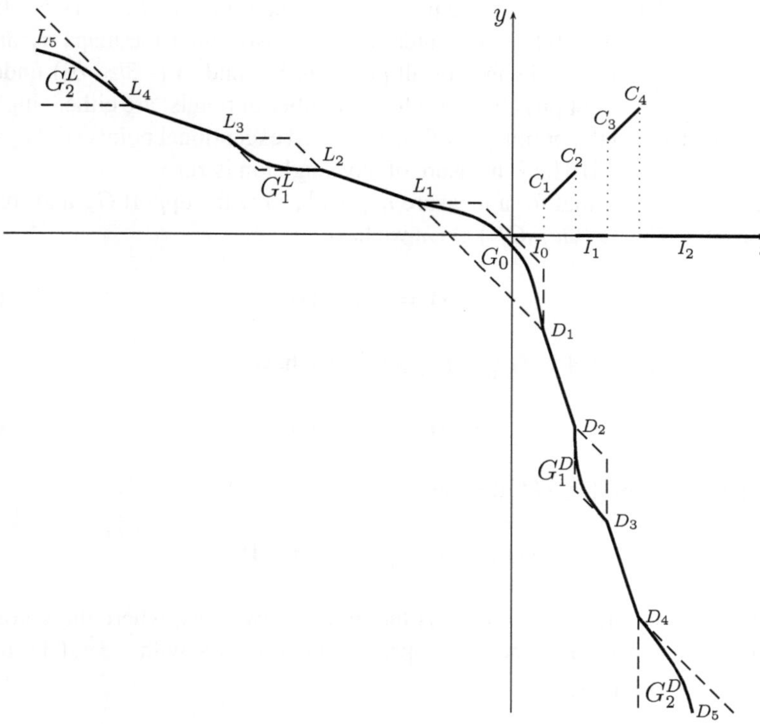

Fig. 5.1 The set $G_{\mathcal{I}}$ corresponding to the system of intervals $\mathcal{I} = \{I_0, I_1, I_3\}$

The dashed lines in Fig. 5.1 are either parallel to one of the coordinate axes or make an angle $\pi/4$ with it. Condition (5.6) in Definition 5.3, as applied to the system of intervals $\mathcal{I} = \{I_0,\ I_1,\ I_2\}$, is equivalent to the following geometric condition: the sets G_0, G_1^D, and G_1^L lie in the corresponding dashed quadrangles, and the sets G_2^D and G_2^L lie in the quadrangles bounded by the corresponding dashed lines.

Note that the system $\mathcal{I}$ under consideration is admissible for the measure λ under consideration. Indeed, condition (5.4) follows from the inequality (5.9); (5.5) is a consequence of the continuity of the curve D_5L_5; and (5.6) follows from the aforementioned geometric condition.

The following lemma is at the same the definition of the measure $\eta_{\mathcal{I}}$ corresponding to an admissible system $\mathcal{I}$.

Lemma 5.1. *For each admissible system $\mathcal{I} = \{I_i\}$ there exists a unique measure $\eta_{\mathcal{I}} \in \Gamma(\lambda, \lambda)$ such that spt $\eta_{\mathcal{I}} = G_{\mathcal{I}}$.*

Proof. We define the maps π_x and π_y from $G_{\mathcal{I}}$ to $\mathbb{R}$ by the formulae $\pi_x(x, y) = x$ and $\pi_y(x, y) = y$. Thus, π_x and π_y are the orthogonal projections of the set $G_{\mathcal{I}}$ on the x- and y-axes, respectively.

Note the following facts. (i) The preimages of the semiaxis $\mathbb{R}_- = (-\infty, 0]$ and (in the case where $0 \in \{i\}$) of the interval $(-3b_0, b_0)$ under the maps π_x and π_y belong to G^-. (ii) The preimages of all points in $\mathbb{R}_-$ and in $(-3b_0, b_0)$ under the maps π_x and π_y, except possibly for a finite number of points, are either singletons or an empty set, and the preimages of the indicated exceptional points (if they exist) are two-point sets. (iii) The λ-measure of any singleton is zero.

These facts imply that for a measure $\eta \in \Gamma(\lambda, \lambda)$ with support $G_{\mathcal{I}}$ and for each Borel set $A \subset G^D_{(0)} \cup \left(\cup_{i\neq 0} G^D_i\right) \cup G_0$ we have

$$\eta(A) = \lambda(\pi_y(A)), \tag{5.10}$$

and for each Borel set $A \subset G^L_{(0)} \cup \left(\cup_{i\neq 0} G^L_i\right)$ we have

$$\eta(A) = \lambda(\pi_x(A)). \tag{5.11}$$

Further, for each $A \subset G^+$ we have

$$\lambda(\pi_x(A)) = \eta(\pi_x^{-1} \circ \pi_x(A)).$$

The set $\pi_x^{-1} \circ \pi_x(A) = A \cup \pi^D A$ is the union of two sets, where the second set $\pi^D A$ is contained in $G^D_{(0)}$ and its π_y-projection coincides with $-3\pi_x(A)$, that is, $\pi_y(\pi^D A) = -3\pi_x(A)$.

Thus, we have

$$\eta(\pi_x^{-1} \circ \pi_x(A)) = \eta(A) + \eta(\pi^D A) = \eta(A) + \lambda(-3\pi_x(A)),$$

and therefore

$$\eta(A) = \lambda(\pi_x(A)) - \lambda(-3\pi_x(A)) = \tilde{\lambda}(\pi_x(A)). \tag{5.12}$$

Taking into account that λ_+ is the upper variation of the signed measure $\tilde{\lambda}$ and using the inclusion $\pi_x(A) \subset \mathbb{R}_+ \setminus (\cup_i I_i)$ and the relation (5.4), we obtain

$$\eta(A) = \lambda_+(\pi_x(A)). \tag{5.13}$$

Thus, we have shown that a measure $\eta \in \Gamma(\lambda, \lambda)$ that satisfies $\operatorname{spt} \eta = G_{\mathcal{I}}$ must satisfy relations (5.10), (5.11), or (5.13) for each Borel set A lying in one of the sets: $G^D_{(0)} \cup \left(\cup_{i\neq 0} G^D_i\right) \cup G_0$, $G^L_{(0)} \cup \left(\cup_{i\neq 0} G^L_i\right)$, or G^+, respectively (note that $G_{\mathcal{I}}$ is the union of these sets). Hence, there exists at most one such measure.

It remains to prove that the measure η defined by (5.10), (5.11), and (5.13) belongs to $\Gamma(\lambda, \lambda)$ and its support is $G_{\mathcal{I}}$. At the moment we can only assert that $\operatorname{spt} \eta \subset G_{\mathcal{I}}$.

(a) We are going to prove that for each $B \subset \mathbb{R}$,

$$\lambda(B) = \eta(\pi_x^{-1}(B)), \quad \lambda(B) = \eta(\pi_y^{-1}(B)); \tag{5.14}$$

this will imply that $\eta \in \Gamma(\lambda, \lambda)$. We shall prove the first equality in (5.14); the second one is proved similarly.

Consider separately four cases, where B belongs (i) to the π_x-projection of $G^L_{(0)} \cup (\cup_{i \neq 0} G^L_i)$, which is the half-line $(-\infty, -3b_0]$ (if $0 \notin \{i\}$, set $b_0 = 0$); (ii) to the π_x-projection of G_0, that is, the interval $(-3b_0, -b_0)$ (in the case where $0 \in \{i\}$); (iii) to the interval I_i ($i \neq 0$); and (iv) to the set $\mathbb{R}_+ \setminus (\cup_i I_i)$.

(i) Taking into account the definition of $G_{\mathcal{I}}$, we see that the preimage of a point x in $(-\infty, -3b_0]$ under π_x is nonempty if and only if $x \in \operatorname{spt} \lambda$. It follows that for any $B \subset (-\infty, -3b_0]$,

$$\pi_x \circ \pi_x^{-1}(B) = B \cap \operatorname{spt} \lambda. \tag{5.15}$$

holds. Using (5.11) we have

$$\lambda(\pi_x \circ \pi_x^{-1}(B)) = \eta(\pi_x^{-1}(B));$$

therefore

$$\lambda(B) = \lambda(B \cap \operatorname{spt} \lambda) = \eta(\pi_x^{-1}(B)).$$

(ii) Assume that $0 \in \{i\}$. It follows from the definition of G_0 that for a Borel set $A \subset G_0$ one has

$$\lambda(\pi_x(A)) = \lambda(\pi_y(A)).$$

Further, for $B \subset (-3b_0, b_0)$, (5.15) holds. Bearing in mind (5.10), we conclude that

$$\lambda(B) = \lambda(B \cap \operatorname{spt} \lambda) = \lambda(\pi_x \circ \pi_x^{-1}(B)) = \lambda(\pi_y \circ \pi_x^{-1}(B)) = \eta(\pi_x^{-1}(B)).$$

Exactly the same argument is valid in case (iii).

(iv) For $B \subset \mathbb{R}_+ \setminus (\cup_i I_i)$ we have

$$\pi_x^{-1}(B) = A_+ + A_-,$$

where

$$A_+ = \{(x, x) : x \in B\} \subset G^+ \quad \text{and}$$
$$A_- = \{(x, -3x) : x \in B \text{ and } -3x \in \operatorname{spt} \lambda\} \subset G^D_{(0)}.$$

By (5.12), we have

$$\eta(A_+) = \lambda(B) - \lambda(-3B),$$

and by (5.10), we have

$$\eta(A_-) = \lambda(\pi_y(A_-)) = \lambda(-3B \cap \operatorname{spt} \lambda) = \lambda(-3B).$$

Therefore,

$$\eta(\pi_x^{-1}(B)) = \eta(A_+) + \eta(A_-) = \lambda(B).$$

Thus, Eq. (5.14), and therefore the inclusion $\eta \in \Gamma(\lambda, \lambda)$, are proved.

Let us finally show that

$$\operatorname{spt}\eta = G_{\mathcal{I}}. \tag{5.16}$$

By the definition of η and by (5.10), (5.11), and (5.13), we have $\operatorname{spt}\eta \subset G_{\mathcal{I}}$. On the other hand, for a point $z \in G^D_{(0)} \cup \left(\cup_{i\neq 0} G^D_i\right) \cup G_0$, $\pi_y(z) \in \operatorname{spt}\lambda$ holds true; therefore, by (5.10) we have $z \in \operatorname{spt}\eta$. Further, for a point $z \in G^L_{(0)} \cup \left(\cup_{i\neq 0} G^L_i\right)$, $\pi_x(z) \in \operatorname{spt}\lambda$ holds true; therefore, by (5.11), we have $z \in \operatorname{spt}\eta$. Finally, for a point $z \in G^+$ by (5.4), $\pi_x(z) \in \operatorname{spt}\lambda_+$ holds true; therefore, by (5.13), we have $z \in \operatorname{spt}\eta$. Thus, the reverse inclusion $G_{\mathcal{I}} \subset \operatorname{spt}\eta$ is also true, and thus (5.16) is proved. □

The following Theorem 5.1 and Corollary 5.1 are the main results of this chapter. Next we state Corollary 5.2, which will be used in Chap. 6 to solve problems of optimal mean resistance.

Theorem 5.1. *The support of $\eta \in \Gamma(\lambda,\lambda)$ is f-monotone if and only if*

$$\eta = \eta_{\mathcal{I}}$$

for some admissible system of intervals $\mathcal{I}$.

Corollary 5.1. *Let η_* be an optimal measure; then $\eta_* = \eta_{\mathcal{I}}$ for some admissible system of intervals $\mathcal{I}$.*

Remark 5.2. Assume that $\operatorname{spt}\lambda_+$ is the disjoint union of m segments; then the number of intervals I_i, $i \neq 0$, in an admissible system is at most m, and the equality $\tilde{\lambda}(I_i) = 0$ in (5.5) describes a relation between the left and right endpoints of the interval I_i, $i \neq 0, r$, and uniquely fixes the left endpoint of I_r. Thus, by Theorem 5.1, the family of f-monotone sets is the union of finitely many k-parameter families with $k \leq m-1$.

Corollary 5.2. *Assume that $\operatorname{spt}\lambda \subset [-3w, w]$, $\operatorname{spt}\lambda_- = [0, c]$, and $\operatorname{spt}\lambda_+ = [c, w]$ for some $0 < c < w$ (an example of a signed measure $\tilde{\lambda}$ satisfying these conditions is displayed in Fig. 5.2). Then the optimal measure η_* belongs to a one-parameter family $\eta_t \in \Gamma_{\lambda,\lambda}$, $c \leq t \leq w$, where the parameter t satisfies an additional condition to be stated below. Each measure η_t is uniquely defined by its support* $\operatorname{spt}\eta_t = G^+ \cup G^L \cup G^D \cup G_0$. *Here* $G^+ = \{(x,x) : x \in [t,w]\}$, $G^D = \{(x,y) : y = -3x,\ x \in [t,w]$ and $y \in \operatorname{spt}\lambda\}$ *(except for the case $t = w$, where G^+ and G^D are empty sets), G^L is symmetric to G^D relative to the line $x = y$, and* $G_0 = \{(x,y) : x, y \in \operatorname{spt}\lambda \cap (-3t, t)$ and $\lambda((x,t)) = \lambda((-3t, y))\}$. *The additional condition on t means that G_0 lies in the band $-2t \leq x + y \leq 0$. This condition can also be written analytically:*

$$\lambda((-3t, -2t - x)) \leq \lambda((x, t)) \leq \lambda((-3t, -x)) \quad \textit{for each } x \in (-3t, t). \tag{5.17}$$

Let a function $g_t(x) : (-3t, t) \to \mathbb{R}$ be defined by $g_t(x) = \min\{y : \lambda((-3t, x)) = \lambda((y, t))\}$*; then the value of the functional on η_t is*

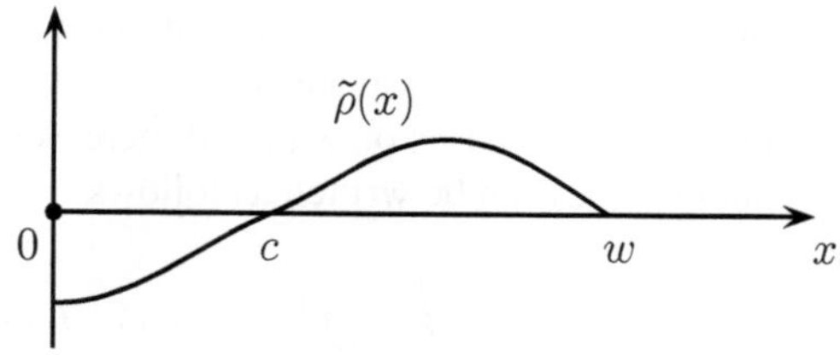

Fig. 5.2 Signed measure $\tilde{\lambda}$ satisfying the conditions of Corollary 5.2; the density of $\tilde{\rho}$ is plotted here

$$\mathcal{F}(\eta_t) = \int_t^w f(2x)\, d(\lambda(x) - \lambda(-3x))$$

$$+ \int_{-3t}^t f(x + g_t(x))\, d\lambda(x) + 2 \int_t^w f(-2x)\, d\lambda(-3x). \tag{5.18}$$

Proof. It follows from (5.4) and (5.5) that $\mathbb{R}_+ \setminus (\cup_i I_i) \subset [c, w]$ and no interval of the system is contained in $[c, w]$. Further, one of the intervals is semi-infinite, and its left endpoint is not contained in $[c, w]$. This implies that an admissible system of intervals must be either $\mathcal{I}_t = \{[0, t), (w, +\infty)\}$, $c \le t < w$, or $\mathcal{I}_w = \{\mathbb{R}^+\}$. The additional condition on the parameter t follows from (5.6).

By Corollary 5.1, an optimal measure belongs to the family of measures $\eta_t = \eta_{\mathcal{I}_t}$ with support $G_{\mathcal{I}_t}$. The description of this support is a direct consequence of the definition of $G_{\mathcal{I}}$. The measure η_t is defined by

$$\eta_t(A) = \tilde{\lambda}(\{x : t \le x \le w,\ (x, x) \in A\}) + \lambda(\{x : -3t < x < t,\ (x, g(x)) \in A\})$$
$$+\lambda(\{-3x : t \le x \le w,\ (x, -3x) \in A\}) + \lambda(\{-3y : t \le y \le w,\ (-3y, y) \in A\})$$

for each Borel set $A \subset \mathbb{R}^2$; this implies formula (5.18). □

5.2 Proof of Theorem 5.1

The proof of the theorem amounts to the chain of Lemmas 5.2–5.9. The main difficulty is to prove that the set spt $\eta \cap \{x+y > 0\}$ lies on the line $x = y$. A number of lemmas up to Lemma 5.8 inclusive is dedicated to the proof of this. Further specification of the set spt η is relatively simple and is contained in Lemma 5.9.

Lemma 5.2. *A set X is f-monotone if and only if*

$$(x_1 - x_2)(y_1 - y_2)(x_1 + y_1 + x_2 + y_2) \ge 0 \tag{5.19}$$

for any pair (x_1, y_1), (x_2, y_2) in X.

Proof. For $x_1 = x_2$ we have both relations (5.2) and (5.19); therefore, it is sufficient to consider the case $x_1 \ne x_2$.

Bearing in mind that f is a continuous function, concave on $\mathbb{R}_+$, and convex on $\mathbb{R}_- = (-\infty, 0]$, we see that f' is defined everywhere except at countably many points, decreases on $\mathbb{R}_+$, and increases on $\mathbb{R}_-$, and f is its primitive. Hence inequality (5.2) can be written as follows:

$$\int_{y_1}^{y_2} \big(f'(x_2 + \theta) - f'(x_1 + \theta)\big)\, d\theta \leq 0. \tag{5.20}$$

After the change of variable $\xi = \theta + (x_1 + x_2)/2$, setting

$$\Delta x = \frac{x_2 - x_1}{2}, \quad \Delta y = \frac{y_2 - y_1}{2}, \quad \sigma = \frac{x_1 + y_1 + x_2 + y_2}{2} \quad \text{and}$$

$$h(\xi) = \frac{f'(\xi + \Delta x) - f'(\xi - \Delta x)}{\Delta x},$$

we can write (5.20) as follows:

$$\Delta x \cdot \int_{\sigma - \Delta y}^{\sigma + \Delta y} h(\xi)\, d\xi \leq 0. \tag{5.21}$$

Bearing in mind that $h(\xi)$ is an odd function, negative for $\xi > 0$, it is easy to see that (5.21) holds if and only if $\Delta x\, \Delta y\, \sigma \geq 0$. □

Remark 5.3. If $\eta \in \Gamma(\lambda, \lambda)$, then $\operatorname{spt} \eta$ contains no isolated points.

Remark 5.4. Let $\eta \in \Gamma(\lambda, \lambda)$. Then $x \in \operatorname{spt} \lambda$ if and only if there exists y such that $(x, y) \in \operatorname{spt} \eta$.

Lemma 5.3. *Let $\mathcal{I}$ be an admissible system of intervals; then the set $\operatorname{spt} \eta_{\mathcal{I}}$ is f-monotone.*

Proof. It is sufficient to verify that inequality (5.19) holds for arbitrary points (x_1, y_1) and (x_2, y_2) in $G^+ \cup G^-$.

It follows from the definition of the set G^+ that if both points lie in G^+, then $(x_1 - x_2)(y_1 - y_2) \geq 0$, $x_1 + y_1 \geq 0$ and $x_2 + y_2 \geq 0$. From the definition of G^- and Remark 5.1 we see that if both points lie in G^-, then $(x_1 - x_2)(y_1 - y_2) \leq 0$, $x_1 + y_1 \leq 0$, and $x_2 + y_2 \leq 0$. Therefore, in both cases we have (5.19).

It remains to consider the case $(x_1, y_1) \in G^+$, $(x_2, y_2) \in G^-$. Two cases are possible: either (i) $(x_2, y_2) \in G_{(0)}$ or (ii) $(x_2, y_2) \in G_i$ for some i.

(i) $(x_2, y_2) \in G_{(0)} = G^D_{(0)} \cup G^L_{(0)}$. Let $(x_2, y_2) \in G^D_{(0)}$; the case $(x_2, y_2) \in G^L_{(0)}$ is treated in a similar way. We have $x_1 = y_1 \geq 0$, $y_2 = -3x_2 \leq 0$; therefore, $(x_1 - x_2)(y_1 - y_2)(x_1 + y_1 + x_2 + y_2) = 2(x_1 - x_2)^2(x_1 + 3x_2) \geq 0$.

(ii) $(x_2, y_2) \in G_i$. In this case, either $(x_2, y_2) \in G^D_i$ $(i \neq 0)$, or $(x_2, y_2) \in G^L_i$ $(i \neq 0)$, or $(x_2, y_2) \in G_0$. Let $(x_2, y_2) \in G^D_i$, $i \neq 0$; the other cases are treated in a similar way.

Due to Remark 5.3, there exists a sequence of points $(x^{(k)}, y^{(k)}) \in \operatorname{spt} \eta$ such that $\lim_{k\to\infty}(x^{(k)}, y^{(k)}) = (x_2, y_2)$ and $(x^{(k)}, y^{(k)}) \neq (x_2, y_2)$. Starting from some number k, the points belong to G_i^D. It follows from the definition of G_i^D (5.7) and condition (5.6) that for the corresponding values of k,

$$\lambda((-3b_i, -2b_i - x^{(k)})) \leq \lambda((-3b_i, y^{(k)})) \leq \lambda((-3b_i, -2a_i - x^{(k)})). \quad (5.22)$$

It follows from (5.22) that

$$-2b_i - x_2 \leq y_2 \leq -2a_i - x_2. \quad (5.23)$$

Indeed, assuming that $y_2 < -2b_i - x_2$ we can find points $(x^{(k)}, y^{(k)})$ and $(x^{(j)}, y^{(j)})$ such that $y^{(k)} \neq y^{(j)}$ and $\max\{y^{(k)}, y^{(j)}\} < \min\{-2b_i - x^{(k)}, -2b_i - x^{(j)}\}$. By (5.22), we obtain

$$\lambda((y^{(k)}, -2b_i - x^{(k)})) = 0 = \lambda((y^{(j)}, -2b_i - x^{(j)})),$$

so that one of the points $y^{(k)}$ and $y^{(j)}$ lies in the interior of an interval of λ-measure zero and therefore does not belong to $\operatorname{spt} \lambda$. This contradiction proves the first inequality in (5.23); the proof for the second inequality is similar.

We now have $x_1 = y_1 > y_2$ and $x_1 \notin I_i$; therefore, $x_1 \leq a_i$ or $x_1 \geq b_i$. In the first case we have $x_1 < x_2$, and taking into account the second inequality in (5.23) we obtain $x_1 + y_1 + x_2 + y_2 = 2x_1 + x_2 + y_2 \leq 2a_i + x_2 + y_2 \leq 0$. In the second case we have $x_1 > x_2$, and taking into account the first inequality in (5.23) we obtain $x_1 + y_1 + x_2 + y_2 \geq 2b_i + x_2 + y_2 \geq 0$. In both cases inequality (5.19) is proved. □

Definition 5.4. Let J_1 and J_2 be open intervals. We say that a set G is *X-like of format* $J_1 \times J_2$ if

$$G \cap (J_1 \times \mathbb{R}) \subset J_1 \times \bar{J}_2 \quad \text{and} \quad G \cap (\mathbb{R} \times J_2) \subset \bar{J}_1 \times J_2.$$

Note that a set can be simultaneously X-like of several formats.

In Fig. 5.3, each subset of the shaded domain is X-like of format $J_1 \times J_2$.

Lemma 5.4. *If* $\eta \in \Gamma(\lambda, \lambda)$ *and* $\operatorname{spt} \eta$ *is X-like of format* $J_1 \times J_2$*, then* $\lambda(J_1) = \lambda(J_2)$.

Proof. We have $\lambda(J_1) = \eta(J_1 \times \mathbb{R}) = \eta(J_1 \times J_2) = \eta(\mathbb{R} \times J_2) = \lambda(J_2)$. □

Definition 5.5. We set

$$\operatorname{spt}^+ \eta := \operatorname{spt} \eta \cap \{x + y > 0\} \quad \text{and} \quad \operatorname{spt}^- \eta := \operatorname{spt} \eta \cap \{x + y \leq 0\}.$$

Remark 5.5. For each pair (x_1, y_1), (x_2, y_2) in $\operatorname{spt}^+ \eta$ we have

$$(x_1 - x_2)(y_1 - y_2) \geq 0, \quad (5.24)$$

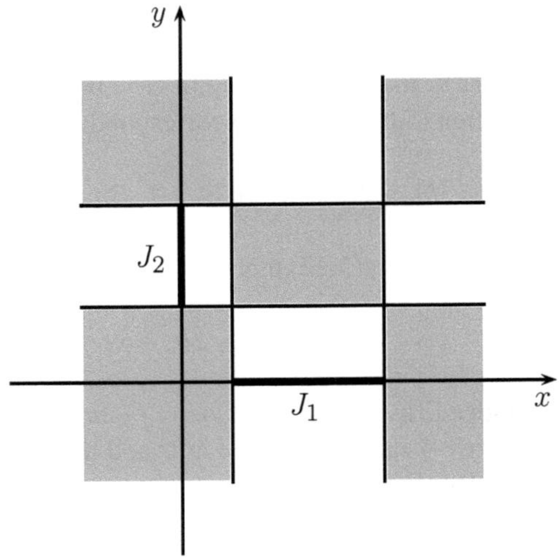

Fig. 5.3 An X-like set of format $J_1 \times J_2$

and for each pair of points in $\text{spt}^-\eta$ we have

$$(x_1 - x_2)(y_1 - y_2) \leq 0. \tag{5.25}$$

Indeed, if both points belong to $\text{spt}^+\eta$, then (5.24) is a consequence of Lemma 5.2 and the inequality $x_1 + y_1 + x_2 + y_2 > 0$. On the other hand, if both points belong to $\text{spt}^-\eta$, two cases are possible: (i) $x_1 + y_1 + x_2 + y_2 < 0$ or (ii) $x_1 + y_1 + x_2 + y_2 = 0$. In case (i), inequality (5.25) is obvious, whereas in case (ii), we have $x_1 + y_1 = 0$, $x_2 + y_2 = 0$, and therefore $x_1 - x_2 = -(y_1 - y_2)$, so again we have inequality (5.25).

In Lemmas 5.5–5.9, the measure η belongs to $\Gamma(\lambda, \lambda)$ and its support $\text{spt}\,\eta$ is f-regular.

Lemma 5.5. *For each x, each of the sets* $\text{spt}^+\eta \cap (\{x\} \times \mathbb{R})$, $\text{spt}^+\eta \cap (\mathbb{R} \times \{x\})$, $\text{spt}^-\eta \cap (\{x\} \times \mathbb{R})$, *and* $\text{spt}^-\eta \cap (\mathbb{R} \times \{x\})$ *contains at most two point.*

Proof. We shall prove this for $\text{spt}^+\eta \cap (\{x\} \times \mathbb{R})$; the proofs for the other sets are similar. It is sufficient to show that if (x_1, y_1), $(x_1, y_2) \in \text{spt}^+\eta$ and $y_1 < y_2$, then $\{x_1\} \times (y_1, y_2)$ is disjoint from $\text{spt}^+\eta$.

Consider a point (x, y) satisfying the conditions $x + y > 0$, $y_1 < y < y_2$, and $x \neq x_1$. We have $x_1 + y_1 + x + y > 0$, $x_1 + y_2 + x + y > 0$, $(y_1 - y)(y_2 - y) < 0$; therefore, at least one of the quantities $(x_1 - x)(y_1 - y)(x_1 + y_1 + x + y)$, $(x_1 - x)(y_2 - y)(x_1 + y_2 + x + y)$ is negative. Hence inequality (5.19) is not fulfilled for one of the pairs $\{(x_1, y_1), (x, y)\}$ and $\{(x_1, y_2), (x, y)\}$. Using Lemma 5.2 and bearing in mind that $\text{spt}\,\eta$ is f-monotone, we get that $(x, y) \notin \text{spt}\,\eta$.

Thus, we have $\eta(\{(x, y) : x + y > 0,\ y_1 < y < y_2,\ x \neq x_1\}) = 0$. On the other hand, $\eta(\{(x, y) : x = x_1\}) = \lambda(\{x_1\}) = 0$. This implies that the η-measure of the open set $A = (\mathbb{R} \times (y_1, y_2)) \cap \{x + y > 0\}$ is zero, $\eta(A) = 0$, and since the set $\{x_1\} \times (y_1, y_2)$ lies in A, it does not intersect $\text{spt}\,\eta$. □

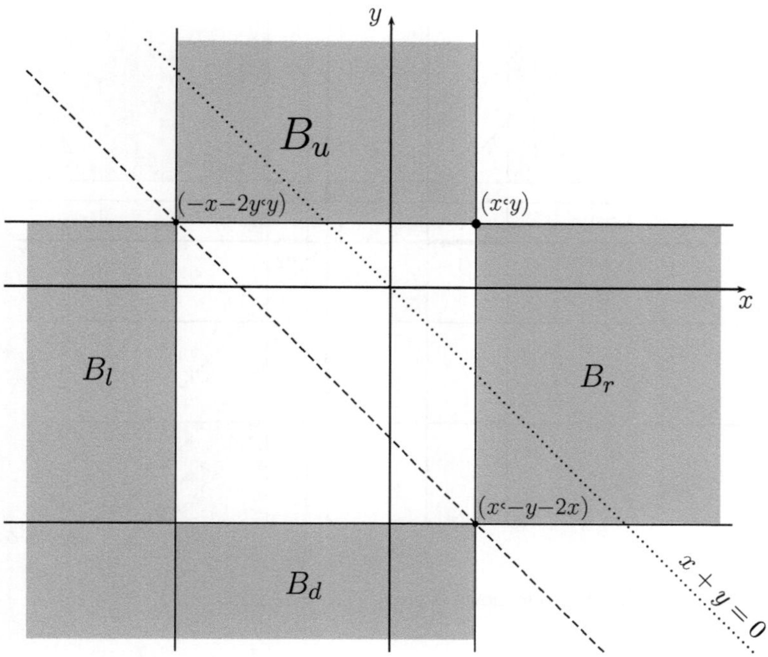

Fig. 5.4 $A(x, y)$ is the complement of the sets B_r, B_u, B_l, B_d and is shown in *white*

Lemma 5.6. *(a) If* $(x, y) \in spt^+\eta$, *then*

$$\lambda((-x-2y,\, x)) = \lambda((-y-2x,\, y)).$$

(b) If $(x_1, y_1), (x_2, y_2) \in spt^+\eta$, $x_1 < x_2$, *and* $y_1 < y_2$, *then*

$$\lambda((x_1,\, x_2)) \geq \lambda((-y_2 - 2x_2,\, -y_1 - 2x_1)),$$
$$\lambda((y_1,\, y_2)) \geq \lambda((-x_2 - 2y_2,\, -x_1 - 2y_1)).$$

Proof. (a) We set

$$B_r = (x, +\infty) \times (-y-2x,\, y), \qquad B_u = (-x-2y,\, x) \times (y, +\infty),$$
$$B_l = (-\infty, -x-2y) \times (-\infty, y), \quad B_d = (-\infty, x) \times (-\infty, -y-2x)$$

(Fig. 5.4). For a point $(x', y') \in B_r \cup B_u \cup B_l \cup B_d$ we can prove that

$$(x'-x)(y'-y)(x'+y'+x+y) < 0$$

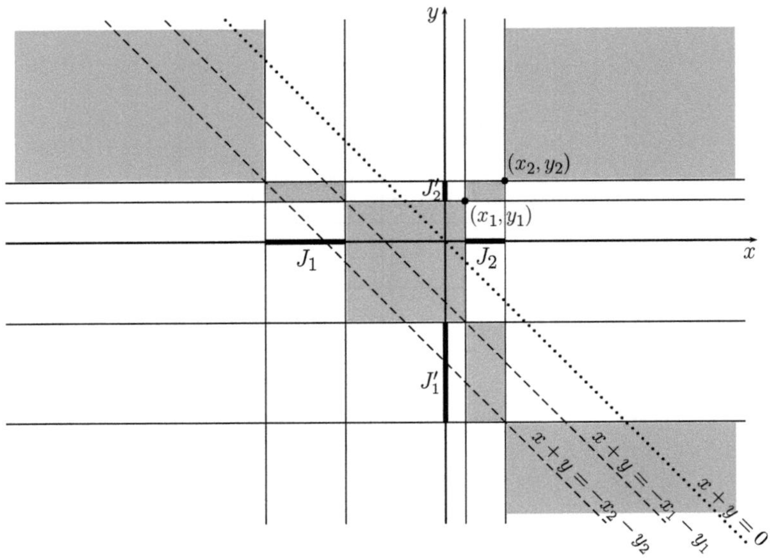

Fig. 5.5 Set A containing spt η is shown in *gray*

by the successive consideration of the cases $(x', y') \in B_j$, $j = r, u, l, d$. Hence by Lemma 5.2 and our assumption that $(x, y) \in \operatorname{spt}\eta$ and $\operatorname{spt}\eta$ is f-monotone, we obtain that $(x', y') \notin \operatorname{spt}\eta$. That is, $\operatorname{spt}\eta \subset A(x, y) := \mathbb{R}^2 \setminus (B_r \cup B_u \cup B_l \cup B_d)$. The set $A(x, y)$ is X-like of format $(-x-2y,\ x) \times (-y-2x,\ y)$; therefore, $\operatorname{spt}\eta$ is also X-like of this format, and taking Lemma 5.4 into account we obtain assertion (a).

(b) It follows from the proof of (a) that $\operatorname{spt}\eta \subset A := A(x_1, y_1) \cap A(x_2, y_2)$. Let

$$J_1 = (-x_2 - 2y_2,\ -x_1 - 2y_1),\ J_2 = (x_1,\ x_2),$$
$$J'_1 = (-y_2 - 2x_2,\ -y_1 - 2x_1),\ J'_2 = (y_1,\ y_2)$$

(Fig. 5.5). One can immediately verify that

$$A \cap (J_1 \times \mathbb{R}) = J_1 \times J'_2, \quad A \cap (J_2 \times \mathbb{R}) = J_2 \times (J'_1 \cup J'_2),$$
$$A \cap (\mathbb{R} \times J'_1) = J_2 \times J'_1, \quad A \cap (\mathbb{R} \times J'_2) = (J_1 \cup J_2) \times J'_2.$$

Hence

$$\lambda(J_2) = \eta(J_2 \times \mathbb{R}) = \eta(J_2 \times (J'_1 \cup J'_2)) \geq \eta(J_2 \times J'_1) = \eta(\mathbb{R} \times J'_1) = \lambda(J'_1),$$
$$\lambda(J'_2) = \eta(\mathbb{R} \times J'_2) = \eta((J_1 \cup J_2) \times J'_2) \geq \eta(J_1 \times J'_2) = \eta(J_1 \times \mathbb{R}) = \lambda(J_1).$$

We see that $\lambda(J_2) \geq \lambda(J'_1)$ and $\lambda(J'_2) \geq \lambda(J_1)$, as required. □

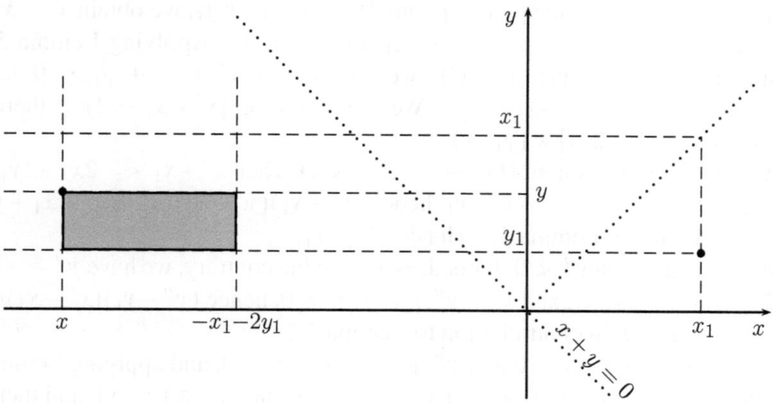

Fig. 5.6 Illustration to Lemma 5.7. The rectangle $(x, -x_1 - 2y_1) \times (y_1, y)$ is shown in *gray*

Lemma 5.7. *Let $(x_1, y_1) \in spt^+\eta$. If $x_1 > y_1$, then there exists a point (x_2, y_2) in $spt^+\eta$ such that $y_2 \in (y_1, x_1)$, whereas if $x_1 < y_1$, then there exists a point $(x_2, y_2) \in spt^+\eta$ such that $x_2 \in (x_1, y_1)$.*

Proof. We shall prove the first assertion of the lemma. The proof of the second is perfectly similar.

Assume the contrary: $x_1 > y_1$ and

$$\operatorname{spt}\eta \cap (\mathbb{R} \times (y_1, x_1)) \subset \operatorname{spt}^-\eta. \tag{5.26}$$

First we assume that $\operatorname{spt}\lambda \cap (y_1, x_1)$ is nonempty. Indeed, using Remark 5.3, we choose a sequence of points $(x_i, y_i) \neq (x_1, y_1)$ in $\operatorname{spt}^+\eta$, $i \geq 2$, converging to (x_1, y_1). Using (5.26), we conclude that $y_i \leq y_1$ starting from some value i; therefore, $x_i \leq x_1$. Using Lemma 5.5, choose this sequence such that $y_i < y_1$ and $x_i < x_1$. This immediately yields that

$$\lambda((x_1 - \varepsilon, x_1)) > 0 \quad \text{for all } \varepsilon > 0. \tag{5.27}$$

Let $y \in \operatorname{spt}\lambda \cap (y_1, x_1)$; then there exists x such that $(x, y) \in \operatorname{spt}\eta \cap (\mathbb{R} \times (y_1, x_1))$. By (5.26) we obtain $x + y \leq 0$; therefore, $x \leq -y < -y_1 < x_1$. Applying Lemma 5.2 to the pair of points $\{(x, y), (x_1, y_1)\}$ we obtain

$$x + y + x_1 + y_1 \leq 0. \tag{5.28}$$

Hence it follows from the assumptions $x_1 + y_1 > 0$ and $y > y_1$ that $x + y < 0$ and $x < -x_1 - 2y_1$ (Fig. 5.6).

We claim that the set $\operatorname{spt}\eta$ is X-like of format $(x, -x_1 - 2y_1) \times (y_1, y)$. Indeed, by (5.26), each point $(x', y') \in \operatorname{spt}\eta \cap (\mathbb{R} \times (y_1, y))$ lies in $\operatorname{spt}^-\eta$; therefore, $x' + y' \leq 0$, so that $x + y + x' + y' < 0$. In addition, we have $y' - y < 0$.

Applying Lemma 5.2 to the pair of points $\{(x, y), (x', y')\}$, we obtain $x' - x \geq 0$. Next, we have $x' \leq -y' < -y_1 < x_1$ and $y' > y_1$. Applying Lemma 5.2 to the pair of points $\{(x_1, y_1), (x', y')\}$ we obtain $x' + y' + x_1 + y_1 \leq 0$, so that $x' \leq -x_1 - y_1 - y' < -x_1 - 2y_1$. We see that $x' \in [x, -x_1 - 2y_1]$; therefore, $(x', y') \in [x, -x_1 - 2y_1] \times (y_1, y)$.

Now let $(x'', y'') \in \operatorname{spt} \eta \cap ((x, -x_1 - 2y_1) \times \mathbb{R})$. Then $x'' - x_1 < -2x_1 - 2y_1 < 0$. If $y'' < y_1$, then $x'' + y'' < -x_1 - y_1$, hence $(x'' - x_1)(y'' - y_1)(x'' + y'' + x_1 + y_1) < 0$, in contradiction to Lemma 5.2. Hence $y'' \geq y_1$.

We claim that $x'' + y'' < 0$. Indeed, assuming the contrary, we have $y'' \geq -x'' > x_1 + 2y_1 > y_1$, $x'' < x_1$, and $x'' + y'' + x_1 + y_1 > 0$, hence $(y'' - y_1)(x'' - x_1)(x'' + y'' + x_1 + y_1) < 0$, in contradiction to Lemma 5.2.

Thus, we have $x'' - x > 0$ and $x'' + y'' + x + y < 0$, and applying Lemma 5.2 to the pair $\{(x, y), (x'', y'')\}$ we get $y'' - y < 0$. Hence $y'' \in [y_1, y]$, and therefore $(x'', y'') \in (x, -x_1 - 2y_1) \times [y_1, y]$. We have thus shown that $\operatorname{spt} \eta$ is an X-like set, and therefore

$$\lambda((x, -x_1 - 2y_1)) = \lambda((y_1, y)). \tag{5.29}$$

Further, by Lemma 5.6(a) we obtain $\lambda((-x_1 - 2y_1, x_1)) = \lambda((-y_1 - 2x_1, y_1))$, so that

$$\lambda((-y_1 - 2x_1, -x_1 - 2y_1)) = \lambda((y_1, x_1)). \tag{5.30}$$

By (5.28)–(5.30), we obtain the chain of relations

$$\begin{aligned}
\lambda((y, x_1)) &= \lambda((y_1, x_1)) - \lambda((y_1, y)) \\
&= \lambda((-y_1 - 2x_1, -x_1 - 2y_1)) - \lambda((x, -x_1 - 2y_1)) \leq \\
&\leq \lambda((-y_1 - 2x_1, -x_1 - 2y_1)) - \lambda((-x_1 - y_1 - y, -x_1 - 2y_1)) \\
&= \lambda((-y_1 - 2x_1, -x_1 - y_1 - y));
\end{aligned}$$

therefore,

$$\lambda((y, x_1)) \leq \lambda((-y_1 - 2x_1, -x_1 - y_1 - y)). \tag{5.31}$$

Let us return to the sequence x_i. Without loss of generality we assume that $x_i \in (y_1, x_1)$. By Lemma 5.6(b),

$$\lambda((x_i, x_1)) \geq \lambda((-y_1 - 2x_1, -y_i - 2x_i)). \tag{5.32}$$

On the other hand, substituting x_i for y in (5.31), we obtain

$$\lambda((x_i, x_1)) \leq \lambda((-y_1 - 2x_1, -x_1 - y_1 - x_i)). \tag{5.33}$$

It follows by (5.32) and (5.33) that for each i,

$$\begin{aligned}
&\lambda((-x_1 - y_1 - x_i, -y_i - 2x_i)) = \\
&\quad = \lambda((-y_1 - 2x_1, -y_i - 2x_i)) - \lambda((-y_1 - 2x_1, -x_1 - y_1 - x_i)) \leq 0;
\end{aligned}$$

therefore, the open set $\mathcal{A} = \cup_i (-x_1 - y_1 - x_i, -y_i - 2x_i)$ has measure zero. Hence

$$\mathcal{A} \subset \mathbb{R} \setminus \operatorname{spt} \lambda \quad \text{and} \quad \inf \mathcal{A} = -y_1 - 2x_1. \tag{5.34}$$

By assumption $\mathbb{R} \setminus \operatorname{spt} \lambda$ is the union of finitely many intervals; therefore, we conclude from (5.34) that

$$\lambda((-y_1 - 2x_1, -y_1 - 2x_1 + a)) = 0$$

for some $a > 0$. Using (5.33) again and bearing in mind that $\lim_{i \to \infty} x_i = x_1$ we see that $\lambda((x_i, x_1)) = 0$ for some i, in contradiction to (5.27). □

Lemma 5.8. *$\operatorname{spt}^+ \eta$ lies on the line $\{x = y\}$.*

Proof. Assume the contrary and consider a point $(x_1, y_1) \in \operatorname{spt}^+\eta$, $x_1 \neq y_1$. Without loss of generality we assume that $x_1 > y_1$. We now construct recursively a sequence of points $(x_i, y_i) \in \operatorname{spt}^+\eta$, $i \geq 1$ such that $x_i > y_i$. Assume that we have defined the point (x_i, y_i) for some i; then by definition

$$y_{i+1} := \sup\{y \in (y_i, x_i) : \text{ there exists } x \text{ such that } (x, y) \in \operatorname{spt}^+\eta\}.$$

The set over which the supremum is taken is nonempty in view of Lemma 5.7. Let $x_{i+1} := \sup\{x : (x, y_{i+1}) \in \operatorname{spt}^+\eta\}$. Since both points (x_i, y_i) and (x_{i+1}, y_{i+1}) lie in $\operatorname{spt}^+\eta$ and $y_{i+1} \geq y_i$, then $x_{i+1} \geq x_i$. Assume that $x_{i+1} = x_i$; then by Lemma 5.7 there exists a point $(x, y) \in \operatorname{spt}^+\eta$ such that $y \in (y_{i+1}, x_i)$, in contradiction to the choice of y_{i+1}. Thus, we have $x_{i+1} > x_i$.

The sequences $\{x_i\}$ and $\{y_i\}$ so defined are increasing and bounded; therefore, there exist limits $x_\infty = \lim_{i \to \infty} x_i$, $y_\infty = \lim_{i \to \infty} y_i$. Taking into account the definition of y_{i+1} and y_{i+2} and the inequality $y_{i+2} > y_{i+1}$ we obtain

$$y_{i+2} \geq x_i. \tag{5.35}$$

We see that $y_{i+1} \leq x_i \leq y_{i+2}$, so that $x_\infty = y_\infty$.

Now, taking into account the facts that x_∞ is the limit of an increasing sequence of points $x_i \in \operatorname{spt} \lambda$ and $\operatorname{spt} \lambda$ consists of finitely many intervals, we conclude that

$$[x_\infty - \varepsilon, x_\infty] \subset \operatorname{spt} \lambda$$

for some $\varepsilon > 0$. Hence for sufficiently large i we have $\lambda((y_i, x_i)) > 0$. By assertion (a) of Lemma 5.6 we have $\lambda((-x_i - 2y_i, x_i)) = \lambda((-y_i - 2x_i, y_i))$, hence

$$\lambda((-y_i - 2x_i, -x_i - 2y_i)) = \lambda((y_i, x_i)) > 0. \tag{5.36}$$

Both sequences $-y_i - 2x_i$ and $-x_i - 2y_i$ decrease and converge to $-3x_\infty$; again using the fact that $\operatorname{spt}\lambda$ consists of finitely many intervals, we conclude that for some $\omega > 0$,

$$[-3x_\infty, -3x_\infty + \omega] \subset \operatorname{spt}\lambda. \tag{5.37}$$

By assertion (b) of Lemma 5.6,

$$\lambda((y_i, y_{i+1})) \geq \lambda((-x_{i+1} - 2y_{i+1}, -x_i - 2y_i)); \tag{5.38}$$

in addition, by the inequality $y_i < y_{i+1} \leq x_i$ we obtain

$$\lambda((y_i, x_i)) \geq \lambda((y_i, y_{i+1})). \tag{5.39}$$

It follows from (5.36), (5.38), and (5.39) that

$$\lambda((-y_i - 2x_i, -x_i - 2y_i)) \geq \lambda((-x_{i+1} - 2y_{i+1}, -x_i - 2y_i));$$

therefore, by (5.37) for sufficiently large i,

$$-y_i - 2x_i \leq -x_{i+1} - 2y_{i+1}. \tag{5.40}$$

It follows from (5.40) that

$$\sum_1^\infty (x_i - y_i) \geq \sum_1^\infty (x_{i+1} + 2y_{i+1} - x_i - 2y_i) = 3x_\infty - x_1 - 2y_1. \tag{5.41}$$

On the other hand, setting $x_0 := y_1$ and using (5.35) we obtain

$$x_\infty - y_1 = \sum_1^\infty (y_{i+1} - y_i) \geq \sum_1^\infty \left(\frac{y_{i+1} - x_{i-1}}{2} + x_{i-1} - y_i \right) =$$

$$= \sum_1^\infty \frac{y_{i+1} - x_{i-1}}{2} + \sum_1^\infty \frac{x_i - y_{i+1}}{2} + \sum_1^\infty \frac{x_{i-1} - y_i}{2} = \frac{1}{2} \sum_1^\infty (x_i - y_i). \tag{5.42}$$

It follows by (5.41) and (5.42) that $3x_\infty - x_1 - 2y_1 \leq 2(x_\infty - y_1)$; therefore, $x_\infty \leq x_1$, in contradiction to the fact that the sequence $\{x_i\}$ is increasing. □

Lemma 5.9. *Each measure η with f-monotone support has the form $\eta = \eta_{\mathcal{I}}$, where $\mathcal{I}$ is an admissible system of intervals.*

Proof. By Lemma 5.8, we have that $\operatorname{spt}^+\eta$ lies on the half-line $\{x = y > 0\}$; in addition, the closure $\overline{\operatorname{spt}^+\eta}$ coincides with either $\operatorname{spt}^+\eta$ or $\operatorname{spt}^+\eta \cup \{(0,0)\}$. The set $\overline{\operatorname{spt}^+\eta}$ can be represented in the following form:

$$\overline{\operatorname{spt}^+\eta} = \{(x,x) : x \in \mathbb{R}_+ \setminus (\cup_i I_i)\},$$

where the I_i make up a countable family of disjoint subintervals in $\mathbb{R}_+$. This family contains at most one semiopen interval of the form $I_0 = [0, b_0)$; the other intervals are open, $I_i = (a_i, b_i)$, $0 < a_i < b_i \le +\infty$, $i \neq 0$. Thus, the index set $\{i\}$ may or may not contain the symbol 0. The set $\text{spt}^+\eta$ contains no isolated points; therefore, $\bar{I}_i \cap \bar{I}_j = \emptyset$ for $i \neq j$.

We shall show that system $\mathcal{I}$ is an admissible system of intervals and satisfies condition (5.4), and additionally $\eta = \eta_{\mathcal{I}}$. Before that we prove several auxiliary results.

(i) If $x \in \mathbb{R}_+ \setminus (\cup_i \bar{I}_i)$, then $(\{x\} \times \mathbb{R}) \cap \text{spt}^-\eta \subset \{(x, -3x)\}$. Thus, $(\{x\} \times \mathbb{R}) \cap \text{spt}^-\eta$ is either a singleton or an empty set.

(ii) If $i \neq 0$, then $\lambda((a_i, b_i)) = \lambda((-3b_i, -3a_i))$.

(iii) If $\tilde{a}_i < x < b_i$, where $\tilde{a}_i = \begin{cases} a_i, & i \neq 0, \\ -3b_i, & i = 0, \end{cases}$ and $(x, y) \in \text{spt}^-\eta$, then

$$\lambda((x, b_i)) = \lambda((-3b_i, y)).$$

(iv) If $x = a_i$, $i \neq 0$, or $x = b_i$, then $(\{x\} \times \mathbb{R}) \cap \text{spt}^-\eta \subset \{(x, -3x)\}$.

(v) If $0 \in \{i\}$, then

$$((-\infty, -3b_0] \times (-\infty, -3b_0]) \cap \text{spt}^-\eta = \emptyset, \tag{5.43}$$

whereas if $0 \notin \{i\}$, then

$$((-\infty, 0] \times (-\infty, 0]) \cap \text{spt}^-\eta \subset \{(0, 0)\}. \tag{5.44}$$

In the latter case, $(0, 0)$ lies in G^+.

(i) Let $x \in \mathbb{R}_+ \setminus (\cup_i \bar{I}_i)$; then there exist sequences $\{x'_i\}$, $\{x''_i\} \subset \mathbb{R}_+ \setminus (\cup_i I_i)$ such that $x'_i < x < x''_i$ and $\lim_{i\to\infty} x'_i = x = \lim_{i\to\infty} x''_i$. By the definition of $\mathbb{R}_+ \setminus (\cup_i I_i)$, the points (x'_i, x'_i) and (x''_i, x''_i) belong to $\text{spt}^+\eta$. Let $(x, y) \in \text{spt}^-\eta$; then $y \le -x < 0$, and therefore $y < x'_i$ and $y < x''_i$. Applying Lemma 5.2 to the pair of points (x, y), (x'_i, x'_i) and using the inequalities $x > x'_i$ and $y < x'_i$ we obtain $x + y + 2x'_i \le 0$, whereas applying Lemma 5.2 to the pair of points (x, y), (x''_i, x''_i) and using the inequalities $x < x''_i$ and $y < x''_i$ we obtain $x + y + 2x''_i \ge 0$. Thus, we have

$$-x - 2x''_i \le y \le -x - 2x'_i.$$

Passing here to the limit as $i \to \infty$ we obtain $y = -3x$, which proves (i).

(ii) Let $(x, y) \in \text{spt}\,\eta$, $0 < a_i < x < b_i$; then, by the definition of I_i, we obtain $(x, y) \in \text{spt}^-\eta$. We have $y \le -x < 0$; therefore, $y < a_i < b_i$. Applying Lemma 5.2 to the pair of points (x, y), (a_i, a_i) and using the inequalities $x > a_i$ and $y < a_i$ we obtain $x + y + 2a_i \le 0$, and applying Lemma 5.2 to (x, y), (b_i, b_i) and using the inequalities $x < b_i$ and $y < b_i$ we obtain $x+y+2b_i \ge 0$.

Hence $-3b_i < y < -3a_i$. We have thus proved that $(I_i \times \mathbb{R}) \cap \operatorname{spt} \eta \subset I_i \times (-3I_i)$.

Assume now that $-3b_i < y < -3a_i$. If $x < a_i$, then $(x - a_i)(y - a_i)(x + y + 2a_i) < 0$, whereas if $x > b_i$, then $(x - b_i)(y - b_i)(x + y + 2b_i) < 0$; hence $(x, y) \notin \operatorname{spt} \eta$ in both cases. We thus see that $(\mathbb{R} \times (-3I_i)) \cap \operatorname{spt} \eta \subset I_i \times (-3I_i)$. This shows that $\operatorname{spt} \eta$ is X-like of format $I_i \times (-3I_i)$; therefore, by Lemma 5.4, $\lambda(I_i) = \lambda(-3I_i)$.

(iii) Assume that $\tilde{a}_i < x < b_i$ and $(x, y) \in \operatorname{spt}^- \eta$; then $y \le b_i$. Indeed, otherwise for $i \ne 0$ we have $x + y > a_i + b_i > 0$, in contradiction to the inclusion $(x, y) \in \operatorname{spt}^- \eta$, whereas for $i = 0$ we have (x, y), $(b_0, b_0) \in \operatorname{spt} \eta$ and $(x - b_0)(y - b_0)(x + y + 2b_0) < 0$, in contradiction to Lemma 5.2.

Consider now a point $(x', y') \in \operatorname{spt} \eta$ such that $-3b_i < y' < y$ and either $x' > b_i$ or $x' < x$. If $x' > b_i$, then

$$(x' - b_i)(y' - b_i)(x' + y' + 2b_i) < 0,$$

whereas if $x' < x$, then

$$(x' - x)(y' - y)(x' + y' + x + y) < 0.$$

These inequalities contradict the result of Lemma 5.2 and the inclusion (x, y), (b_i, b_i), and $(x', y') \in \operatorname{spt} \eta$; hence we can conclude that

$$\operatorname{spt} \eta \cap (\mathbb{R} \times (-3b_i,\ y)) \subset [x,\ b_i] \times (-3b_i,\ y).$$

Consider now a point $(x'', y'') \in \operatorname{spt} \eta$ such that $x < x'' < b_i$ and either $y'' < -3b_i$ or $y'' > y$. By the definition of intervals I_i, we obtain $(x'', y'') \in \operatorname{spt}^- \eta$. Now, if $y'' < -3b_i$, then

$$(x'' - b_i)(y'' - b_i)(x'' + y'' + 2b_i) < 0,$$

whereas if $y'' > y$, then

$$(x'' - x)(y'' - y)(x'' + y'' + x + y) < 0.$$

Both inequalities contradict Lemma 5.2 and the inclusions (x, y), (b_i, b_i), $(x'', y'') \in \operatorname{spt} \eta$. Hence

$$\operatorname{spt} \eta \cap ((x,\ b_i) \times \mathbb{R}) \subset (x,\ b_i) \times [-3b_i,\ y].$$

We see that $\operatorname{spt} \eta$ is X-like of format $(x,\ b_i) \times (-3b_i,\ y)$; therefore, by Lemma 5.4, $\lambda((x,\ b_i)) = \lambda((-3b_i,\ y))$.

(iv) Let $x = a_i$, $i \ne 0$, or $x = b_i$, $(x, y) \in \operatorname{spt}^- \eta$. By Remark 5.3, there exists a sequence of points $(x_n, y_n) \in \operatorname{spt} \eta$ convergent to (x, y). We select it so that

$x_n \notin \{a_i, b_i\}$. Using (i)–(iii) we now see that $y_i + 3x_i \to 0$ as $i \to \infty$; therefore, $y = -3x$.

(v) Assume that $0 \in \{i\}$, $x \le -3b_0$, $y \le -3b_0$. Then we have $(b_0, b_0) \in \operatorname{spt}\eta$, $(x - b_0)(y - b_0)(x + y + 2b_0) < 0$; therefore, $(x, y) \notin \operatorname{spt}\eta$.

Assume that $0 \notin \{i\}$, $x \le 0$, $y \le 0$, $(x, y) \ne (0, 0)$. We can find a point $(x', x') \in \operatorname{spt}^+\eta$ such that $0 < x' < -(x + y)/2$. Then $(x - x')(y - x')(x + y + 2x') < 0$; therefore, $(x, y) \notin \operatorname{spt}\eta$. In this case, 0 lies in $\mathbb{R}_+ \setminus (\cup_i I_i)$; therefore, $(0, 0)$ lies in G^+. This proves assertion (v).

It follows by (i) and (iv) that

$$((\mathbb{R}_+ \setminus (\cup_i I_i)) \times \mathbb{R}) \cap \operatorname{spt}^-\eta \subset \{(x, y) : y = -3x,\ x \in \mathbb{R}_+ \setminus (\cup_i I_i)\}, \quad (5.45)$$

and it follows by (iii) that

$$(\tilde{I}_i \times \mathbb{R}) \cap \operatorname{spt}^-\eta \subset \{(x, y) : x \in \tilde{I}_i,\ \lambda((x, b_i)) = \lambda((-3b_i, y))\}, \quad (5.46)$$

where $\tilde{I}_i = \begin{cases} I_i & \text{if } i \ne 0 \\ (-3b_0, b_0) & \text{if } i = 0 \end{cases}$.

In a similar way we prove that

$$(\mathbb{R} \times (\mathbb{R}_+ \setminus (\cup_i I_i))) \cap \operatorname{spt}^-\eta \subset \{(x, y) : x = -3y,\ y \in \mathbb{R}_+ \setminus (\cup_i I_i)\}, \quad (5.47)$$

$$(\mathbb{R} \times \tilde{I}_i) \cap \operatorname{spt}^-\eta \subset \{(x, y) : y \in \tilde{I}_i,\ \lambda((y, b_i)) = \lambda((-3b_i, x))\}. \quad (5.48)$$

By Remark 5.3, $\operatorname{spt}\eta$ contains no isolated points; therefore, all isolated points of $\operatorname{spt}^-\eta$ lie in $\overline{\operatorname{spt}^+\eta}$. Hence $\operatorname{spt}^-\eta$ can have only a unique isolated point $(0, 0)$, and only if it is also contained in $\overline{\operatorname{spt}^+\eta}$. Furthermore, taking into account relations (5.43)–(5.48) and Remark 5.4 we see that

$$\operatorname{spt}^-\eta \subset G^- \cup (\{(0, 0)\} \cap G^+),$$

and by the definition of system $\mathcal{I}$ we have

$$\overline{\operatorname{spt}^+\eta} = G^+. \quad (5.49)$$

This proves that

$$\operatorname{spt}\eta \subset G_{\mathcal{I}}. \quad (5.50)$$

To prove the reverse inclusion designate $\tilde{b} = \begin{cases} b_0, \text{ if } 0 \in \{i\}, \\ 0, \ \text{ if } 0 \notin \{i\}, \end{cases}$ and note that

$$G^- \subset (\mathbb{R} \times (-\infty, \tilde{b})) \cup ((-\infty, \tilde{b}) \times \mathbb{R}) \cup \{(0, 0)\}. \quad (5.51)$$

We claim that

$$G^- \cap \{y < \tilde{b}\} \subset \operatorname{spt}\eta, \quad G^- \cap \{x < \tilde{b}\} \subset \operatorname{spt}\eta. \quad (5.52)$$

Let y^α be the values of y such that the intersection $(\mathbb{R} \times \{y\}) \cap G^-$ consists of two points, which we denote by (x_1^α, y^α) and (x_2^α, y^α), $x_1^\alpha < x_2^\alpha$. We have $\lambda((x_1^\alpha, x_2^\alpha)) = 0$; therefore, x_1^α and x_2^α belong to the finite set $\partial(\operatorname{spt}\lambda)$ and the number of points y^α is also finite. For y distinct from y^α the set $(\mathbb{R} \times \{y\}) \cap G^-$ contains at most one point.

Assume that $(x, y) \in G^-$, $y < \tilde{b}$, $y \notin \{y^\alpha\}$. By (5.50) we have

$$(\mathbb{R} \times \{y\}) \cap \operatorname{spt}\eta \subset (\mathbb{R} \times \{y\}) \cap G_{\mathcal{I}} = (\mathbb{R} \times \{y\}) \cap G^- = \{(x, y)\}.$$

By the definition of G^-, we have $y \in \operatorname{spt}\lambda$; therefore, by Remark 5.4, we obtain $(\mathbb{R} \times \{y\}) \cap \operatorname{spt}\eta \neq \emptyset$. Hence $(x, y) \in \operatorname{spt}\eta$. Further, if $y = y^\alpha$, then (x, y) is a limit point of $\operatorname{spt}\eta$, which therefore also belongs to $\operatorname{spt}\eta$.

We have thus proved the first relation in (5.52); the second is proved in a similar way.

Finally, assume that $(0, 0) \in G^-$; then $(0, 0)$ is a limit of G^- and, therefore, also of $\operatorname{spt}\eta$. Since $\operatorname{spt}\eta$ is closed, we conclude that $(0, 0)$ lies in $\operatorname{spt}\eta$. Thus, we have proved the following fact:

$$\text{if } (0, 0) \in G^-, \quad \text{then } (0, 0) \in \operatorname{spt}\eta. \tag{5.53}$$

By (5.51)–(5.53) we have

$$G^- \subset \operatorname{spt}\eta.$$

It now follows by (5.49) and the closedness of $\operatorname{spt}\eta$ that

$$\overline{G^+} \subset \operatorname{spt}\eta.$$

Thus, $G_{\mathcal{I}} \subset \operatorname{spt}\eta$, and therefore by (5.50),

$$G_{\mathcal{I}} = \operatorname{spt}\eta.$$

Using Lemma 5.1 we now see that $\eta = \eta_{\mathcal{I}}$.

It remains to prove inclusion (5.4) and verify that system $\mathcal{I}$ satisfies conditions (5.5) and (5.6) and is therefore admissible.

Recall that λ_+ is the upper variation of the signed measure $\tilde{\lambda}$ defined by $\tilde{\lambda}(Q) = \lambda(Q) - \lambda(-3Q)$.

Consider an arbitrary point $x \in (0, +\infty) \setminus (\cup_i I_i)$ and an open set $\tilde{\mathcal{O}}$ containing x. Select an open interval $\mathcal{O}$ contained in $\tilde{\mathcal{O}}$ and also containing x such that its endpoints lie in $(0, +\infty) \setminus (\cup_i I_i)$. Such a selection is always possible; otherwise x is the right endpoint of one interval and the left endpoint of another, which is impossible.

Next we have

$$\lambda(\mathcal{O}) = \eta(\mathcal{O} \times \mathbb{R}) = \eta((\mathcal{O} \times \mathbb{R}) \cap G^+) + \eta((\mathcal{O} \times \mathbb{R}) \cap G^-). \tag{5.54}$$

The first term in the right-hand side of (5.54) contains the point $(x, x) \in \operatorname{spt} \eta$, and therefore is positive, while the second term is $\lambda(-3\mathcal{O})$. It follows that $\lambda(\mathcal{O}) - \lambda(-3\mathcal{O}) > 0$, hence $x \in \operatorname{spt} \lambda_+$.

It remains to consider the case $0 \in \mathbb{R}_+ \setminus (\cup_i I_i)$. Then 0 is a limit point of $(0, +\infty) \setminus (\cup_i I_i)$; therefore, it also is a limit point of $\operatorname{spt} \lambda_+$. By the closedness of $\operatorname{spt} \lambda_+$, we see that $0 \in \operatorname{spt} \lambda_+$. Inclusion (5.4) is thus proved.

Further, the first relation in (5.5) coincides with claim (ii). It remains to prove inequalities (5.6). Recall that we defined the intervals $\tilde{I}_i$ below formula (5.46). Let $x \in \tilde{I}_i \cap \operatorname{spt} \lambda$. We find y such that $(x, y) \in \operatorname{spt}^- \eta$. If $i \neq 0$, then $(a_i, a_i) \in \operatorname{spt} \eta$ and $x - a_i > 0$, $y - a_i < 0$, $x + y + 2a_i \leq 0$ by Lemma 5.2. On the other hand, if $i = 0$, then $x + y \leq 0$. Next, $(b_i, b_i) \in \operatorname{spt} \eta$ and $x - b_i < 0$, $y - b_i < 0$; therefore, $x + y + 2b_i \geq 0$. We thus obtain

$$-x - 2b_i \leq y \leq -x - 2a_i. \tag{5.55}$$

From (5.55) and (iii) we deduce (5.6).

Consider now an arbitrary point $x \in \tilde{I}_i$ and let $x' = \inf\{\xi \geq x : \xi \in \operatorname{spt} \lambda\}$. If $x' < b_i$, then the second inequality in (5.6) for x' is already proved, so

$$\lambda((x, b_i)) = \lambda((x', b_i)) \leq \lambda((-3b_i, -2a_i - x')) \leq \lambda((-3b_i, -2a_i - x)).$$

On the other hand, if $x' = b_i$, then $\lambda((x, b_i)) = 0$. This proves the second inequality in (5.6) for arbitrary x.

The first inequality in (5.6) is trivial for $i = r$. For $i \neq r$, 0 we set $x'' = \sup\{\xi \leq x : \xi \in \operatorname{spt} \lambda\}$. If $x'' > a_i$, then

$$\lambda((x, b_i)) = \lambda((x'', b_i)) \geq \lambda((-3b_i, -2b_i - x'')) \geq \lambda((-3b_i, -2b_i - x)).$$

On the other hand, if $x'' = a_i$, then

$$\lambda((x, b_i)) = \lambda((a_i, b_i)) = \lambda((-3b_i, -3a_i)) \geq \lambda((-3b_i, -2b_i - x)).$$

In the case $i = 0$, the reasoning is similar.

Thus the first inequality in (5.6) is proved for arbitrary x. Hence $\mathcal{I}$ is an arbitrary system of intervals. □

The result of Theorem 5.1 is an immediate consequence of Lemmas 5.3 and 5.9.

5.3 Examples

We now apply our results to several special cases.

Example 3. Assume that the inequality

$$\lambda(B) \geq \lambda(-3B) \tag{5.56}$$

holds for any Borel set $B \subset \mathbb{R}$. This implies that $\tilde{\lambda} = \lambda_+$ and $\lambda_- = 0$. In particular, if the measure λ has a density ρ, then (5.56) is equivalent to

$$\rho(x) \geq 3\rho(3x).$$

Then there exists a unique admissible system of intervals determined by $\mathbb{R}_+ \setminus (\cup_i I_i) = \operatorname{spt} \lambda_+$; therefore, there exists a unique measure with f-monotone support $\eta_* = \eta_{\mathcal{I}}$, which is the solution of minimization problem (5.1).

For each $x \in I_i$ we have $\tilde{\lambda}((x, b_i)) = 0$, hence $\lambda((x, b_i)) = \lambda((-3b_i, -3x))$. It follows that all points of the sets G_i^D, not only of the set $G_{(0)}^D$, lie on the line $y = -3x$. Repeating this argument for points in G_i^L and $G_{(0)}^L$, we see that G^- is contained in the union of two rays: $G^- \subset \{y = -3x \leq 0 \text{ or } x = -3y \leq 0\}$. Using that $\operatorname{spt} \eta_* = G_{\mathcal{I}}$, we conclude that η_* is concentrated on the union of three rays:

$$\operatorname{spt} \eta_* \subset \{(x, y) : x = y \geq 0, \text{ or } y = -3x \leq 0, \text{ or } x = -3y \leq 0\},$$

and additionally

$$\eta_*(\{(x, -3x) : x \in A\}) = \eta_*(\{(-3x, x) : x \in A\}) = \lambda(-3A),$$
$$\eta_*(\{(x, x) : x \in A\}) = \lambda(A) - \lambda(-3A)$$

for a Borel set $A \subset \mathbb{R}_+$. The minimum value of the functional $\mathcal{F}$ is

$$\mathcal{F}(\eta_*) = \int_0^{+\infty} f(2x)\, d(\lambda(x) - \lambda(-3x)) + 2\int_0^{+\infty} f(-2x)\, d\lambda(-3x) =$$
$$= \int_0^{+\infty} f(2x)\, d\lambda(x) + \int_0^{+\infty} (2f(-2x) - f(2x))\, d\lambda(-3x).$$

Making the change of variable $\xi = -3x$ and using that $f(-x) = -f(x)$, we have

$$\mathcal{F}(\eta_*) = \int_0^{+\infty} f(2x)\, d\lambda(x) + \int_{-\infty}^{0} (3f(2\xi/3))\, d\lambda(\xi).$$

Finally, denote $\tilde{f}(x) = \begin{cases} f(x) & \text{if } x \geq 0 \\ 3f(x/3) & \text{if } x < 0 \end{cases}$; then we have

$$\mathcal{F}(\eta_*) = \int_{-\infty}^{+\infty} \tilde{f}(2x)\, d\lambda(x).$$

Example 4. Assume now that the inequality

$$\lambda(B) \leq \lambda(-3B)$$

holds for each Borel set $B \subset \mathbb{R}$. This implies that $\tilde{\lambda} = -\lambda_-$ and $\lambda_+ = 0$. In this case, there exists a unique admissible system of intervals that consists of the unique interval $[0, +\infty)$, and $G_{\mathcal{I}}$ coincides with G_0; therefore,

$$G_{\mathcal{I}} = G_0 = \{(x, y) : x, y \in \operatorname{spt}\lambda, \lambda((x, +\infty)) = \lambda((-\infty, y))\}.$$

Let

$$g(x) := \min\{y : \lambda((-\infty, x)) = \lambda((y, +\infty))\};$$

then G_0 coincides with $\{y = g(x) : x \in \operatorname{spt}\lambda\}$ up to a finite set.

The unique measure with an f-monotone support $\eta_* = \eta_{\mathcal{I}}$ solves minimization problem (5.1). For each Borel set $B \subset \mathbb{R}^2$ we have

$$\eta_*(B) = \lambda(\{x : (x, g(x)) \in B\}),$$

and the minimum value of $\mathcal{F}$ is

$$\mathcal{F}(\eta_*) = \int_{-\infty}^{+\infty} f(x + g(x))\, d\lambda(x).$$

Example 5. Let λ be a measure with density function ρ that is positive and nondecreasing on an interval $[-\alpha, \beta]$ and equal to zero outside this interval; in addition, assume that $\beta > \alpha/3 > 0$. Then the measure $\tilde{\lambda}$ is defined by the density function $\tilde{\rho}(x) = \rho(x) - 3\rho(-3x)$, $x \ge 0$, which is nonpositive on an interval $0 \le x < a_0$ (which may be empty), positive for $a_0 < x \le \beta$, and vanishes for $x > \beta$, where $0 \le a_1 \le a_0 \le \alpha/3$.

Consider now two cases.

(a) $\lambda((-\alpha, 0)) \ge \lambda((0, \beta))$. By (5.4) and (5.5), the endpoints of intervals in an admissible system that are distinct from 0 lie in $[a_0, \beta]$, and none of these intervals lies entirely in $[a_0, \beta]$. In addition, no two endpoints of the intervals are the same. Hence each admissible system of intervals either coincides with $\{\mathbb{R}_+\}$ or has the form $\{I_0, I_r\}$, where $I_0 = [0, b_0)$, $I_r = (a_r, +\infty)$, $a_0 \le b_0 < a_r \le \beta$. Using (5.5) again we obtain $a_r = \beta$.

We claim that each family of the form $\{I_0, I_r\}$, where $I_0 = [0, b_0)$ and $I_r = (\beta, +\infty)$, is admissible. To prove this, it suffices to check relation (5.6) for the interval $(0, b_0)$ (for the interval $(\beta, +\infty)$ it is obvious). Substituting 0 for a_i and $b \in [a_0, \beta) \cup \{+\infty\}$ for b_i in this relation, we obtain

$$\lambda((-3b, -2b - x)) \le \lambda((x, b)) \le \lambda((-3b, -x)), \quad x \in (-3b, b). \quad (5.57)$$

The first inequality in (5.57) is obtained by integrating both sides of the inequality $\rho(\xi-3b) \le \rho(\xi+x)$ over the interval $[0, b-x]$. The right inequality in (5.57) is for $x = 0$ a consequence of the equality

$$\lambda((0, b)) - \lambda((-3b, 0)) = [\lambda((0, \beta)) - \lambda((-3\beta, 0))] - \int_b^\beta \tilde{\rho}(\xi)\, d\xi,$$

where the expression in the square brackets on the right-hand side is negative and the integrand is positive. Hence

$$\lambda((0, b)) \le \lambda((-3b, 0)). \tag{5.58}$$

For $x \in [0, b]$ we have $\lambda((0, x)) \ge \lambda((-x, 0))$; it follows from this inequality and (5.58) that

$$\lambda((x, b)) \le \lambda((-3b, -x)); \tag{5.59}$$

thus, the second inequality in (5.57) also holds for $x \in [0, b]$. Adding the term $\lambda((-x, x))$ to both sides of (5.59), we obtain $\lambda((-x, b)) \le \lambda((-3b, x))$; thus, we conclude that the second inequality in (5.57) is also true for $x \in [-b, 0]$. Finally, for $x \in (-3b, -b)$ this inequality is a consequence of the inclusion $(x, b) \subset (-3b, -x)$.

We thus conclude that the system $\mathcal{I}_t = \{(0, t), (\beta, +\infty)\}$, $t \in [a_0, \beta)$ is admissible; additionally, the trivial system $\mathcal{I}_0 = \{\mathbb{R}_+\}$ may also be admissible. There are no other admissible systems.

We define the function $g(x)$ as in the previous example and set

$$g_t(x) := \min\{y : \lambda((-3t, x)) = \lambda((y, t))\}, \quad x \in (-3t, t).$$

Then $\eta_{\mathcal{I}_0}$ is defined by the formula

$$\eta_{\mathcal{I}_0}(B) = \lambda(\{x : (x, g(x)) \in B\}) \text{ for each Borel set } B \subset \mathbb{R}^2,$$

$$\mathcal{F}(\eta_{\mathcal{I}_0}) = \int_{-\infty}^{+\infty} f(x + g(x))\, d\lambda(x).$$

Then the measure $\eta_{\mathcal{I}_t}$ is defined by

$$\eta_{\mathcal{I}_t}(B) = \lambda(\{x : (x, g_t(x)) \in B\}) \text{ for Borel sets } B \subset (-\infty, t) \times (-\infty, t);$$

$$\eta_{\mathcal{I}_t}(B) = \lambda(\{y : (-y/3, y) \in B\}) \text{ for Borel sets } B \subset [t, +\infty) \times (-\infty, t);$$

$$\eta_{\mathcal{I}_t}(B) = \lambda(\{x : (x, -x/3) \in B\}) \text{ for Borel sets } B \subset (-\infty, t) \times [t, +\infty);$$

$$\eta_{\mathcal{I}_t}(B) = \tilde{\lambda}(\{x : (x, x) \in B\}) \quad \text{for Borel sets } B \subset [t, +\infty) \times [t, +\infty),$$

and

$$\mathcal{F}(\eta_{\mathcal{I}_t}) = \int_{-3t}^{t} f(x + g_t(x))\, d\lambda(x) + \int_{\mathbb{R}\setminus[-3t,t]} \tilde{f}(2x)\, d\lambda(x),$$

where $\tilde{f}$ is as in Example 1.

(b) $\lambda((-\alpha, 0)) < \lambda((0, \beta))$. Reasoning similarly to case (a) we conclude that all admissible systems of intervals have the form $\mathcal{I}_t = \{[0, t), (\beta, +\infty)\}$, $t \in [a_0, b_0]$, with b_0 defined by $\lambda((0, b_0)) = \lambda((-\alpha, 0))$. The measure $\eta_{\mathcal{I}_t}$ and the value $\mathcal{F}(\eta_{\mathcal{I}_t})$ are determined in the same fashion as in case (a).

We thus see that in both cases (a) and (b), the minimization problem reduces to minimizing a function of one variable.

In the special case where $\rho(x) = 1$ for $x \in [-\alpha, \beta]$, we have $a_0 = \alpha/3$; for $\alpha/3 < \beta \le \alpha$ we obtain case (a), whereas for $\beta > \alpha$ we obtain (b); moreover, $b_0 = \alpha$. The corresponding values of $\mathcal{F}$ are as follows:

$$\mathcal{F}(\eta_{\mathcal{I}_0}) = -(\alpha + \beta)\, f(\alpha - \beta);$$

$$\mathcal{F}(\eta_{\mathcal{I}_t}) = \int_t^{\beta} f(2x)\, dx - (\alpha + t)\, f(\alpha - t).$$

5.4 The Problem of Mass Transfer on a Sphere

Here we consider a special problem of mass transportation on a sphere. Let us fix a vector $n \in \mathbb{R}^d$ and define the measures λ_n and λ_{-n} in S^{d-1} by $d\lambda_n(x) = b_d(x \cdot n)_+ dx$ and $d\lambda_{-n}(x) = b_d(x \cdot n)_- dx$. Let $\nu_e \in \Gamma_{\lambda_{-n},\lambda_n}$ be the measure generated by the shift along n. This measure is defined by

$$d\nu_e(x, y) = b_d(x \cdot n)_- \delta(y - x + 2(x \cdot n)n)\, dx\, dy.$$

Consider the following:

Problem. Find

$$\inf_{\nu\in\Gamma_{\lambda_{-n},\lambda_n}} R(\nu), \text{ where } R(\nu) = \frac{d+1}{4} \iint_{(S^{d-1})^2} \frac{1}{2}|x - y|^2\, d\nu(x, y). \tag{5.60}$$

The normalizing factor $(d + 1)/4$ is chosen so that the equality $R(\nu_e) = 1$ is fulfilled.

Since Problem (5.60) (that is, both mass distributions and the cost function) is invariant relative to revolution about n, it can be reduced to a one-dimensional problem, where the parameter is the angle between the radius vector of a point and the axis n. Namely, let λ_d be the measure on $[0, \pi/2]$ with $d\lambda_d(\varphi) = (d-1)\sin^{d-2}\varphi \cos\varphi \, d\varphi$. We will show that Problem (5.60) is equivalent to the following one-dimensional problem: find

$$m_d = \inf_{\eta \in \Gamma_{\lambda_d, \lambda_d}} F(\eta),$$

$$\text{where } F(\eta) = \frac{d+1}{4} \iint_{[0,\pi/2]^2} \left(1 + \cos(\varphi + \varphi^+)\right) d\eta(\varphi, \varphi^+). \tag{5.61}$$

The following reduction lemma holds true.

Lemma 5.10. *We have*

$$\inf_{\nu \in \Gamma_{\lambda_{-n}, \lambda_n}} R(\nu) = \inf_{\eta \in \Gamma_{\lambda_d, \lambda_d}} F(\eta).$$

Remark 5.6. The proof of the lemma indicates a way of constructing the solution ν_* on a sphere, given the solution η_* of the one-dimensional problem.

Proof. Recall that $S_n^{d-1} = \{v \in S^{d-1} : v \cdot n \geq 0\}$. Let the map $\Phi : S_{-n}^{d-1} \times S_n^{d-1} \to [0, \pi/2]^2$ be given by $\Phi(x, y) = (\varphi, \varphi^+)$, where $\varphi = \arccos(-x \cdot n)$ and $\varphi^+ = \arccos(y \cdot n)$. In other words, φ and φ^+ are the angles formed by the radius vectors of x and y with the axis n. Further, we have

$$\frac{1}{2}|x - y|^2 \geq 1 + \cos(\varphi + \varphi^+); \tag{5.62}$$

moreover, the equality in (5.62) is achieved only in the case where vectors x, y, and n are coplanar and n is situated between $-x$ and y. Additionally, we have

$$\Phi^{\#}\nu \in \Gamma_{\lambda_d, \lambda_d}. \tag{5.63}$$

It follows from (5.62) that

$$\iint_{(S^{d-1})^2} \frac{1}{2}|x - y|^2 \, d\nu(x, y) \geq \iint_{[0,\pi/2]^2} \left(1 + \cos(\varphi + \varphi^+)\right) d(\Phi^{\#}\nu)(\varphi, \varphi^+), \tag{5.64}$$

and the equality in (5.64) is achieved only in the case where for all $(x, y) \in \operatorname{spt}\nu$ vectors x, y, and n are coplanar and n is situated between $-x$ and y.

By (5.63) and (5.64), we have that

$$\inf_{\nu \in \Gamma_{\lambda_{-n}, \lambda_n}} R(\nu) \geq m_d. \tag{5.65}$$

Assume that η_* is a solution of Problem (5.61), that is,

$$F(\eta_*) = m_d. \tag{5.66}$$

For a continuous function $f : (S^{d-1})^2 \to \mathbb{R}$ denote

$$\tilde{f}(\varphi, \varphi^+) = \int_{S^{d-1} \cap n^\perp} f(-\cos\varphi \cdot n + \sin\varphi \cdot w,\ \cos\varphi^+ \cdot n + \sin\varphi^+ \cdot w)\, dw \quad (5.67)$$

and set by definition

$$\nu_*(f) := \iint_{[0,\pi/2]^2} \tilde{f}(\varphi, \varphi^+)\, d\eta_*(\varphi, \varphi^+). \quad (5.68)$$

Equality (5.68) defines a nonnegative continuous linear functional ν_* in the space of continuous functions, that is, a measure on $(S^{d-1})^2$. It can be immediately checked that both natural projections of this measure on S^{d-1} are λ_{-n} and λ_n. By (5.67) and (5.68), its support belongs to the set of points (x, y) of the form

$$x = -\cos\varphi \cdot n + \sin\varphi \cdot w, \quad y = \cos\varphi^+ \cdot n + \sin\varphi^+ \cdot w, \quad \varphi,\ \varphi^+ \in [0,\ \pi/2].$$

This set can be otherwise characterized as follows: (x, y) belongs to the set *iff* x, y, and n are coplanar and n lies between $-x$ and y. By the definition of ν_*, we have $\Phi^\# \nu_* = \eta_*$; therefore, by formula (5.64) and the observation after it,

$$R(\nu_*) = F(\eta_*).$$

Therefore, bearing in mind (5.65) and (5.66), we have

$$\inf_{\nu \in \Gamma_{\lambda_{-n}, \lambda_n}} R(\nu) = m_d,$$

so ν_* is optimal. Formulas (5.67) and (5.68) indicate a method of constructing ν_*. □

By Lemma 5.10, it suffices to solve the one-dimensional problem (5.61). After we make the change of variables $x = \pi/4 - \varphi$, $y = \pi/4 - \varphi^+$, it takes the form

$$\inf_{\eta \in \Gamma_{\hat{\lambda}, \hat{\lambda}}} \iint_{\mathbb{R}^2} f(x + y)\, d\eta(x, y), \quad (5.69)$$

where $d\hat{\lambda}(x) = \rho(x)\, dx$ with

$$\rho(x) = (d-1) \sin^{d-2}(\pi/4 - x) \cos(\pi/4 - x) \cdot \chi_{[-\pi/4,\, \pi/4]}(x),$$

and $f(x) = \frac{d+1}{4} \sin x$ for $|x| \le \pi/2$. We define f on $\mathbb{R} \setminus [-\pi/2, \pi/2]$ in such a way that the resulting function is strictly concave on $\mathbb{R}_+$ and odd. Thus, condition (A1) in Sect. 5.1 is satisfied.

The support of $\hat{\lambda}$ is the segment $[-\pi/4,\ \pi/4]$, and the $\hat{\lambda}$-measure of each point is zero; therefore, condition (A2) is also satisfied. Let us show that $\hat{\lambda}$ satisfies the conditions of Corollary 5.2.

Lemma 5.11. *Set $\tilde{\rho}(x) = \rho(x) - 3\rho(-3x)$. Then there exists $c = c_d$ such that $\tilde{\rho}(x) < 0$ for $0 \le x < c$ and $\tilde{\rho}(x) > 0$ for $c < x \le \pi/4$.*

Proof. One clearly has $\tilde{\rho}(x) = \rho(x) > 0$ for $\pi/12 < x \le \pi/4$; therefore, $0 \le c \le \pi/12$. When we make the inverse transformation $\varphi = \pi/4 - x$, the function $\tilde{\rho}(x)$ takes the form

$$\hat{\rho}(\varphi) = \tilde{\rho}(\pi/4-\varphi) = (d-1)(\sin^{d-2}\varphi\cos\varphi + 3\sin^{d-2}3\varphi\cos 3\varphi), \quad \varphi \in [\pi/6,\ \pi/4].$$

Let $\hat{c} = \pi/4 - c$. We will prove that for some $\hat{c} \in [\pi/6,\ \pi/4]$

$$\hat{\rho}(\varphi) > 0 \text{ for } 0 \le \varphi < \hat{c} \text{ and } \hat{\rho}(\varphi) < 0 \text{ for } \hat{c} < \varphi \le \pi/4$$

holds.

After some additional algebra the function $\hat{\rho}$ can be reduced to the form

$$\hat{\rho}(\varphi) = (d-1)\sin^{d-2}\varphi\cos\varphi\,[1 + 3(3 - 4\sin^2\varphi)^{d-2}(1 - 4\sin^2\varphi)].$$

Further, after the change of variable $z = 4\sin^2\varphi - 1 \in [0,\ 1]$ the function in the square brackets takes the form $1 - 3z(2-z)^{d-2}$. In the case $d = 2$ it is positive for $0 \le z < 1/3$ and negative for $1/3 < z \le 1$. In the case $d \ge 3$ it equals 1 for $z = 0$, equals -2 for $z = 1$, decreases on $[0,\ 2/(d-1)]$, and increases on $[2/(d-1),\ 1]$. Hence it has exactly one zero $z_0 \in (0,\ 2/(d-1))$ and is positive for $0 \le z < z_0$ and negative for $z_0 < z \le 1$. In particular, in the case $d = 3$, we have $z_0 = 1 - \sqrt{2/3}$. As we have already seen, in the case $d = 2$, $z_0 = 1/3$. □

Observe that z_0 has the following asymptotics: $z_0 = z_0(d) \approx \frac{1}{3}2^{-d+2}$ as $d \to \infty$.

Using Lemma 5.11 and bearing in mind that $\operatorname{spt}\hat{\lambda} = [-\pi/4,\ \pi/4]$, we conclude that the condition of Corollary 5.2 is satisfied with the parameters

$$c = \frac{\pi}{4} - \arcsin\sqrt{\frac{z_0+1}{4}} \in \left(0,\ \frac{\pi}{12}\right) \quad \text{and} \quad w = \frac{\pi}{4}.$$

In the case $d = 2$, we have $c = \pi/4 - \arcsin\frac{1}{\sqrt{3}}$, and condition (5.17) on the parameter t takes the form $t < \pi/4 - \arcsin(1 - \sqrt{2})$; hence t ranges over the following segment:

$$t \in \left[\frac{\pi}{4} - \arcsin\frac{1}{\sqrt{3}},\ \frac{\pi}{4} - \arcsin(1-\sqrt{2})\right].$$

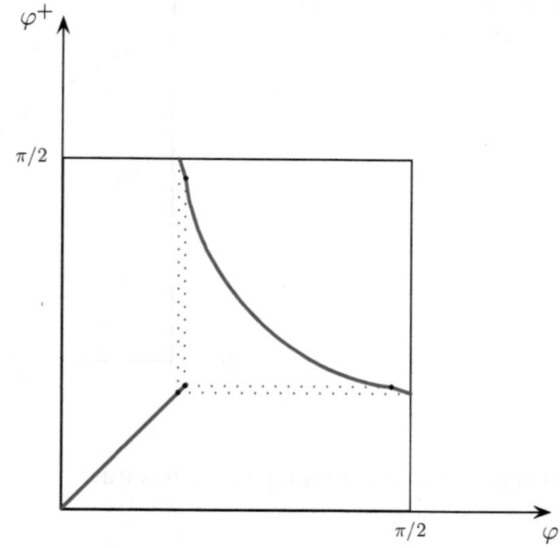

Fig. 5.7 Support of optimal measure in Problem (5.61) for $d = 2$

In the case $d = 3$, we have $c = \pi/4 - \arcsin\sqrt{\frac{1}{2} - \frac{1}{2\sqrt{6}}}$, and condition (5.17) is always satisfied, so that t ranges over the segment

$$t \in \left[\frac{\pi}{4} - \arcsin\sqrt{\frac{1}{2} - \frac{1}{2\sqrt{6}}},\ \frac{\pi}{4}\right].$$

One should distinguish between two cases: (i) $t < \pi/12$ and (ii) $t \geq \pi/12$.

In case (i), the intersection $[-3w, -3t] \cap \operatorname{spt}\hat{\lambda}$ is nonempty; therefore, the sets G^L and G^D are also nonempty. The corresponding measure η_t is not the graph of a function.

In case (ii), the intersection $[-3w, -3t] \cap \operatorname{spt}\hat{\lambda}$ is empty (or in the case $t = \pi/12$, it is a singleton); therefore, $G^L = G^D = \emptyset$ and $\operatorname{spt}\eta_t = G^+ \cup G_0$ is the graph of a piecewise continuous function decreasing on $[-\pi/4, t)$ and increasing on $[t, \pi/4]$.

Using numerical simulation, we select the optimal solution $\eta_* = \eta_{t_*}$ from the one-parameter family of candidates for the solution; the family $\eta_t = \eta_t^d$, and therefore the optimal measure $\eta_* = \eta_*(d) = \eta^d_{t_*(d)}$, depends on the dimension d.

For $d = 2$ we have $t_* = t_*(2) < \pi/12$, hence the optimal measure is not a Monge solution. Making the inverse transformation $\varphi = \pi/4 - x$, $\varphi^+ = \pi/4 - y$, we obtain the measure solving the original one-dimensional Problem (5.61). Its support is depicted in Fig. 5.7.

For $d \geq 3$ it has been checked numerically that $t_* = t_*(d) > \pi/12$; therefore, the optimal measure $\eta_* = \eta_*(d)$ is a Monge solution. We have also numerically found the limiting relation $\lim_{d\to\infty} t_*(d) = \pi/4$.

The quantities m_d for $d = 2, 3, \ldots, 11$ calculated numerically are plotted on the graph in Fig. 5.8. In particular, $m_2 \approx 0.9878$, $m_3 \approx 0.9694$; in addition we assume

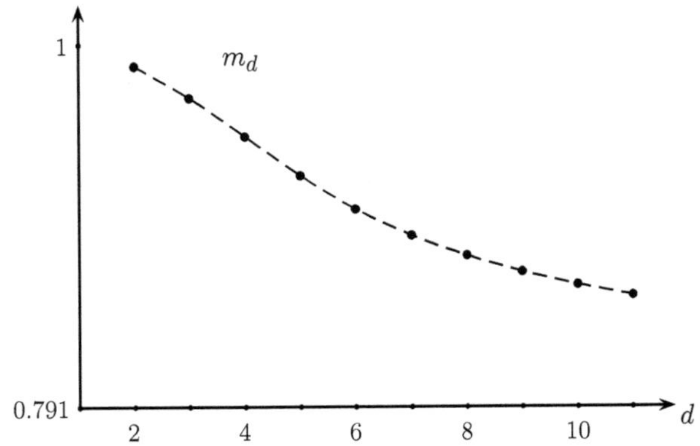

Fig. 5.8 Quantities m_d for several values of d

that the following limiting relation is true:

$$\lim_{d\to\infty} m_d = \int_0^1 \sqrt{\ln z \ln(1-z)}\, dz = 0.791\ldots$$

Numerical simulation related to this chapter was carried out by G. Mishuris. I am very grateful to him for this work.

Chapter 6
Problems on Optimization of Mean Resistance

Here we apply the results of Chaps. 4 and 5 to problems of minimum and maximum resistance in Newtonian aerodynamics.

We recall the general framework for these problems. A body $B \subset \mathbb{R}^d$ is moving translationally with constant velocity in a rarefied medium, and the velocity v of the motion is a random variable uniformly distributed over S^{d-1}. It is convenient to take a reference frame attached to the moving body. In it we have a flow of particles falling on the resting body B. The velocity of the flow is chosen randomly at the initial moment of observation and thereafter remains constant. The problem consists in minimizing or maximizing the expectation of the pressure force in the direction of the flow.

We obtain an equivalent setting if B is moving translationally and at the same time rotates slowly and uniformly about some point (the center of mass, for example) in it. In the reference frame attached to the body we have a resting body and a flow of particles impinging on it with a velocity v, $|v| = 1$. The velocity $v = v(t)$ of the flow changes slowly with time. The condition that the rotation is uniform means that the fraction of time that $v(t)$ belongs to a Borel subset A of S^{d-1} (provided that the period of observation is sufficiently large) is proportional to the relative area of A. In the two-dimensional case $d = 2$, uniform rotation means that the angular velocity of the rotation is constant and small. The problem consists in minimizing or maximizing the resistance of the body in the direction of the motion averaged over the (sufficiently large) period of observations.

Two-dimensional problems are considered in Sect. 6.1 and problems in higher dimensions in Sect. 6.2.

The main results of this chapter can also be found in [45, 50–52].

6.1 Two-Dimensional Case

Here we consider two cases: (a) the flow has internal temperature zero, that is, all the particles have the same velocity v; (b) the particles in the flow perform thermal motion, and the distribution of their velocities is centrally symmetric, with center at v.

A. Plakhov, *Exterior Billiards: Systems with Impacts Outside Bounded Domains*, DOI 10.1007/978-1-4614-4481-7_6,

6.1.1 Resistance in a Medium with Temperature Zero

Scattering of particles by B is conveniently described in terms of the map

$$\xi^+ = \xi^+_{B,\mathrm{Conv}B}(\xi, v), \quad v^+ = v^+_B = v^+_{B,\mathrm{Conv}B}(\xi, v).$$

We recall that we defined this map in Sect. 1.1. Here ξ and ξ^+ are the initial and final positions, and v and v^+ are the initial and final velocities of a scattered particle. We also recall that this map generates a measure ν_B, which is defined in Sect. 4.1.

In the reference frame attached to a slowly rotating body, the momentum that a particle transmits to the body is equal to $v - v^+$ (with suitable normalization). The projection of this momentum on the direction of the translational motion of the body is $(v - v^+) \cdot v$, and the component of the momentum orthogonal to this direction is $-v^+ \cdot v^\perp$, where $v^\perp$ is the vector turned through 90° (say, counterclockwise) relative to v. The mean resistance of the body is the sum of the momenta transmitted to the body over the period of the rotation divided by this period. It is convenient to calculate the projection $R_u(B)$ of the mean resistance on the direction of the motion and the projection $R_u^\perp(B)$ on the orthogonal direction. These are the integrals

$$R_u(B) = \int_{(\partial C \times S^1)_-} (v - v^+_B(\xi, v)) \cdot v \, d\mu(\xi, v),$$

$$R_u^\perp(B) = \int_{(\partial C \times S^1)_-} v^+_B(\xi, v) \cdot v^\perp \, d\mu(\xi, v), \tag{6.1}$$

where $C = \mathrm{Conv}B$. Recall that the measure $\mu = \mu_{\partial C}$ in $\partial C \times S^1$ is defined by $d\mu(\xi, v) = \frac{1}{2} |n(\xi) \cdot v| \, d\xi \, dv$. After the substitution $(\xi, v) \to (v, v^+_B(\xi, v), n(\xi))$ in these integrals, we obtain

$$R_u(B) = \int_{\mathbb{T}^3} (v - v^+) \cdot v \, d\nu_B(v, v^+, n),$$

$$R_u^\perp(B) = -\int_{\mathbb{T}^3} v^+ \cdot v^\perp \, d\nu_B(v, v^+, n).$$

After another substitution $(v, v^+, n) \mapsto (\varphi, \varphi^+) = \varpi(v, v^+, n)$ and use of relation (4.30) and the equalities $-v^+ \cdot v = \cos(\varphi - \varphi^+)$ and $v^+ \cdot v^\perp = \sin(\varphi - \varphi^+)$, we see that

$$R_u(B) = |\partial(\mathrm{Conv}B)| \iint_\square (1 + \cos(\varphi - \varphi^+)) \, d\eta_B(\varphi, \varphi^+), \tag{6.2}$$

$$R_u^\perp(B) = |\partial(\mathrm{Conv}B)| \iint_\square \sin(\varphi - \varphi^+) \, d\eta_B(\varphi, \varphi^+). \tag{6.3}$$

The measure η_B belongs to $\mathcal{M}$, hence it is invariant under the transposition of variables $v \leftrightarrow v^+$. On the other hand, $\sin(\varphi - \varphi^+)$ is antisymmetric under this transposition. Therefore, the integral (6.3) vanishes.

We have arrived at the following important conclusion.

Proposition 6.1. *The mean resistance force points in the direction opposite to the translational motion of the body, and its magnitude is equal to the integral $R_u(B)$ in (6.2).*

Recall that the following problem was stated in Sect. 1.7 of the introduction.

Problem 3. Find $\inf R_u(B)$ in the class

(a) Of connected (*in general* **nonconvex**) *bodies B of fixed area.*
(b) Of **convex** *bodies B of fixed area.*

The solution is given by the following theorem.

Recall that the quantity $m_2 \approx 0.9878$ is defined in Sect. 5.4.

Theorem 6.1.

(a) $\inf\{R_u(B) :$ *B is connected,* $\text{Area}(B) = A\} = \frac{8}{3}\, m_2 \sqrt{\pi A}$.
(b) $\inf\{R_u(B) :$ *B is convex,* $\text{Area}(B) = A\} = \frac{8}{3}\sqrt{\pi A}$.

Before starting the proof of the theorem we shall state and prove the following auxiliary proposition.

Let $S^1_\pm = \{x = (x_1, x_2) \in S^1 : \pm x_2 \geq 0\}$, and let dx be a Lebesgue measure on S^1. The measures in S^1 defined by $d\lambda_+(x) = \frac{1}{2}(x_2)_+\, dx$ and $d\lambda_-(x) = \frac{1}{2}(x_2)_-\, dx$ are denoted, respectively, by λ_+ and λ_-. Recall that $z_\pm = \max\{\pm z, 0\}$ is the positive or negative part of a real number z. Obviously, we have $\operatorname{spt} \lambda_+ = S^1_+$ and $\operatorname{spt} \lambda_- = S^1_-$.

Consider two minimization problems: find

$$\inf_{\eta \in \mathcal{M}} R(\eta), \quad \text{where } R(\eta) = \frac{3}{4} \iint_{\square} (1 + \cos(\varphi - \varphi^+))\, d\eta(\varphi, \varphi^+), \tag{6.4}$$

and find

$$\inf_{\nu \in \Gamma_{\lambda_-,\lambda_+}} F(\nu), \quad \text{where } F(\nu) = \frac{3}{4} \iint_{(S^1)^2} \frac{1}{2}\, |x - y|^2\, d\nu(x, y) \tag{6.5}$$

(the second problem is related to optimal mass transfer on the circle). We now show that they are equivalent.

Using the natural identification of plane vectors and complex numbers, $x = (x_1, x_2) \sim x_1 + i x_2$, define the map $\Psi : S^1_- \times S^1_+ \to \square$ by $\Psi(x, y) = (\text{Arg}(ix), \text{Arg}(-iy))$. Recall that the map $\pi_{\rm d}$ transposes the arguments, $\pi_{\rm d} : (\varphi, \varphi^+) \mapsto (\varphi^+, \varphi)$. For a measure η on $\square$ define the symmetrized measure $\eta_{\rm symm} = \frac{1}{2}(\eta + \pi^\#_{\rm d} \eta)$.

Proposition 6.2. *We have*

$$\inf_{\eta \in \mathcal{M}} R(\eta) = \inf_{\nu \in \Gamma_{\lambda_-,\lambda_+}} F(\nu). \tag{6.6}$$

Moreover, if ν_* is a solution of (6.5), then $\eta_* = (\Psi^\# \nu_*)_{symm}$ is a solution of (6.4).

Proof. Recall that λ is a measure on $[-\pi/2,\ \pi/2]$ defined by $d\lambda(\varphi) = \frac{1}{2}\cos\varphi\, d\varphi$. The map $\Psi^\#$ is a one-to-one correspondence between $\Gamma_{\lambda_-,\lambda_+}$ and $\Gamma_{\lambda,\lambda}$, and the equality $F(\nu) = R(\Psi^\# \nu)$ holds. Therefore, we have

$$\inf_{\eta \in \Gamma_{\lambda,\lambda}} R(\eta) = \inf_{\nu \in \Gamma_{\lambda_-,\lambda_+}} F(\nu), \tag{6.7}$$

and if ν_* minimizes $F(\nu)$, then $\Psi^\# \nu_*$ minimizes $R(\eta)$.

Since $\mathcal{M} \subset \Gamma_{\lambda,\lambda}$, we have

$$\inf_{\eta \in \mathcal{M}} R(\eta) \ge \inf_{\eta \in \Gamma_{\lambda,\lambda}} R(\eta). \tag{6.8}$$

Assume that $\eta \in \Gamma_{\lambda,\lambda}$; then $\eta_{\text{symm}} \in \mathcal{M}$, and by symmetry of the integrand in (6.4) under the transposition $(\varphi, \varphi^+) \mapsto (\varphi^+, \varphi)$ we obtain $R(\eta) = R(\eta_{\text{symm}})$. It follows that the measure $(\Psi^\# \nu_*)_{\text{symm}}$ minimizes $R(\eta)$ in $\mathcal{M}$ and

$$\inf_{\eta \in \mathcal{M}} R(\eta) \le \inf_{\eta \in \Gamma_{\lambda,\lambda}} R(\eta). \tag{6.9}$$

Relation (6.6) is a consequence of (6.7)–(6.9). The proof of Proposition 6.2 is complete. □

Proof. of Theorem 6.1. By (6.2) and (6.4) we have

$$R_u(B) = |\partial(\text{Conv}B)| \cdot \frac{4}{3} R(\eta_B). \tag{6.10}$$

(a) The proof reduces to three steps.

Step 1. Find $\inf |\partial(\text{Conv}B)|$ provided that $\text{Area}(B) = \text{A}$. This is an isoperimetric problem; its solution is a disc of the corresponding area: $\text{Ball}_r(O)$ with $r = \sqrt{A/\pi}$. Its boundary has the length

$$|\partial \text{Ball}_r(O)| = \inf |\partial(\text{Conv}B)| = 2\sqrt{\pi A}.$$

Step 2. Find the minimum $\inf_{\eta \in \mathcal{M}} R(\eta)$ in Problem (6.4). By Proposition 6.2, it reduces to Problem (6.5). The latter, in turn, is the two-dimensional case of the problem on optimal mass transportation on a sphere studied in Sect. 5.4. The optimal measure is determined in that section.

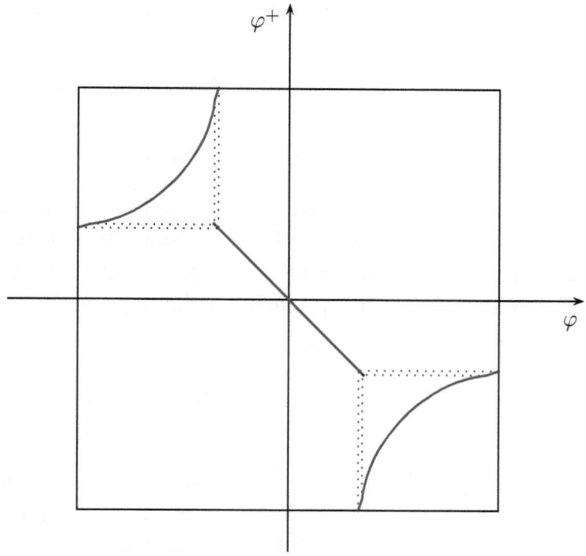

Fig. 6.1 The optimal measure η_*

We have $\inf_{\nu \in \Gamma_{\lambda_-,\lambda_+}} F(\nu) = m_2 \approx 0.98782374$. Using Proposition 6.2, we conclude that $\inf_{\eta \in \mathcal{M}} R(\eta) = m_2$ and find the optimal measure in Problem (6.4). Its support consists of the segment with endpoints $(-\varphi_0, \varphi_0)$ and $(\varphi_0, -\varphi_0)$, $\varphi_0 \approx 0.554$, on a diagonal of the square $\square$, together with two centrally symmetric curves (Fig. 6.1). The lower right curve consists of the segment with endpoints $(\pi/6, -\pi/2)$ and $(\varphi_0, -\pi + 3\varphi_0)$, the segment with endpoints $(\pi/2, -\pi/6)$ and $(\pi - 3\varphi_0, -\varphi_0)$ (these two segments are bounded between the pairs of dotted lines), and the curve $\sin\varphi - \sin\varphi^+ = \sin\varphi_0 + \sin 3\varphi_0$ with $\varphi_0 \le \varphi \le \pi - 3\varphi_0$ joining them. The density of the measure on these curves can be found from conditions (M1) and (M2), which means that $\eta_* \in \mathcal{M}$ (Definition 4.5, Sect. 4.1.1). Note that the support of η_* is not the graph of a function. From the standpoint of mass transfer theory, this means that the corresponding solution is not a Monge solution, whereas from the standpoint of billiards, it means that knowing the incidence angle of a particle is not always enough to predict the direction in which it goes away, and conversely, knowing the angle at which a reflected particle goes away is not always enough to determine the incidence angle.

Note that the part of the support contained in the lower right square in Fig. 6.1 is symmetric to the set (support of the optimal measure) plotted in Fig. 5.7 with respect to the horizontal axis. This symmetry (rather than identity) of the sets is related to the fact that the angle φ^+ is measured in different ways: clockwise relative to the vertical axis in the problem of Sect. 5.4 and counterclockwise in this section.

Step 3. It follows from two previous steps and formula (6.2) that

$$\inf_{\text{Area}(B)=A} R_u(B) \ge 2\sqrt{\pi A} \cdot \frac{4}{3} m_2. \tag{6.11}$$

Now we construct a sequence of bodies B_n such that

$$\lim_{n\to\infty} R_u(B_n) = 2\sqrt{\pi A}\cdot\frac{4}{3}\,m_2; \tag{6.12}$$

by this the equality in (6.11) will be proved.

Using Theorem 4.3, we find for each n a sequence of connected bodies $B_{n,m}$ such that $\mathrm{Ball}_{r-1/n}(0) \subset B_{n,m} \subset \mathrm{Ball}_r(0)$ and the sequence of measures $\eta_{B_{n,m}}$ associated with these bodies converges weakly to η_* as $m \to \infty$. From it we select a diagonal sequence $B_{n,m(n)}$ such that the sequence of corresponding measures $\eta_{B_{n,m(n)}}$ converges to η_*. We have

$$\lim_{n\to\infty} |\partial(\mathrm{Conv}\, B_{n,m(n)})| = 2\sqrt{\pi A} \quad \text{and} \quad \lim_{n\to\infty} \mathrm{Area}(B_{n,m(n)}) = A.$$

Let $\kappa = \sqrt{A/\mathrm{Area}(B_{n,m(n)})}$ and $B_n = \kappa B_{n,m(n)}$. Obviously, the area of each B_n is A, $\lim_{n\to\infty} |\partial(\mathrm{Conv}\, B_n)| = 2\sqrt{\pi A}$ and $\eta_{B_n} = \eta_{B_{n,m(n)}}$; therefore, $\lim_{n\to\infty} R(\eta_{B_n}) = m_2$. Taking into account (6.10), we now have (6.12).

Thus statement (a) of the theorem is proved.

(b) This statement is much easier than (a). The measure η_B associated with a convex body B coincides with the "elastic reflection measure" η_0. Recall that the latter has density $\frac{1}{2}\cos\varphi\,\delta(\varphi+\varphi^+)$. We obtain by direct calculations that $R(\eta_0) = 1$. In view of (6.10) we have $R_u(B) = \frac{4}{3}|\partial B|$, and the problem reduces to the isoperimetric problem of minimizing the perimeter of a convex body with prescribed area. Its solution is a disc $\mathrm{Ball}_r(O)$ of the same radius as in case (a), with perimeter $|\partial(\mathrm{Ball}_r(O))| = 2\sqrt{\pi A}$. Hence $R_u(\mathrm{Ball}_r(O)) = \frac{8}{3}\sqrt{\pi A}$. Statement (b) is thus proved. □

Remark 6.1. Note that we apparently have a great deal of freedom in choosing the hollow for approximating a measure in $\mathcal{M}$. Hollows constructed in accordance with the general scheme of Theorem 4.1 have an intricate shape and can hardly be used in practical problems. On the other hand, we hope that the construction of hollows can be made simpler.

Remark 6.2. It is of interest that the ratio

$$\frac{\textit{(Least resistance in the class of } \textbf{\textit{connected}} \textit{ bodies)}}{\textit{(Least resistance in the class of } \textbf{\textit{convex}} \textit{ bodies)}} = m_2 \approx 0.9878$$

is close to 1, so that the gain we can achieve by extending the class of admissible bodies is just 1.22 % This gain is achieved at the cost of extreme complication of the boundary of optimal body (a "rough disc" with an intricate boundary in place of a disc).

We now consider another problem. Let C_1 and C_2 be bounded convex bodies such that $C_1 \subset C_2$ and $\partial C_1 \cap \partial C_2 = \emptyset$, and consider the class of connected bodies B such that $C_1 \subset B \subset C_2$.

Problem 4. Find

$$r_{\min} = \inf_{C_1 \subset B \subset C_2} R_u(B) \quad \textit{and} \quad r_{\max} = \sup_{C_1 \subset B \subset C_2} R_u(B).$$

The solution is close to the solution of Problem 3(a). We only present the answer. A minimizing sequence of bodies $B_n^{\min}$ is such that the corresponding sequence of measures $\eta_{B_n^{\min}}$ converges weakly to η_* and

$$\cap_m \cup_{n \geq m} B_n^{\min} = C_1.$$

Recall that $R(\eta_0) = 1$ and $R(\eta_*) = m_2$. A maximizing sequence $B_n^{\max}$ is such that $\eta_{B_n^{\max}}$ converges weakly to the retroreflector measure η_{retr} with density $\frac{1}{2} \cos\varphi\, \delta(\varphi - \varphi^+)$ and

$$\cup_m \cap_{n \geq m} B_n^{\max} = C_2.$$

By direct calculation we obtain $R(\eta_{\text{retr}}) = 3/2$. Thus, we have

$$r_{\min} = 0.9878\ldots \cdot R_u(C_1) \quad \text{and} \quad r_{\max} = 1.5 \cdot R_u(C_2).$$

6.1.2 Media with Nonzero Temperature

Here we show that the resistance in media with nonzero temperature is proportional to the resistance in a medium of resting particles, with a proportionality coefficient greater than 1. This coefficient does not depend on the shape of the body and is determined only by the nature of the thermal motion of particles in the medium. Thus, Problems 3 and 4 in the previous item have the same solutions here.

We suppose that the velocity $v + u$ of each particle approaching the body is the sum of the average velocity v of the flow and the velocity u of the thermal motion. The quantities v and u are independent; $v \in S^1$ is uniformly distributed on the unit circle, and the distribution of u is centrally symmetric with density

$$\sigma(u) = \frac{1}{2\pi} f(|u|), \quad \text{where} \quad \int_0^\infty f(r)\, dr = 1.$$

The medium has a finite temperature:

$$\overline{u^2} = \overline{u_f^2} := \int_0^\infty r^2 f(r)\, dr < \infty.$$

An individual particle hits the body with velocity $v + u =: \tilde{u} = r\tilde{v}$ (where $\tilde{v} = \tilde{u}/|\tilde{u}|$ and $r = |\tilde{u}|$) and flies away after several reflections, with velocity $\tilde{u}^+ := rv_B^+(\xi, \tilde{v})$. Here $\xi \in \partial(\operatorname{Conv} B)$ is the point where the particle intersects $\partial(\operatorname{Conv} B)$ for the first time. Vector v can be represented as the sum $v = \kappa\tilde{v} + \tilde{w}$, where $\tilde{w}$ has expectation zero, $\mathbb{E}\tilde{w} = 0$, and the constant $\kappa = \kappa_f$ can be found from the relation

$$\kappa_f = \int_0^\infty \left(\frac{1}{\pi} \int_0^\pi \frac{1 + r_0 \cos\varphi}{\sqrt{1 + 2r_0\cos\varphi + r_0^2}}\, d\varphi \right) f(r)\, dr.$$

In this and the next two paragraphs we multiply and divide two-dimensional vectors as complex numbers. It is easy to see that $\tilde{v}$ and the "twisted" vector $w = \tilde{v}^{-1}\tilde{w}$ are independent and $\mathbb{E}w = 0$. Hence for each vector-valued function g,

$$\mathbb{E}[g(\tilde{v}) \cdot \tilde{v}w] = \mathbb{E}[\tilde{v}^{-1}g(\tilde{v}) \cdot w] = 0. \tag{6.13}$$

Let $\tilde{\sigma}$ be the density of the distribution of the variable $\tilde{u} = v + u$. It is centrally symmetric, so for each $\check{v} \in S^1$

$$2\pi \int_0^\infty r^2\, \tilde{\sigma}(r\check{v})\, r dr = \mathbb{E}|v + u|^2 = 1 + \overline{u_f^2}. \tag{6.14}$$

Consider an infinitesimal arc $d\xi$ on the boundary ∂C and an infinitesimal domain $d\tilde{u}$ in the plane. The number of particles hitting this arc during a small period of time Δt and having velocities in $d\tilde{u}$ is

$$(n(\xi) \cdot \tilde{u})_-\, d\xi\, \tilde{\sigma}(\tilde{u})\, d\tilde{u}\Delta t. \tag{6.15}$$

The momentum that each of these particles transmits to the body is (after suitable normalization) $\tilde{u} - \tilde{u}^+$, and its projection on the direction of v is

$$(\tilde{u} - \tilde{u}^+) \cdot v = r\,(\tilde{v} - v_B^+(\xi, \tilde{v})) \cdot v. \tag{6.16}$$

The force in the direction of the flow produced by these particles is the product of (6.15) and (6.16) divided by Δt, that is,

$$r((\tilde{v} - v_B^+(\xi, \tilde{v})) \cdot v)\, (n(\xi) \cdot \tilde{u})_-\, d\xi\, \tilde{\sigma}(\tilde{u})\, d\tilde{u}.$$

The mean resistance $R_{u,f}(B)$ is the sum of all these forces, that is, the integral of the last expression. Bearing in mind that $v = \kappa\tilde{v} + \tilde{v}w$, $\tilde{u} = r\tilde{v}$, $d\tilde{u} = rdrd\tilde{v}$, denoting the distribution function of w by W, and setting $C = \operatorname{Conv} B$, we obtain

$$R_{u,f}(B) = \int_{\partial C \times \mathbb{R}_+ \times S^1 \times \mathbb{R}^2} r^2(\tilde{v} - v_B^+(\xi, \tilde{v})) \cdot (\kappa\tilde{v} + \tilde{v}w)\, (n(\xi) \cdot \tilde{v})_-\, d\xi\, \pi\tilde{\sigma}(r\tilde{v})\, rdr\, d\tilde{v}\, dW(w). \tag{6.17}$$

By (6.13), the term connected with w in this integral vanishes, so in view of relation (6.14) the substitution

$$d\mu(\xi, \tilde{v}) = \frac{1}{2} |n(\xi) \cdot \tilde{v}| \, d\xi \, d\tilde{v}$$

reduces formula (6.17) to the simpler form

$$R_{u,f}(B) = \kappa_f (1 + \overline{u_f^2}) \int_{(\partial C \times S^1)_-} (\tilde{v} - v_B^+(\xi, \tilde{v})) \cdot \tilde{v} \, d\mu(\xi, \tilde{v}).$$

Comparing this with (6.1) and setting $c_f = \kappa_f(1 + \overline{u_f^2})$, we obtain $R_{u,f}(B) = c_f \, R_u(B)$ as a result. Thus, in a medium with positive temperature the resistance is proportional to that in a medium of resting particles. The proportionality coefficient c_f depends only on the distribution of velocities in the medium.

We assert that $c_f > 1$, that is, the resistance is higher when there is thermal motion of particles. First we note that in the special case $f(r) = \delta(r - r_0)$ we have $\kappa_f = \kappa(r_0)$, where

$$\kappa(r) = \frac{1}{\pi} \int_0^\pi \frac{1 + r\cos\varphi}{\sqrt{1 + 2r\cos\varphi + r^2}} \, d\varphi.$$

The coefficient κ_f of general form admits the following representation in terms of $\kappa(r)$, $r > 0$:

$$\kappa_f = \int_0^\infty \kappa(r) \, f(r) \, dr. \tag{6.18}$$

Note also the following relation:

$$1 + \overline{u_f^2} = \int_0^\infty (1 + r^2) \, f(r) \, dr. \tag{6.19}$$

We have the asymptotic relations $\kappa(r) = 1 - r^2/4 + o(r^2)$ as $r \to 0^+$ and $\kappa(r) = 1/(2r) + o(1/r)$ as $r \to +\infty$. It can be verified numerically that $\kappa(r)\,(1 + r^2) > 1$ for each $r > 0$. Hence from relations (6.18) and (6.19) we see that

$$\begin{aligned} \kappa_f(1 + \overline{u_f^2}) &= \int_0^\infty \int_0^\infty \frac{1}{2} \left[\kappa(r)(1 + s^2) + \kappa(s)(1 + r^2)\right] f(r) f(s) \, dr ds \\ &> \int_0^\infty \int_0^\infty f(r) f(s) \, dr ds = 1. \end{aligned}$$

We now consider briefly the asymptotic behavior of c_f in the case where the distribution function depends on a parameter: $f_\varepsilon(r) = \varepsilon^{-1}\phi(\varepsilon^{-1} r)$, where $\phi(0) = 0$ and $\phi'(0) > 0$. Here ε^2 plays the role of temperature. In the zero-temperature limit

$$c_f = 1 + \text{const}_1 \cdot \varepsilon^2 + o(\varepsilon^2) \quad \text{as} \quad \varepsilon \to 0^+,$$

whereas in the high-temperature limit

$$c_f = \text{const}_2 \cdot \varepsilon + o(\varepsilon) \quad \text{as} \quad \varepsilon \to \infty.$$

Here

$$\text{const}_1 = \frac{3}{4}\int_0^\infty r^2\phi(r)\,dr, \qquad \text{const}_2 = \frac{1}{2}\int_0^\infty r\phi(r)\,dr \cdot \int_0^\infty r^2\phi(r)\,dr.$$

6.2 Case of Higher Dimension

Here we consider the problem of optimizing the mean resistance of rough bodies in a medium of resting particles. The dimension d will be arbitrary.

The mean resistance of the body B, as in the two-dimensional case, can be expressed by the integral

$$R_u(B) = a_d \iint_{(\partial C \times S^{d-1})_-} (v - v_B^+(\xi, v)) \cdot v \, d\mu(\xi, v),$$

where C is an ambient convex body, $B \subset C$, and the normalizing coefficient a_d is specified below. Taking into account that $(v - v^+) \cdot v = \frac{1}{2}|v - v^+|^2$, after the substitution

$$(\xi, v) \mapsto (v, v_B^+(\xi, v), n(\xi)),$$

we obtain

$$R_u(B) = a_d \iiint_{(S^{d-1})^3} \frac{1}{2}|v - v^+|^2 \, d\nu_{B,C}(v, v^+, n).$$

Now let $\mathcal{B}$ be a rough body obtained by grooving C. By definition, its resistance $R_u(\mathcal{B})$ is the limit of the resistances of a sequence of bodies B_n representing it. Since the measures $\nu_{B_n,C}$ converge weakly to $\nu_{\mathcal{B}}$, the mean resistance of $\mathcal{B}$ is

$$R_u(\mathcal{B}) = a_d \iiint_{(S^{d-1})^3} \frac{1}{2}|v - v^+|^2 \, d\nu_{\mathcal{B}}(v, v^+, n). \tag{6.20}$$

The constant a_d is determined from the condition $R_u(C) = |\partial C|$, that is, the resistance of the original convex body C is $|\partial C|$. Substituting ν_C^{e} for $\nu_{\mathcal{B}}$ into (6.20) and using (4.38) we obtain

$$1 = \frac{1}{|\partial C|} R_u(C) = a_d \frac{1}{|\partial C|} \iiint_{(S^{d-1})^3} \frac{1}{2}|v - v^+|^2 \, d\nu_C^{\mathrm{e}}(v, v^+, n) =$$

$$= a_d \frac{1}{|\partial C|} \int_{S^{d-1}} d\tau_C(n) \int_{S^{d-1}} 2(v \cdot n)_-^3 \, b_d \, dv = a_d \, b_d \int_{S^{d-1}} 2(v \cdot n)_-^3 dv.$$

We have

$$\int_{S^{d-1}} 2(v\cdot n)_-^3\,dv = |S^{d-2}|\int_0^1 2(1-r^2)r^{d-2}\,dr = 2\pi^{(d-1)/2}/\Gamma\left(\frac{d+3}{2}\right)$$

and

$$b_d = \Gamma\left(\frac{d+1}{2}\right)\pi^{(1-d)/2};$$

therefore,

$$a_d = (d+1)/4; \qquad (6.21)$$

in particular, $a_2 = \frac{3}{4}$ and $a_3 = 1$.

Recall the problem on optimization of resistance in the class of rough bodies obtained by grooving a convex body C stated in Sect. 1.7.

Problem 5. Find

(a) $\frac{1}{|\partial C|}\sup\{R_u(\mathcal{B}) : \mathcal{B} \text{ is a grooving of } C\}$;

(b) $\frac{1}{|\partial C|}\inf\{R_u(\mathcal{B}) : \mathcal{B} \text{ is a grooving of } C\}$.

The solution of this problem is provided by the following theorem.

Theorem 6.2. *We have*

(a) $\frac{1}{|\partial C|}\sup_{\mathcal{B}} R_u(\mathcal{B}) = \frac{d+1}{2}$,
(b) $\frac{1}{|\partial C|}\inf_{\mathcal{B}} R_u(\mathcal{B}) = m_d$,

where sup *and* inf *are taken over all rough bodies $\mathcal{B}$ obtained by grooving a convex body C and the coefficients m_d are defined in Sect. 5.4.*

Proof. Using Theorem 4.5 and formulas (6.20) and (6.21) we can represent Problem 5 in the following form: find

$$\begin{Bmatrix}\sup\\ \inf\end{Bmatrix}_{\nu\in\Gamma_C} \mathcal{C}(\nu), \quad \text{where } \mathcal{C}(\nu) = \frac{d+1}{4}\iiint_{(S^{d-1})^3}\frac{1}{2}|v-v^+|^2\,d\nu(v,v^+,n). \qquad (6.22)$$

In turn, a simple trick reduces this problem to a problem of mass transfer on the sphere. Namely, set $e = (1,0,\ldots,0)$ and consider a piecewise smooth family of rotations $W_n : \mathbb{R}^3 \to \mathbb{R}^3$ such that $W_n(n) = e$. For instance, if $n \neq \pm e$, then we can define W_n as the rotation about $n \times e$ taking n to e, while the rotation $W_{\pm e}$ is defined somehow.

Define the map $G : (S^{d-1})^3 \to (S^{d-1})^2$ by $G(v,v^+,n) = (W_n(v), W_n(v^+), e)$ and note that the inclusion $\nu \in \Gamma_C$ yields $G^\#\nu \in \mathcal{M}_e$. Since for $V = W_n(v)$ and $V^+ = W_n(v^+)$ we have $|V-V^+| = |v-v^+|$, the integral in (6.22) can be written as $\mathcal{C}(\nu) = R(G^\#\nu)$, where

$$R(\nu) = \frac{d+1}{4}\iint_{(S^{d-1})^2}\frac{1}{2}|V-V^+|^2\,d\nu(V,V^+). \qquad (6.23)$$

Since the map $G^{\#} : \Gamma_C \to \mathcal{M}_e$ is surjective, Problem (6.22) can be represented in an equivalent form:

$$\begin{Bmatrix} \sup \\ \inf \end{Bmatrix}_{\nu \in \mathcal{M}_e} R(\nu). \tag{6.24}$$

Finally, note that the relation $\nu \mapsto \nu_{\text{symm}} = \frac{1}{2}\,(\nu + \pi_{\text{ad}}^{\#}\nu)$ defines a surjective map from $\Gamma_{\lambda_{-e},\lambda_e}$ to $\mathcal{M}_e$, and taking into account that $(v, v^+, n) \mapsto \frac{1}{2}\,|v - v^+|^2$ is invariant under the map π_{ad}, we have $R(\nu) = R(\nu_{\text{symm}})$. It follows that Problem (6.24) is equivalent to the problem in a wider space of measures:

$$\begin{Bmatrix} \sup \\ \inf \end{Bmatrix}_{\nu \in \Gamma_{\lambda_{-e},\lambda_e}} R(\nu). \tag{6.25}$$

(a) The optimal transport plan in the maximization problem in (6.25) takes v to $v^+ = -v$. The cost function at this pair of points equals $\frac{1}{2}\,|v - v^+|^2 = 2$. This transport plan is realized by means of the retroreflector measure ν_{retr}^e defined by $d\nu_{\text{retr}}^e(v, v^+) = \delta(v + v^+)\, d\lambda_{-e}(v)$. The maximum value of R is

$$\sup_{\nu \in \Gamma_{\lambda_{-e},\lambda_e}} R(\nu) = R(\nu_{\text{retr}}^e) = \frac{d+1}{4} \cdot 2 = (d+1)/2.$$

This completes the proof of (a).

(b) The minimization problem in (6.25) is precisely Problem (5.60) on mass transfer on a sphere considered in Sect. 5.4. As proved there, the minimum value in this problem is m_d. This completes the proof of (b). □

In conclusion of this chapter, consider several examples of two-dimensional rough bodies, where the boundary of each set B_m representing the rough body $\mathcal{B}$ is obtained by repetition of a single hollow (Ω_m, I_m). Clearly, the size of the hollow vanishes as $m \to \infty$. We assume that η_{Ω_m, I_m} weakly converges, $\lim_{m\to\infty} \eta_{\Omega_m, I_m} = \eta$. Then the resistance of $\mathcal{B}$ takes an especially simple form:

$$\frac{1}{|\partial C|} R_u(\mathcal{B}) = R(\eta).$$

In the examples below, we utilize the results of Sect. 4.1.2.

(a) $\mathcal{B}$ coincides with C (Fig. 6.2a). Then

$$\frac{1}{|\partial C|} R_u(C) = 1.$$

(b) The hollow is a right isosceles triangle (Fig. 6.2b). Then we have

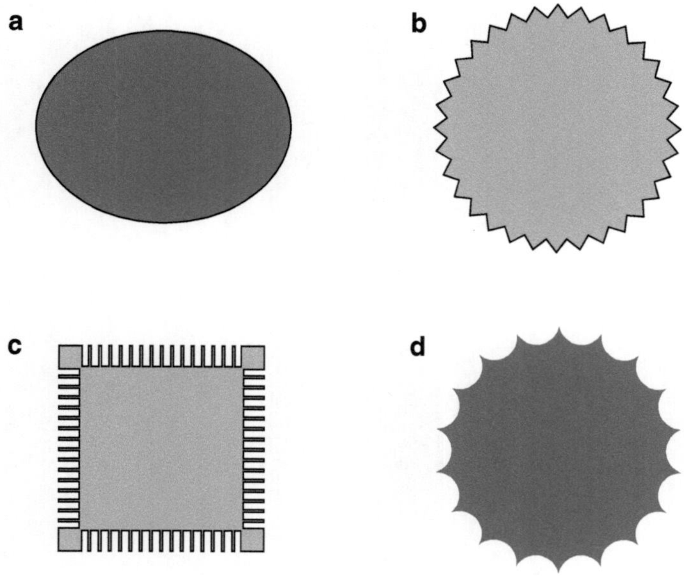

Fig. 6.2 (**a**) A convex body (no hollows). Rough bodies with special shapes of hollows: (**b**) right isosceles triangles; (**c**) rectangles; (**d**) semicircles

$$\frac{1}{|\partial C|} R_u(\mathcal{B}) = \sqrt{2} \approx 1.414.$$

(c) The hollow is a rectangle $\Pi_m = a_m \times b_m$, with $\lim_{m\to\infty} a_m/b_m = 0$ (Fig. 6.2c) or an isosceles triangle with the angle at the apex going to 0. In both cases we have

$$\frac{1}{|\partial C|} R_u(\mathcal{B}) = 1.25.$$

(d) The hollow is a semicircle (Fig. 6.2d). We have

$$\frac{1}{|\partial C|} R_u(\mathcal{B}) = \frac{3\pi}{8} \approx 1.178.$$

(e) The hollow is helmet-shaped (Fig. 9.14c). Then

$$\frac{1}{|\partial C|} R_u(\mathcal{B}) \approx 1.4965.$$

Chapter 7
The Magnus Effect and the Dynamics of a Rough Disc

In this chapter we are concerned with the Magnus effect: the phenomenon governing the deflection of the trajectory of a spinning body (for example, a golf ball or a soccer ball). Surprisingly enough, in highly rarefied media (on Mars or in the thin atmosphere at a height corresponding to low Earth orbits: between 100 and 1,000 km) the *inverse* effect takes place; this means that the trajectory deflection has opposite signs in sparse and in dense media.

There is a vast literature devoted to the Magnus effect, motivated by sports and technology applications (see, e.g., [38, 62, 64]). The inverse effect is also well known to physicists; study of this phenomenon has become increasingly important nowadays because of potential applications to the aerodynamics of artificial satellites [6,7,27,74,75]. Theoretical studies on the inverse Magnus effect are based on models of *nonelastic* reflections of medium particles from convex bodies.

At present all theoretical works (see, however, [58, 59]) ignore the connection between the shape of roughness (which is always present on the body surface) and the Magnus effect. The kind of roughness (that is, the shape of, e.g., microscopic dimples, hollows, gullies) depends on the body material; the surface may also be artificially roughened. Due to the roughness, particles bounce off the body surface in directions other than that prescribed by the visible orientation of the surface and may also have multiple reflections.

We believe that the roughness of a body surface should be taken into account when modeling the Magnus effect. We propose a new approach based on examining the shape of the hollows and their contribution to the effect. We restrict ourselves to the case of a two-dimensional rough disc and (as everywhere in this book) assume that all reflections of the particles with the body are elastic.

This approach meets evident difficulties: there is a huge variety of shapes governing roughness. The way to get rid of these difficulties is based on the characterization of scattering laws given in Chap. 4. Using this characterization, we can determine the set of scattering laws for *all possible* shapes of roughness and then calculate the range of forces and of the moments of force acting on the disc

A. Plakhov, *Exterior Billiards: Systems with Impacts Outside Bounded Domains*, DOI 10.1007/978-1-4614-4481-7_7,

for a fixed angular velocity. In addition, in this chapter we calculate the forces and trajectories in several special cases of roughness.

Another novelty of our approach consists in applying optimal mass transportation (OMT) to the study of the Magnus effect. A sort of *vector-valued* OMT problem naturally appears and is examined here. To the best of our knowledge, this kind of generalization of OMT has never been considered before.

The results of this chapter were first published in [58, 59].

7.1 Description of the Effect and Statement of the Problem

7.1.1 Statement of the Problem for a Rough Disc

We now proceed to a detailed description of the problem. A spinning two-dimensional body moves through a homogeneous medium in a plane. The medium is extremely rarefied, so that the free path length of particles is much larger than the body's size. In this case the interaction of the body with the medium can be described in terms of *free molecular flow*, where point particles fall on the body's surface and each particle interacts with the body but not with the other particles. There is no gravitation force. The particles of the medium originally stay at rest; that is, the absolute temperature of the medium equals zero. In a frame of reference moving forward together with the body, we have a parallel flow of particles falling on the resting body.

Neglecting the angular momentum of particles, each particle is identified with a mass point that approaches the body, makes several (maybe none) elastic collisions with its surface, and goes away.

The body under consideration is a *rough disc* $\mathcal{B}$ obtained by grooving a disc $C = C_r$ of radius r. The roughness is supposed to be uniform along the boundary, that is, the scattering law is the same at all points of the body boundary and is defined by a certain measure $\eta \in \mathcal{M}$. (Recall that $\eta = \eta_{\mathcal{B}}$ is defined on the square $\square = [-\pi/2, \pi/2] \times [-\pi/2, \pi/2]$ with coordinates φ and φ^+.) More precisely, we assume that the scattering law on $\mathcal{B}$ is

$$\nu_{\mathcal{B}} = \sigma^{\#}(\eta \otimes 2\pi r\, u) \in \Gamma_C. \tag{7.1}$$

Here u is the uniform probability measure on S^1; therefore, $2\pi r\, u$ is the surface measure of a disc of radius r. The map σ is defined by (4.14) (Sect. 4.1.1). The body $\mathcal{B}$ is called a *rough disc*.

Remark 7.1. Note in passing that in the case of nonuniform roughness, that is, if the shape of dimples varies along the boundary, periodic oscillations of the disc along the trajectory may occur, the period being equal to the period of one turn of the disc. The "averaged" trajectory, however, coincides with the trajectory of the uniformly rough disc, where the roughness is obtained by "averaging" the

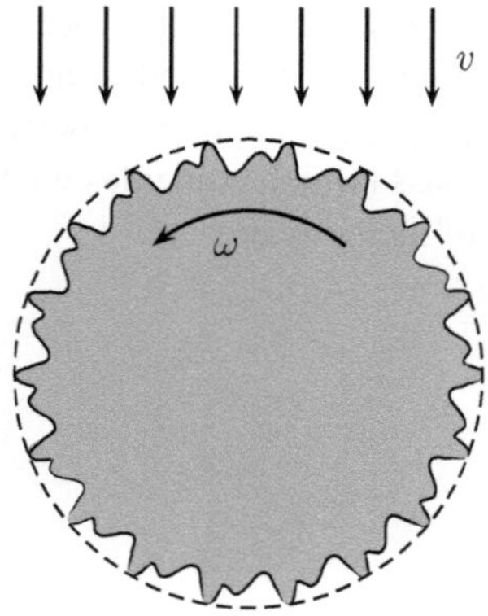

Fig. 7.1 A rotating rough disc in a parallel flow of particles

original one. The mathematical procedure of averaging the roughness (in the general case of two-dimensional convex body C) is as follows. Let $\nu_{\mathcal{B}}$ be the scattering law on $\mathcal{B}$; then $\eta = \varpi^{\#}\nu_{\mathcal{B}}$ defines the "reduced" scattering law. Recall that the map $\varpi : (v, v^+, n) \mapsto (\varphi, \varphi^+)$ is defined in Sect. 4.1.5 by (4.29.) The measure η can be treated as a scattering law at a point obtained by averaging over the boundary of $\mathcal{B}$. Then the averaged roughness is $\sigma^{\#}(\eta \otimes \tau_C)$.

Consider a sequence of bodies B_m representing $\mathcal{B}$. There is a certain mass distribution in each B_m. We assume that the mass in B_m is distributed in such a way that the total mass M is constant and the center of mass coincides with the geometric center O of the disc. In addition, the moment of inertia I_m of B_m relative to O converges to a positive quantity I as $m \to \infty$. We have $I \le Mr^2$. The quantity $\beta = Mr^2/I$ is the inverse specific moment of inertia; we have $1 \le \beta < +\infty$. In the sequel, we pay special attention to the following two particular cases: (a) $\beta = 1$, that is, the mass of the disc is concentrated on its boundary; and (b) $\beta = 2$, that is, the mass is distributed uniformly in the disc.

We have a parallel flow of particles with density ρ and velocity $-\vec{v}$ falling on a body B_m rotating about a fixed center O with angular velocity ω (Fig. 7.1). Let $\varphi = \varphi(t)$ be the rotation angle at moment t (measured counterclockwise), so that $d\varphi/dt = \omega$. Let $\vec{R}_m(B_m, \varphi, \omega, \vec{v})$ be the resistance force acting on the body, and let $R_{I,m}(B_m, \varphi, \omega, \vec{v})$ be the moment of this force. Below we see that the quantities $\vec{R}_m(B_m, \varphi, \omega, \vec{v})$ and $R_{I,m}(B_m, \varphi, \omega, \vec{v})$ converge $m \to \infty$, and the limits

$$\lim_{m\to\infty} \vec{R}_m(B_m, \varphi, \omega, \vec{v}) = \vec{R}(\mathcal{B}, \omega, \vec{v}) \quad \text{and} \quad \lim_{m\to\infty} R_{I,m}(B_m, \varphi, \omega, \vec{v}) = R_I(\mathcal{B}, \omega, \vec{v})$$

do not depend on the sequence B_m representing the body, and the convergence is uniform on compacts.

Definition 7.1. The limiting functions $\vec{R}(\mathcal{B}, \omega, \vec{v})$ and $R_I(\mathcal{B}, \omega, \vec{v})$ are called the force of the flow pressure and the moment of the force acting on the rough disc rotating with angular velocity ω.

The body B_m is placed in a medium with density ρ. At the initial moment of time $t = 0$ the body starts moving. The rotation angle of the body at moment t is $\varphi_m(t)$, the angular velocity is $\omega_m(t) = d\varphi_m/dt$, and the velocity of the center of mass (which coincides with its geometric center) is $\vec{v}_m(t)$. The equations of motion of the body are

$$M\,\frac{d\vec{v}}{dt} = \vec{R}_m(B_m, \varphi, \omega, \vec{v}),$$

$$I_m\,\frac{d\omega}{dt} = R_{I,m}(B_m, \omega, \vec{v}),$$

$$\frac{d\varphi}{dt} = \omega.$$

We assume that there exist the limits $\lim_{m\to\infty}\varphi_m(0) =: \varphi(0)$, $\lim_{m\to\infty}\omega_m(0) =: \omega(0)$, $\lim_{m\to\infty}\vec{v}_m(0) =: \vec{v}(0)$. Then for all $t > 0$ we have the convergence $\lim_{m\to\infty}\varphi_m(t) =: \varphi(t)$, $\lim_{m\to\infty}\omega_m(t) =: \omega(t)$, $\lim_{m\to\infty}\vec{v}_m(t) =: \vec{v}(t)$. The functions $\omega(t)$ and $\vec{v}(t)$ are defined by the equations

$$M\,\frac{d\vec{v}}{dt} = \vec{R}(B, \omega, \vec{v}), \tag{7.2}$$

$$I\,\frac{d\omega}{dt} = R_I(B, \omega, \vec{v}), \tag{7.3}$$

and $\varphi(t)$ is obtained by integrating $\omega(t)$.

Equations (7.2) and (7.3) are naturally interpreted as equations of dynamics of a rough disc, and the functions $\varphi(t)$, $\omega(t)$, and $\vec{v}(t)$ as, respectively, the rotation angle, angular velocity, and velocity of the center of mass at time t.

We consider the following problems:

(A) Determine the force acting on a spinning disc, find the moment of this force, and investigate their dependence on the roughness.
(B) Analyze the motion of a rough disc in a medium, that is, study the behavior of the functions $\omega(t)$ and $\vec{v}(t)$.

Problem (A) is primary with respect to (B). In this chapter we will devote our attention primarily to Problem (A), having just touched upon Problem (B), where we will restrict ourselves to deducing equations of motion and solving these equations in several simple cases.

7.1.2 Summary of the Remainder of the Chapter

We shall see subsequently that, generally speaking, the force $\vec{R}(\mathcal{B}, \omega, \vec{v})$ is not collinear to the velocity $\vec{v}$.

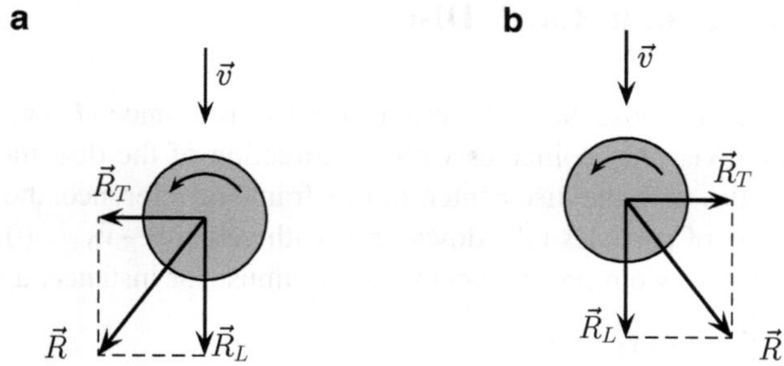

Fig. 7.2 **(a)** Proper Magnus effect **(b)** Inverse Magnus effect

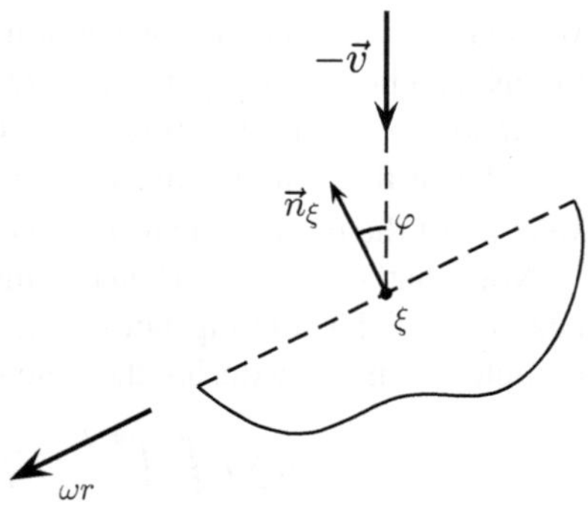

Fig. 7.3 A particle falling in a hollow

If a transversal component of the resistance force appears, resulting in deflection of a body's trajectory, then we encounter the (proper or inverse) Magnus effect. If the direction of the transversal component coincides with the instantaneous velocity of the front point of the body, then a *proper* Magnus effect takes place. If these directions are opposite, then an *inverse* Magnus effect occurs (see Fig. 7.2a, b).

The limiting case of slow rotation was studied in the previous chapter. In that case, the mean resistance force is parallel to the direction of the body's motion, and therefore the Magnus effect does not appear.

In Sect. 7.2 we will define the set of all possible values of $\vec{R}$, when ω and $\vec{v}$ are fixed and $\nu_{\mathcal{B}}$ takes all admissible values in Γ_C. In other words, we will answer the following question: what is the range of values of the force acting on a rough disc? This problem is formulated in terms of a special vector-valued Monge–Kantorovich problem and is solved numerically in Sect. 7.3 for several fixed values of the parameter $\gamma = \omega r/v$. Further, we calculate the functions $\vec{R}(\gamma)$ and $R_I(\gamma)$ for some special values of $\nu \in \Gamma_C$ (and thus for some special kinds of rough bodies). In Sect. 7.4, we deduce the equations of dynamics in a convenient form and solve them in several simple particular cases. Finally, in Sect. 7.5, a comparison of our results with previous works (by physicists) on the inverse Magnus effect in rarefied media is given.

7.2 Resistance of a Rough Disc

Denote $v = |\vec{v}|$ and choose the (noninertial) frame of reference Ox_1x_2 such that the direction of the axis Ox_2 coincides with the direction of the disc motion and the origin O coincides with the disc center. In this frame of reference, the disc stays at rest, and the flow of particles falls down on it with velocity $-\vec{v}_0 = (0;\ -v)^T$. Here and in what follows, we represent vectors as columns; for instance, a vector $\vec{x}$ will be denoted by $\begin{bmatrix} x_1 \\ x_2 \end{bmatrix}$ or $(x_1;\ x_2)^T$.

Let us calculate the force $\vec{R}$ of the medium resistance and the moment of this force R_I with respect to O. To that end, first we consider the prelimit body B_m. Without loss of generality we assume that all the hollows of the body are identical. Let us fix a hollow and parameterize its inlet by the variable ξ varying from 0 to 1.

Denote by φ the rotation angle of the hollow [that is, the external normal at the inlet of the hollow equals $\vec{n}_\xi = (-\sin\varphi;\ \cos\varphi)^T$], and by $\vec{v}^{+}_{(m)}(\xi,\varphi)$ the final velocity of the particle entering the hollow at point ξ with velocity $-\vec{v}_0$ (Fig. 7.3).

Note that $\Delta t = 2\pi/(\omega m)$ is the minimal time period between two identical positions of the rotating body B_m. Then the momentum imparted to B_m by the particles of the flow during the time interval Δt equals

$$2r\rho v\Delta t \int_0^1 \int_{-\pi/2}^{\pi/2} \left(-\vec{v}_0 - \vec{v}^{+}_{(m)}(\xi,\varphi)\right) \frac{1}{2} \cos\varphi \, d\varphi \, d\xi. \tag{7.4}$$

Consider the frame of reference $\tilde{O}\tilde{x}_1\tilde{x}_2$ having its center at the midpoint of the inlet I of the hollow, the axis $\tilde{O}\tilde{x}_1$ being parallel to I, and $\tilde{O}\tilde{x}_2$, codirectional with $\vec{n}_\xi$. That is, the frame of reference rotates jointly with the segment I. The change of variables from $\vec{x} = (x_1;\ x_2)^T$ to $\vec{\tilde{x}} = (\tilde{x}_1;\ \tilde{x}_2)^T$ and the inverse one are given by

$$\vec{\tilde{x}} = A_{-\omega t}\vec{x} - r\cos(\pi/m)\, e_{\pi/2} \quad \text{and} \quad \vec{x} = A_{\omega t}\vec{\tilde{x}} + r\cos(\pi/m)\, e_{\pi/2+\omega t},$$

where $A_\phi = \begin{pmatrix} \cos\phi & -\sin\phi \\ \sin\phi & \cos\phi \end{pmatrix}$ and $\vec{e}_\phi = \begin{bmatrix} \cos\phi \\ \sin\phi \end{bmatrix}$.

Suppose now that $\vec{x}(t)$ and $\vec{\tilde{x}}(t)$ are the coordinates of a moving point in the initial and rotating frames of reference, respectively, and let $\vec{v} = (v_1;\ v_2)^T = d\vec{x}/dt$ and $\vec{\tilde{v}} = (\tilde{v}_1;\ \tilde{v}_2)^T = d\vec{\tilde{x}}/dt$. Then

$$\vec{\tilde{v}} = A_{-\omega t}\vec{v} - \omega A_{\pi/2-\omega t}\vec{x} \quad \text{and} \quad \vec{v} = A_{\omega t}\vec{\tilde{v}} + \omega A_{\pi/2+\omega t}\vec{\tilde{x}} - \omega r\cos(\pi/m)\,\vec{e}_{\omega t}. \tag{7.5}$$

We apply formulas (7.5) to the velocity of the particle at the two moments of its intersection with I. At the first moment, it holds that $\omega t = \varphi$ and $\vec{x} = re_{\pi/2+\varphi} + o(1)$ as $m \to \infty$. [Here and in what follows, the estimates $o(1)$ are not necessarily uniform with respect to ξ and φ.] Then the incidence velocity $-\vec{v}_0$ takes the form

$$-\vec{\tilde{v}}_0 = v\,(\gamma - \sin\varphi;\ -\cos\varphi)^T + o(1) = -v\varrho\,(-\sin x;\ \cos x)^T + o(1), \tag{7.6}$$

where $\gamma = \omega r/v$ and

$$\varrho = \varrho(\varphi, \gamma) = \sqrt{\gamma^2 - 2\gamma \sin\varphi + 1}, \quad x = x(\varphi, \gamma) = \arcsin \frac{\gamma - \sin\varphi}{\varrho(\varphi, \gamma)}. \tag{7.7}$$

As $m \to \infty$, the time spent by the particle in the hollow tends to zero; therefore, the rotating frame of reference can be considered "approximately inertial" during that time, and the velocity at the second point of the intersection given by

$$\vec{\dot{v}}^{+} = v\varrho\,(-\sin y;\, \cos y)^T + o(1), \quad \text{where} \quad y = y(\xi, \varphi, \gamma) = \varphi^{+}(\xi, x(\varphi, \gamma)).$$

[Here $(\xi, \varphi) \mapsto (\xi^{+}(\xi, \varphi),\ \varphi^{+}(\xi, \varphi))$ denote the map generated by the hollow.] Applying the second formula in (7.5) and taking into account that $\vec{\dot{x}} = o(1)$ and $\omega t = \varphi + o(1)$, we find the velocity in the initial frame of reference:

$$\vec{v}^{+} = \vec{v}^{+}_{(m)}(\xi, \varphi, \gamma) = v\varrho A_\varphi(-\sin y;\, \cos y)^T - v\gamma\, e_\varphi + o(1) = v^{+}(\xi, \varphi, \gamma) + o(1),$$

where

$$v^{+}(\xi, \varphi, \gamma) = v \begin{bmatrix} -\varrho \sin(\varphi + y) - \gamma \cos\varphi \\ \varrho \cos(\varphi + y) - \gamma \sin\varphi \end{bmatrix}. \tag{7.8}$$

Letting $m \to \infty$ in formula (7.4) for the imparted momentum and dividing it by Δt, we obtain the following formula for the force of resistance acting on a disc:

$$\vec{R} = \begin{bmatrix} R_T \\ R_L \end{bmatrix} = r\rho v \int_0^1 \int_{-\pi/2}^{\pi/2} (-\vec{v}_0 - \vec{v}^{+}(\xi, \varphi, \gamma)) \cos\varphi \, d\xi d\varphi. \tag{7.9}$$

The angular momentum transmitted to B_m by an individual particle equals $rv\varrho(\sin x + \sin y) + o(1)$ times the mass of the particle. Summing up the angular momenta over all incident particles and passing to the limit $m \to \infty$, one finds the moment of the resistance force acting on the disc:

$$R_I = r^2\rho v \int_0^1 \int_{-\pi/2}^{\pi/2} v\varrho(\varphi, \gamma)\, (\sin x(\varphi, \gamma) + \sin y(\xi, \varphi, \gamma)) \cos\varphi \, d\xi \, d\varphi. \tag{7.10}$$

Theorem 7.1. *The resistance and the moment of resistance of a rough disc of radius r moving through a rarefied medium are equal to*

$$\vec{R} = \frac{8}{3} r\rho v^2 \cdot \vec{R}[\eta, \gamma], \tag{7.11}$$

$$R_I = \frac{8}{3} r^2 \rho v^2 \cdot R_I[\eta, \gamma]. \tag{7.12}$$

Here ρ is the medium density, v the velocity of translation, ω the angular velocity, $\gamma = \omega r/v$, and η the scattering law for this disc, and the dimensionless values $\vec{R}[\eta, \gamma]$ and $R_I[\eta, \gamma]$ are given by the integral formulas

$$\vec{R}[\eta, \gamma] = \begin{bmatrix} R_T[\eta, \gamma] \\ R_L[\eta, \gamma] \end{bmatrix} = \iint_{\square} \vec{c}(x, y, \gamma)\, d\eta(x, y), \tag{7.13}$$

$$R_I[\nu, \gamma] = \iint_{\square} c_I(x, y, \gamma)\, d\eta(x, y), \tag{7.14}$$

with the functions $\vec{c}$ are c_I being given by the relations (7.15)–(7.22). Recall that $\square = [-\pi/2, \pi/2] \times [-\pi/2, \pi/2]$. We also use the notation $\zeta = \zeta(x, \gamma) = \arcsin\sqrt{1 - \gamma^2 \cos^2 x}$ and $x_0 = x_0(\gamma) = \arccos\frac{1}{\gamma}$; χ stands for the characteristic function.

(a) If $0 < \gamma \le 1$, then

$$\vec{c}(x, y, \gamma) = \frac{3}{2}\frac{(\gamma \sin x + \sin\zeta)^3}{\sin\zeta}\cos\frac{x-y}{2}\begin{bmatrix} \cos\left(\zeta + \frac{x-y}{2}\right) \\ -\sin\left(\zeta + \frac{x-y}{2}\right) \end{bmatrix}, \tag{7.15}$$

$$c_I(x, y, \gamma) = -\frac{3}{4}\frac{(\gamma \sin x + \sin\zeta)^3}{\sin\zeta}(\sin x + \sin y), \tag{7.16}$$

and in particular,

$$\vec{c}(x, y, 1) = 6\sin^2 x \begin{bmatrix} \cos(2x - y) + \cos x \\ -\sin(2x - y) - \sin x \end{bmatrix} \chi_{x \ge 0}(x, y), \tag{7.17}$$

$$c_I(x, y, 1) = -6\sin^2 x\,(\sin x + \sin y)\,\chi_{x \ge 0}(x, y). \tag{7.18}$$

In the limiting case $\gamma \to 0^+$, one has

$$\vec{c}(x, y, \gamma) = -\frac{3}{4}\begin{bmatrix} \sin(x - y) \\ 1 + \cos(x - y) \end{bmatrix} + O(\gamma), \tag{7.19}$$

$$c_I(x, y) = \frac{9\gamma}{4}\sin x\,(\sin x + \sin y) + O(\gamma^2). \tag{7.20}$$

(b) If $\gamma > 1$, then

$$\vec{c}(x, y, \gamma) = \frac{3\cos\frac{x-y}{2}}{\sin\zeta}\Bigg\{(\gamma^3 \sin^3 x + 3\gamma \sin x \sin^2\zeta)\cos\zeta \begin{bmatrix} \cos\frac{x-y}{2} \\ -\sin\frac{x-y}{2} \end{bmatrix} -$$

$$- (3\gamma^2 \sin^2 x \sin\zeta + \sin^3\zeta)\sin\zeta \begin{bmatrix} \sin\frac{x-y}{2} \\ \cos\frac{x-y}{2} \end{bmatrix}\Bigg\}\chi_{x \ge x_0}(x, y), \tag{7.21}$$

$$c_I(x, y, \gamma) = -\frac{3}{2}\frac{\gamma^3 \sin^3 x + 3\gamma \sin x \sin^2\zeta}{\sin\zeta}(\sin x + \sin y)\,\chi_{x \ge x_0}(x, y). \tag{7.22}$$

Proof. The theorem will be proved separately for cases $\gamma = 1$, $0 < \gamma < 1$, and $\gamma > 1$.

Case. $\boldsymbol{\gamma = 1}$. We have

$$x = x(\varphi, 1) = \arcsin\sqrt{(1 - \sin\varphi)/2} = \pi/4 - \varphi/2,$$

and so the function $\varphi \mapsto x(\varphi, 1)$ is a bijection between the intervals $[-\pi/2, \pi/2]$ and $[0, \pi/2]$. Further, one has

$$\varrho = \varrho(\varphi, 1) = \sqrt{2(1 - \sin\varphi)} = 2\sin x, \quad \cos\varphi = \sin 2x,$$

and we get from Eq. (7.8) that

$$\vec{v}^{+} = v \begin{bmatrix} -2\sin x\, \cos(2x - y) - \sin 2x \\ 2\sin x\, \sin(2x - y) - \cos 2x \end{bmatrix},$$

whence

$$-\vec{v}_0 - \vec{v}^{+} = 2v\sin x \begin{bmatrix} \cos(2x - y) + \cos x \\ -\sin(2x - y) - \sin x \end{bmatrix}.$$

Making the change of variables $\{\xi, \varphi\} \to \{\xi, x\}$ in the integral in Eq. (7.9) and using (7.11), one gets

$$\vec{R}[\eta, 1] = 3 \int_0^1 \int_0^{\pi/2} \sin^2 x \begin{bmatrix} \cos(2x - y) + \cos x \\ -\sin(2x - y) - \sin x \end{bmatrix} \cos x \, d\xi \, dx.$$

In this integral y is a function of ξ and x, $y = \varphi^{+}(\xi, x)$. Changing the variables once again, $\{\xi, x\} \to \{x, y\}$, and taking into account that $\frac{1}{2}\cos x \, d\xi \, dx = d\eta(x, y)$, we obtain

$$\vec{R}[\eta, 1] = 6 \iint_{\square\blacksquare} \sin^2 x \begin{bmatrix} \cos(2x - y) + \cos x \\ -\sin(2x - y) - \sin x \end{bmatrix} d\eta(x, y). \qquad (7.23)$$

Here the symbol $\square\blacksquare$ stays for the rectangle $x \in [0, \pi/2]$, $y \in [-\pi/2, \pi/2]$.

The moment of the resistance force is calculated analogously, resulting in

$$R_I[\eta, 1] = -3 \int_0^1 \int_0^{\pi/2} \sin^2 x \, (\sin x + \sin y) \cos x \, d\xi \, dx =$$

$$= -6 \iint_{\square\blacksquare} \sin^2 x \, (\sin x + \sin y) \, d\eta(x, y). \qquad (7.24)$$

Case. $\mathbf{0 < \gamma < 1}$. The second relation in (7.7) implies that for a fixed value of γ, $x = x(\varphi, \gamma)$ is a monotonically decreasing function of φ that varies from $\pi/2$ to $-\pi/2$ as φ changes from $-\pi/2$ to $\pi/2$. From formula (7.6) and the first relation in (7.7) we have

$$\sin\varphi = \gamma\cos^2 x - \sin x\sqrt{1-\gamma^2\cos^2 x}, \quad \cos\varphi = \cos x\,(\gamma\sin x + \sqrt{1-\gamma^2\cos^2 x}),$$
$$\varrho = \gamma\sin x + \sqrt{1-\gamma^2\cos^2 x}.$$

Recall that

$$\zeta = \zeta(x,\gamma) = \arcsin\sqrt{1-\gamma^2\cos^2 x}. \tag{7.25}$$

We have

$$\cos\zeta = \gamma\cos x, \quad x+\zeta = \pi/2 - \varphi, \quad \zeta\in[\arccos\gamma,\, \pi/2],$$

and taking into account (7.8) we get

$$-\vec{v}_0 - \vec{v}^+ = v(\gamma\sin x + \sin\zeta)\cdot 2\cos\frac{x-y}{2}\begin{bmatrix} \cos\left(\zeta+\frac{x-y}{2}\right) \\ -\sin\left(\zeta+\frac{x-y}{2}\right)\end{bmatrix},$$
$$\frac{\cos\varphi}{\cos x} = \gamma\sin x + \sin\zeta = \varrho,$$
$$\frac{d\varphi}{dx} = -1 - \frac{d\zeta}{dx} = -\frac{\gamma\sin x + \sin\zeta}{\sin\zeta}.$$

Using the obtained formulas, making the change of variables $\{\xi,\varphi\}\to\{\xi,x\}$ in the integral (7.9) and taking into account (7.11), one obtains

$$\vec{R}[\eta,\gamma] = \frac{3}{4}\int_0^1\int_{-\pi/2}^{\pi/2}\frac{(\gamma\sin x+\sin\zeta)^3}{\sin\zeta}\cos\frac{x-y}{2}\begin{bmatrix} \cos\left(\zeta+\frac{x-y}{2}\right) \\ -\sin\left(\zeta+\frac{x-y}{2}\right)\end{bmatrix}\cos x\, d\xi\, dx.$$

Finally, the change of variables $\{\xi,x\}\to\{x,y\}$ results in

$$\vec{R}[\eta,\gamma] = \frac{3}{2}\iint_{\square}\frac{(\gamma\sin x+\sin\zeta)^3}{\sin\zeta}\cos\frac{x-y}{2}\begin{bmatrix} \cos\left(\zeta+\frac{x-y}{2}\right) \\ -\sin\left(\zeta+\frac{x-y}{2}\right)\end{bmatrix} d\eta(x,y). \tag{7.26}$$

Recall that the symbol $\square$ denotes the square $[-\pi/2,\,\pi/2]\times[-\pi/2,\,\pi/2]$ and $\zeta = \zeta(x,\gamma)$.

In a similar way, from (7.10) one obtains

$$R_I = -\frac{3}{8}\int_0^1\int_{-\pi/2}^{\pi/2}\frac{(\gamma\sin x+\sin\zeta)^3}{\sin\zeta}(\sin x + \sin y)\cos x\, d\xi\, dx;$$

whence

$$R_I[\eta,\gamma] = -\frac{3}{4}\iint\limits_{\square} \frac{(\gamma\sin x + \sin\zeta)^3}{\sin\zeta}\,(\sin x + \sin y)\; d\eta(x,y). \tag{7.27}$$

Formulas (7.23) and (7.24) are the particular cases of (7.26) and (7.27) for $\gamma = 1$. This can be easily verified taking into account that $\zeta(x,1) = |x|$.

Case. $\boldsymbol{\gamma > 1}$. In this case, the function $x = x(\varphi,\gamma)$ of Eq. (7.7) is no longer injection. When φ varies from $-\pi/2$ to $\varphi_0 = \varphi_0(\gamma) := \arcsin\frac{1}{\gamma}$, the value of x monotonically decreases from $\pi/2$ to $x_0 = x_0(\gamma) := \arccos\frac{1}{\gamma}$, and when φ varies from φ_0 to $\pi/2$, x monotonically increases from x_0 to $\pi/2$. Denote by $\varphi_- := \varphi_-(x,\gamma)$ and $\varphi_+ := \varphi_+(x,\gamma)$ the functions inverse to $x(\varphi,\gamma)$ on the intervals $[-\pi/2, \varphi_0]$ and $[\varphi_0, \pi/2]$, respectively. Then one has

$$\sin\varphi_\pm = \gamma\cos^2 x \pm \sin x\sqrt{1-\gamma^2\cos^2 x}.$$

Here and in what follows the "+" and "−" signs are related to the functions φ_+ and φ_-, respectively. The functions φ_+, φ_-, and $\zeta = \zeta(x,\gamma)$ in (7.25) satisfy the relations

$$\pi/2 - \varphi_+ = x - \zeta, \qquad \pi/2 - \varphi_- = x + \zeta.$$

The function ζ is defined for $x \in [x_0, \pi/2]$ and monotonically increases from 0 to $\pi/2$ when x changes in the interval $[x_0, \pi/2]$.

After some algebra [and using the shorthand notation $\varrho_\pm = \varrho(\varphi_\pm(x,\gamma),\gamma)$, $\vec v^{\,+}_\pm = \vec v^{\,+}(\xi,\varphi_\pm(x,\gamma),\gamma)$, $y = y(\xi,\varphi_\pm(x,\gamma),\gamma) = \varphi_i^+(\xi,x)$] one obtains

$$\begin{aligned}
\frac{\cos\varphi_\pm}{\cos x} &= \frac{\sin(x\mp\zeta)}{\cos x} = \gamma\sin x \mp \sin\zeta;\\
\pm\frac{d\varphi_\pm}{dx} &= \frac{d\zeta}{dx} \mp 1 = \frac{\gamma\sin x\mp\sin\zeta}{\sin\zeta};\\
\varrho_\pm &= \gamma\sin x \mp \sin\zeta;\\
-\vec v_0 - \vec v^{\,+}_\pm &= v\,(\gamma\sin x \mp \sin\zeta)\cdot 2\cos\frac{x-y}{2}\begin{bmatrix}\cos\left(\frac{x-y}{2}\mp\zeta\right)\\ -\sin\left(\frac{x-y}{2}\mp\zeta\right)\end{bmatrix}.
\end{aligned}$$

The resistance force takes the form

$$\vec R[\eta,\gamma] = \vec R_- + \vec R_+,$$

where

$$\begin{aligned}
\vec R_\pm &= \frac{3}{8}\int_0^1\!\int_{x_0}^{\pi/2} (-\vec v_0 - \vec v^{\,+}_\pm)\,\frac{\cos\varphi_\pm}{\cos x}\left(\pm\frac{d\varphi_\pm}{dx}\right)\cos x\; d\xi\, dx =\\
&= \frac{3}{4}\int_0^1\!\int_{x_0}^{\pi/2} \frac{(\gamma\sin x\mp\sin\zeta)^3}{\sin\zeta}\,\cos\frac{x-y}{2}\begin{bmatrix}\cos\left(\frac{x-y}{2}\mp\zeta\right)\\ -\sin\left(\frac{x-y}{2}\mp\zeta\right)\end{bmatrix}\cos x\; d\xi\, dx.
\end{aligned}$$

Summing the integrals $\vec{R}_-$ and $\vec{R}_+$ and making the change of variables, one obtains

$$\vec{R}[\eta,\gamma] = 3 \iint_{\square\!\blacksquare} \frac{\cos\frac{x-y}{2}}{\sin\zeta} \left\{ (\gamma^3 \sin^3 x + 3\gamma \sin x \sin^2\zeta) \cos\zeta \begin{bmatrix} \cos\frac{x-y}{2} \\ -\sin\frac{x-y}{2} \end{bmatrix} - \right.$$

$$\left. - (3\gamma^2 \sin^2 x \sin\zeta + \sin^3\zeta) \sin\zeta \begin{bmatrix} \sin\frac{x-y}{2} \\ \cos\frac{x-y}{2} \end{bmatrix} \right\} d\eta(x,y). \qquad (7.28)$$

Here the symbol $\square\!\blacksquare$ stands for the rectangle $[x_0,\ \pi/2] \times [-\pi/2,\ \pi/2]$.

The moment of the resistance force is calculated analogously. One has $R_I[\eta,\gamma] = R_{I-} + R_{I+}$, where

$$R_{I\pm} = -\frac{3}{8} \int_0^1 \int_{x_0}^{\pi/2} \varrho_\pm \frac{\cos\varphi_\pm}{\cos x} \left(\pm \frac{d\varphi_\pm}{dx} \right) (\sin x + \sin y) \cos x \, d\xi \, dx =$$

$$= -\frac{3}{8} \int_0^1 \int_{x_0}^{\pi/2} \frac{(\gamma \sin x \mp \sin\zeta)^3}{\sin\zeta} (\sin x + \sin y) \cos x \, d\xi \, dx.$$

Therefore,

$$R_I[\eta,\gamma] = -\frac{3}{4} \int_0^1 \int_{x_0}^{\pi/2} \frac{\gamma^3 \sin^3 x + 3\gamma \sin x \sin^2\zeta}{\sin\zeta} (\sin x + \sin y) \cos x \, d\xi \, dx.$$

Making the change of variables, we have

$$R_I[\eta,\gamma] = -\frac{3}{2} \iint_{\square\!\blacksquare} \frac{\gamma^3 \sin^3 x + 3\gamma \sin x \sin^2\zeta}{\sin\zeta} (\sin x + \sin y) \, d\eta(x,y). \qquad (7.29)$$

The proof of Theorem 7.1 is complete. □

7.3 Magnus Effect

We are primarily concerned here with determining the two-dimensional set of admissible normalized forces $\mathcal{R}_\gamma := \{\vec{R}[\eta,\gamma] : \eta \in \mathcal{M}\}$.

Recall that according to the characterization Theorem 4.5, for each $\vec{R} \in \mathcal{R}_\gamma$ there exists a suitable rough disc that experiences the force $\vec{R}$ when moving at the relative angular velocity γ. However, this theorem gives us no idea as to what the corresponding shape of roughness looks like. It may well be too complicated to appear in nature or be fabricated. Therefore, it makes sense to describe subsets of $\mathcal{R}_\gamma$ generated by simple shapes. In this section, we present subsets generated by triangular hollows. In addition, we calculate analytically the resistance force and its moment for several simple shapes (e.g., rectangle, right isosceles triangle).

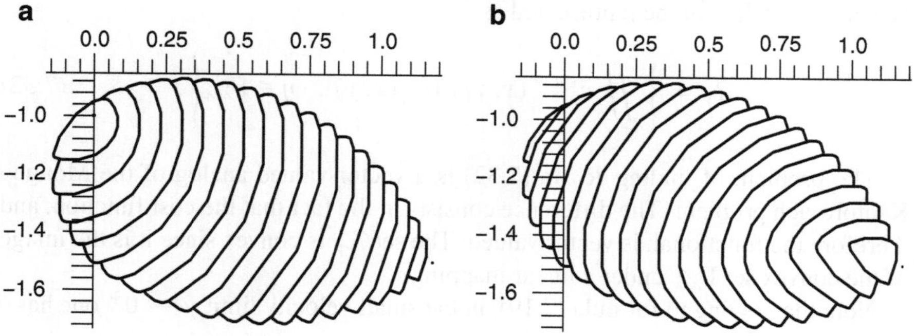

Fig. 7.4 "Level sets" $\mathcal{R}_{1,c} = \{$ all possible values of $\vec{R}[\eta, 1]$, with $R_I[\eta, 1] = c\}$ are shown for 21 values of c, from left to right: $c = 0, -0.075, -0.15, -0.225, \ldots, -1.425, -1.5$. (**a**) "View from above" and (**b**) view from below on these sets

7.3.1 Vector-Valued Monge–Kantorovich Problem

Here we determine the set of all possible resistance forces that can act on a rough disc, with fixed angular velocity. The force is scaled so that the resistance of the "ordinary circle" equals $(0; -1)^T$. The problem is as follows: given γ, find the two-dimensional set

$$\mathcal{R}_\gamma = \{\vec{R}[\eta, \gamma] : \eta \in \mathcal{M}\}. \tag{7.30}$$

This can be viewed as a restriction of the following more general problem: find the three-dimensional set

$$\{(\vec{R}[\eta, \gamma];\ R_I[\eta, \gamma]) : \eta \in \mathcal{M}\}. \tag{7.31}$$

Problem (7.31) is more important, but also more time-consuming, and is mainly postponed to the future. The only exception is the case $\gamma = 1$, where several "level sets" $\mathcal{R}_{1,c} = \{\vec{R}[\eta, 1] : \eta \in \mathcal{M},\ R_I[\eta, 1] = c\}$ are numerically found. These curves are depicted in Fig. 7.4, suggesting what the corresponding three-dimensional set might look like. In this case, $R_I[\eta, 1]$ varies between -1.5 and 0, and the level sets are found for 21 values $c = -1.5, -1.425, -1.35, \ldots, -0.15, -0.075, 0$.

Note that the functional $\vec{R}$, defined on the set $\mathcal{M}$ by (7.13), will not change if the integrand $\vec{c}$ is replaced with the symmetrized function $\vec{c}^{\,\text{symm}}(x, y, \gamma) = \frac{1}{2}(\vec{c}(x, y, \gamma) + \vec{c}(y, x, \gamma))$:

$$\vec{R}[\eta, \gamma] = \iint_{\square} \vec{c}^{\,\text{symm}}(x, y, \gamma)\, d\eta(x, y).$$

Recall that $\Gamma_{\lambda,\lambda}$ is the set of measures η on the square $\square$ that satisfy condition (M1) in Definition 4.5, that is, the set of measures with both marginals equal to λ. For a measure $\eta \in \Gamma_{\lambda,\lambda}$ define the symmetrized measure $\eta^{\text{symm}} = \frac{1}{2}(\eta + \pi_d^{\#}\eta) \in \mathcal{M}$; then we have

$$\iint_{\square} \vec{c}^{\,\text{symm}}\, d\eta = \iint_{\square} \vec{c}^{\,\text{symm}}\, d\eta^{\text{symm}}.$$

Hence the set $\mathcal{R}_\gamma$ can be represented as

$$\mathcal{R}_\gamma = \left\{ \iint_{\square} \vec{c}^{\,\text{symm}}(x, y, \gamma)\, d\eta(x, y) : \ \eta \in \Gamma_{\lambda,\lambda} \right\}. \tag{7.32}$$

The problem of finding $\mathcal{R}_\gamma$ in (7.32) is a vector-valued analog of the Monge–Kantorovich problem. The difference consists in the fact that the cost function, and therefore the functional, is vector valued. The set $\mathcal{R}_\gamma$ is convex since it is the image of the convex set $\Gamma_{\lambda,\lambda}$ under a linear mapping.

Note that, owing to formula (7.19), in the small velocity limit $\gamma \to 0^+$ one has

$$\vec{c}^{\,\text{symm}}(x, y, 0^+) = \frac{3}{4}(1 + \cos(x - y))(0;\ 1)^T;$$

therefore, the problem of finding $\mathcal{R}_{0+}$ amounts to minimizing and maximizing the integral

$$\frac{3}{4} \iint_{\square} (1 + \cos(x - y))\, d\eta(x, y)$$

over all $\eta \in \Gamma_{\lambda,\lambda}$. This special Monge–Kantorovich problem was solved in Sect. 5; the minimal and maximal values of the integral were found to be $m_2 \approx 0.9878$ and 1.5.

In Figs. 7.5 and 7.6 we present numerical solutions of this problem for the values $\gamma = 0.1,\ 0.3$, and 1, as well as the analytical solution for $\gamma = 0^+$. The case of larger γ requires more involved calculation and therefore is postponed to the future. The method of solution is the following: for n equidistant vectors $\vec{e}_i,\ i = 1, \dots, n$ on S^1, we find the solution of the Monge–Kantorovich problem $\inf(\vec{R}[\eta, \gamma] \cdot \vec{e}_i) =: r_i$. This problem is reduced to the transport problem of linear programming and is solved numerically.[1]

Next, the intersection of the half-planes $\vec{r} \cdot \vec{e}_i \geq r_i$ is built. It is a convex polygon approximating the required set $\mathcal{R}_\gamma$, and the approximation accuracy increases as n increases. The value $n = 100$ was used in our calculations.

Figure 7.5 shows the sets $\mathcal{R}_\gamma$ for $\gamma = 0^+$, 0.1, 0.3, and 1. The set $\mathcal{R}_{0+}$ is the vertical segment $\{0\} \times [-1.5,\ -0.9878]$, $\mathcal{R}_{0.1}$ is the thin set with white interior, and $\mathcal{R}_{0.3}$ is the set with a gray interior. The largest set is $\mathcal{R}_1$.

In Fig. 7.6 the same sets are shown in more detail. In Fig. 7.6b–d, additionally, we present the regions corresponding to all possible triangular hollows, with the angles being multiples of 5°. These regions are colored gray. For $\gamma = 0^+$ the corresponding region is the vertical interval $\{0\} \times [-1.42,\ -1]$ marked by a (slightly shifted) dashed line in Fig. 7.6a.

The part of the set $\mathcal{R}_\gamma$ situated to the left of the vertical axis corresponds to resistance forces producing the *proper* Magnus effect. The part of $\mathcal{R}_\gamma$ to the right of this axis is related to forces that cause the *inverse* Magnus effect. We see that

---

[1] All the computational tests were performed on a PC with a Pentium IV, 2.0 GHz, and 512 MB RAM and using the optimization package Xpress-IVE, Version 1.19.00, with the modeler MOSEL.

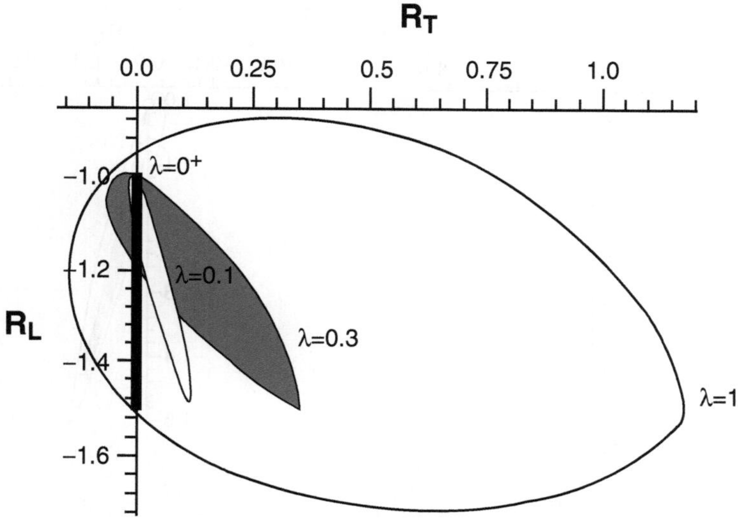

Fig. 7.5 Convex sets $\mathcal{R}_\gamma$ with $\gamma = 0^+$, 0.1, 0.3, and 1. The set $\mathcal{R}_{0^+}$ is the vertical segment (*boldface*) with endpoints $(0, -0.9878\ldots)$ and $(0, -1.5)$

most of the set (in the case $\gamma = 1$, approximately 93.6 % of the area) is situated to the right of the axis. This suggests that the inverse effect is a more common phenomenon than the proper one. Actually, although Theorem 7.1 guarantees that shapes of roughness generate exactly the set $\mathcal{R}_\gamma$, we have never encountered a shape producing the *proper* Magnus effect (and thus corresponding to a point to the left of the vertical axis).

7.3.2 Special Cases of Rough Discs

We present here only the final expressions for the forces and their moments calculated by formulas (7.15)–(7.22); the calculation details are omitted.

1. *Circle (no cavities).* The measure $\eta = \eta_0$ corresponding to the circle is given by $d\eta_0(x, y) = \frac{1}{2}\cos x \cdot \delta(x + y)$. One has $R_T[\eta_0, \gamma] = R_I[\eta_0, \gamma] = 0$ and $R_L[\eta_0, \gamma] = -1$. Thus, as one might expect, the resistance does not depend on the angular velocity and is collinear to the body's velocity. There is no Magnus effect in this case.
2. *Retroreflector.* There exists a unique measure $\eta = \eta_{\text{retr}} \in \mathcal{M}$ supported on the diagonal $x = y$; its density equals $d\eta_{\text{retr}}(x, y) = \frac{1}{2}\cos x \cdot \delta(x - y)$. By Theorem 4.5, there exists a rough disc with this scattering law. One has $R_T[\eta_{\text{retr}}, \gamma] = 3\pi\gamma/8$, $R_L[\eta_{\text{retr}}, \gamma] = -3/2$, and $R_I[\eta_{\text{retr}}, \gamma] = -3\gamma/2$. Thus, the longitudinal component of the resistance force does not depend on the angular velocity γ, whereas the transversal component and the moment of this force are proportional to γ.

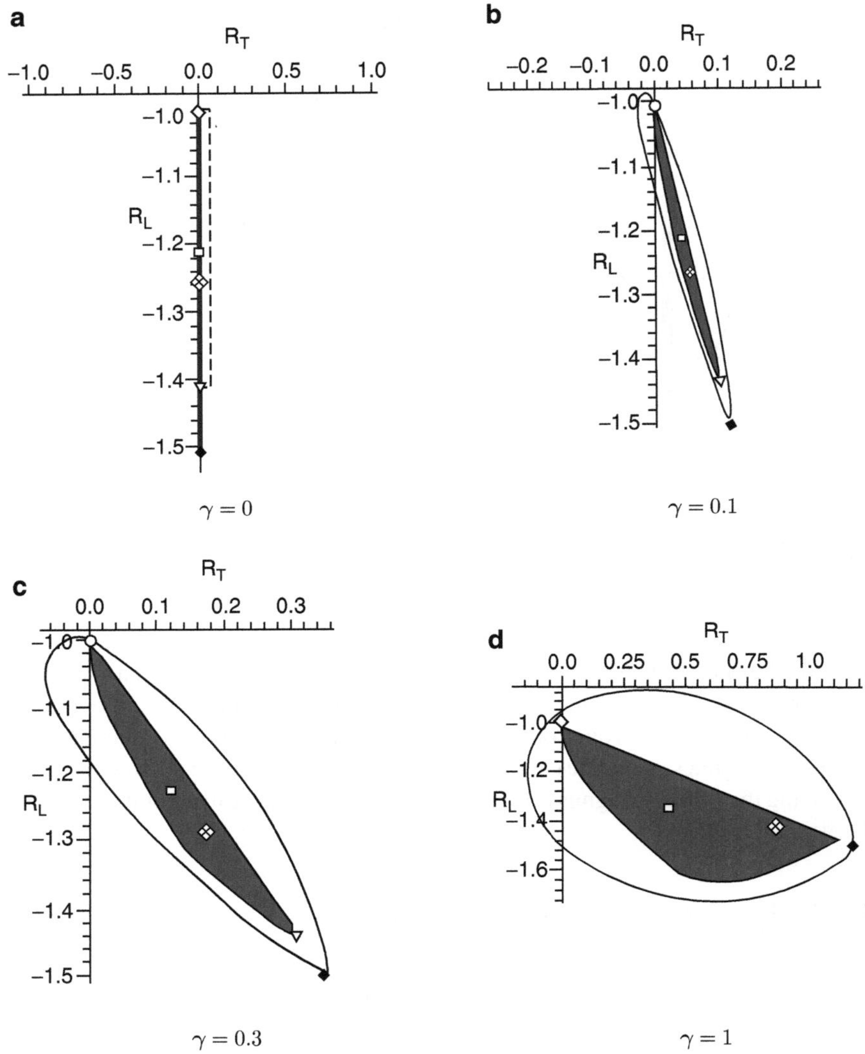

Fig. 7.6 The sets $\mathcal{R}_\gamma$ with (**a**) $\gamma = 0^+$, (**b**) 0.1, (**c**) 0.3, and (**d**) 1 are shown here separately. The values $R[\nu, \gamma]$ with $\nu = \nu_0,\ \nu_*,\ \nu_{\text{rect}},\ \nu_\nabla,\ \nu_\otimes$ are indicated by the symbols *open circle*, *filled circle*, *open square*, *down triangle*, *otimes* respectively. (**a**) The region generated by triangular hollows is marked by a (slightly shifted) *vertical dashed line*. It is the segment with the endpoints $(0, -1)$ and $(0, -1.42)$. (**b**)–(**d**) The regions generated by triangular hollows are painted over

3. *Rectangular hollows*. The rough disc $\mathcal{B}$ is represented by the sequence of sets B_m that are regular m-gons with m congruent rectangles removed (Fig. 7.7a). The width of the rectangles is much smaller than their height, (*width*)/(*side of rectangle*) $= 1/m$. A smaller side of each rectangle is contained in a side of the polygon, in addition $|$*side rectangle*$|/|$*side of polygon*$| = 1 - 1/m$.

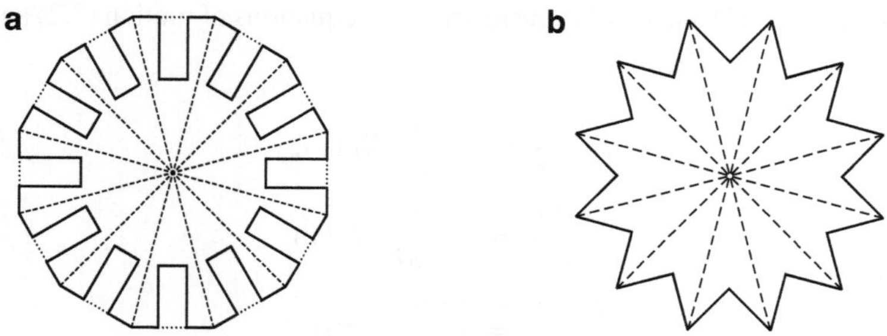

Fig. 7.7 **(a)** Rough disc with rectangular hollows **(b)** Rough disc with triangular hollows

Then $\eta = \eta_{\text{rect}} := \frac{1}{2}(\eta_{\text{retr}} + \eta_0)$. One can easily calculate that $R_T[\eta_{\text{rect}}, \gamma] = 3\pi\gamma/16$, $R_L[\eta_{\text{rect}}, \gamma] = -1.25$, and $R_I[\eta_{\text{rect}}, \gamma] = -3\gamma/4$.

4. *Triangular hollows*. The sets B_m representing the rough disc $\mathcal{B}$ are regular m-gons with m right isosceles triangles removed (Fig. 7.7b). Then the measure $\eta =: \eta_\nabla$ has the following support (which looks like an inclined letter **H**; see also Fig. 4.2b):

$$\{x + y = -\pi/2 : x \in [-\pi/2, 0]\} \cup \{y = x : x \in [-\pi/4, \pi/4]\}$$
$$\times \cup \{x + y = \pi/2 : x \in [0, \pi/2]\}.$$

The density of this measure is given by (4.11). One has $\vec{R}[\eta_\nabla, 0^+] = (0; -\sqrt{2})^T$ and $R_I[\eta_\nabla, 0^+] = 0$; $\vec{R}[\eta_\nabla, 1] = (1/4 + 3\pi/16; 3\pi/16 - 2)^T$. The remaining values are still unknown.

5. *Cavity realizing the product measure*. Consider the measure $\eta = \eta_\otimes$ with density $d\eta_\otimes(x, y) = \frac{1}{4}\cos x \cos y\, dx\, dy$. Evidently in this case $\eta_\otimes \in \mathcal{M}$. The angles of incidence and of reflection are statistically independent; that is, at the moment when the particle leaves the hollow, it completely "forgets" its initial velocity. Here we have $R_T[\eta_\otimes, \gamma] = (10\gamma + \gamma^3)\pi/80$ for $0 < \gamma \le 1$; $R_L[\eta_\otimes, 1] = -3/4 - \pi/5 \approx -1.378$; and $R_I[\eta_\otimes, \gamma] = -3\gamma/4$ for any γ. The remaining values are unknown.

The points in Fig. 7.6a–d corresponding to cases 1–5 are indicated by special symbols: η_0 by a circle, η_* by a diamond, η_{rect} by an open square, η_∇ by a triangle, and $\eta_\otimes$ by a circumscribed cross.

7.4 Dynamics of a Rough Disc

The motion of a spinning rough disc $\mathcal{B}$ is determined by the values $R_T[\eta_\mathcal{B}, \gamma]$, $R_L[\eta_\mathcal{B}, \gamma]$, and $R_I[\eta_\mathcal{B}, \gamma]$. For the sake of brevity, below we omit the fixed argument $\eta = \eta_\mathcal{B}$ and write $R(\gamma)$ instead of $R[\eta, \gamma]$. Recall that the absolute value of the disc velocity is denoted by $v = |\vec{v}|$ and the angular velocity is $\omega = \gamma v/r$. Denote by θ the angle the velocity makes with a fixed direction in an inertial frame of reference.

Using Eqs. (7.11) and (7.12), one rewrites the equations of motion (7.2) and (7.3) in the form

$$\frac{dv}{dt} = \frac{8r\rho v^2}{3M} R_L(\gamma), \tag{7.33}$$

$$\frac{d\theta}{dt} = -\frac{8r\rho v}{3M} R_T(\gamma), \tag{7.34}$$

$$\frac{d(\gamma v)}{dt} = \frac{8r^3\rho v^2}{3I} l R_I(\gamma). \tag{7.35}$$

Recall that $\beta = Mr^2/I$ is the inverse relative moment of inertia. In particular, if the mass is concentrated near the disc boundary, we have $\beta = 1$, and if the mass is distributed uniformly in the disc, then $\beta = 2$. In the intermediate case, when the mass is distributed in an arbitrary (generally speaking, nonuniform) fashion in the disc, one has $\beta \geq 1$.

With the change of variables $d\tau = (8r\rho v/3M)\,dt$, Eqs. (7.33)–(7.35) are transformed into

$$\frac{d\gamma}{d\tau} = \beta R_I(\gamma) - \gamma R_L(\gamma), \tag{7.36}$$

$$\frac{dv}{d\tau} = v\,R_L(\gamma), \tag{7.37}$$

$$\frac{d\theta}{d\tau} = -R_T(\gamma). \tag{7.38}$$

Denote by s the path length of the disc; thus, $ds/dt = v$. One readily finds that s is proportional to τ, $s = (3M/8r\rho)\,\tau$. Below we solve the system of Eqs. (7.36)–(7.38) for cases 1–3 considered in Sect. 7.3.2. Next, we determine the dynamics numerically for some triangular hollows.

1. *Circle*. One has $d\gamma/d\tau = -\gamma$, $dv/d\tau = -v$, and $d\theta/dt = 0$; therefore the circle moves in a straightforward manner. Solving these equations, one gets that its center moves according to $\vec{x}(t) = (3M/8r\rho)\ln(t - t_0)\vec{e} + \vec{x}_0$, where $\vec{e} \in S^1$ and $\vec{x}_0 \in \mathbb{R}^2$ are constants. Thus, having started the motion at some moment, the circle passes a half-line during infinite time. This equation also implies that the motion cannot be extended to all $t \in \mathbb{R}$.
2. *Retroreflector*. Here system (7.36)–(7.38) takes the form

$$d\gamma/d\tau = -3\gamma(\beta - 1)/2,\ \ dv/d\tau = -3v/2,\ \ d\theta/d\tau = -3\pi\gamma/8. \tag{7.39}$$

   In the case $\beta = 1$, one evidently has $\gamma = \text{const}$. The disc moves along a circumference of radius $M/(\pi r\rho\gamma)$ in the direction opposite to the angular velocity of rotation: if the disc rotates counterclockwise, then its center moves clockwise along the circumference. The radius of the circumference is proportional to the disc mass and inversely proportional to the relative angular velocity. The path length is proportional to the logarithm of time, $s(t) = \frac{M}{4r\rho}\ln(t - t_0)$.

In the case $\beta > 1$, we have $s(t) = \frac{M}{4r\rho}\ln(t - t_0)$, $\theta = \theta_0 + \text{const}\cdot\exp(-(\beta - 1)\frac{4r\rho}{M}s)$, and $\gamma = \frac{4}{\pi}(\beta - 1)(\theta - \theta_0)$. The path length once again depends logarithmically on the time, the relative angular velocity γ converges to zero, and the direction θ converges to a limiting value θ_0; thus, the values γ and θ are exponentially decreasing functions of the path length and are inversely proportional to the $(\beta - 1)$th degree of the time passed since a fixed moment. The trajectory of motion is a semibounded curve approaching an asymptote as $t \to +\infty$.

3. *Rectangular cavity*. Equations of motion (7.36)–(7.38) in this case take the form

$$d\gamma/d\tau = -3\gamma(\beta - 5/3)/4, \quad dv/d\tau = -5v/4, \quad d\theta/d\tau = -3\pi\gamma/16.$$

Solving these equations one obtains $\tau = \frac{4}{5}\ln(t - t_0)$, $v = v_0\, e^{-5\tau/4}$, $\gamma = \gamma_0\, e^{3\tau(5/3-\beta)/4}$, $\theta = \theta_0 + \frac{\pi\gamma_0}{4(\beta-5/3)}\, e^{3\tau(5/3-\beta)/4}$. Thus, the path depends on t logarithmically, and the relative angular velocity and the rotation angle are proportional to $(t - t_0)^{1-3\beta/5}$ and to $\exp\left(\frac{2r\rho(5-3\beta)}{3M}\, s\right)$.

If $\beta < 5/3$, then γ and θ tend to infinity and the trajectory of the disc center is a converging spiral. In the case $\beta > 5/3$, γ converges to zero, θ converges to a constant value, and the trajectory is a semibounded curve approaching an asymptote as $t \to +\infty$. In the case $\beta = 5/3$, γ is constant, and the trajectory is a circumference of radius $2M/(\pi r\rho\gamma)$.

Finally, we examine numerically some triangular hollows. It is helpful to denote $g(\gamma) = \gamma R_L(\gamma)/R_I(\gamma)$ and rewrite equation (7.36) in the form

$$\frac{d\gamma}{d\tau} = -R_I(\gamma)(g(\gamma) - \beta). \tag{7.40}$$

In Fig. 7.8a, the function $g(\gamma)$ is shown for two cases where the hollow is an isosceles triangle with the angles (i) 30°, 120°, 30° and (ii) 60°, 60°, 60°. We see that $g(\gamma)$ monotonically increases in case (i) and has three intervals of monotonicity in case (ii). In both cases, $R_I(\gamma) < 0$. This implies, in case (i), that the disc trajectory is a converging spiral, if $\beta < 1.5$, and may take the form of a converging spiral or a curve approaching a straight line, depending on the initial conditions, if $\beta > 1.5$.

The disc behavior is even richer in case (ii) of the equilateral triangle. If $1.38 < \beta < 1.49$, then three kinds of asymptotic behavior may be realized, depending on the initial conditions: (I) the trajectory is a converging spiral, (II) the trajectory approaches a circumference, and (III) the trajectory approaches a straight line (Fig. 7.9). If $1.16 < \beta < 1.38$, only two asymptotic behaviors of types (I) and (II) are possible; if $\beta > 1.49$, then the possible behaviors are (I) and (III); and if $\beta < 1.16$, the asymptotic behavior is always (I).

In the case of triangular cavities, as our numerical evidence shows, the function $g(\gamma)$ monotonically increases for γ sufficiently large and $\lim_{\gamma\to+\infty} g(\gamma) = +\infty$. This implies that the trajectory is a converging spiral for appropriate initial conditions (namely, if the initial angular velocity is large enough). If, in addition, β is large enough (that is, the mass of the disc is concentrated near the center), the trajectory may also be a curve approaching a straight line. If the function g has

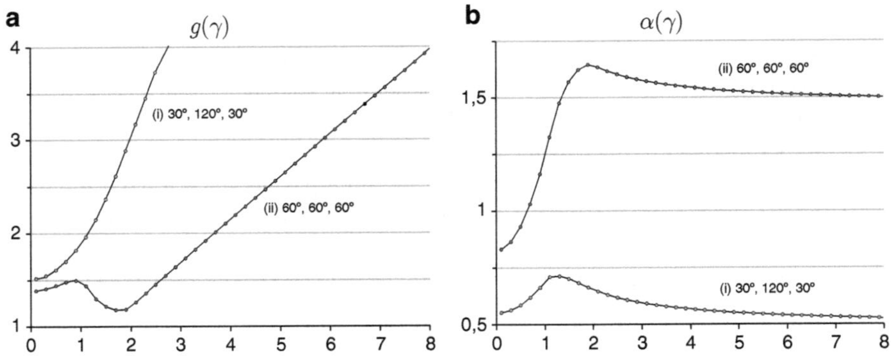

Fig. 7.8 The functions (**a**) $g(\gamma) = \gamma R_L(\gamma)/R_I(\gamma)$ and (**b**) $\alpha(\gamma) = 4R_T(\gamma)/\gamma$ are plotted for the triangular hollows with the angles (**i**) 30°, 120°, 30° and (**ii**) 60°, 60°, 60°

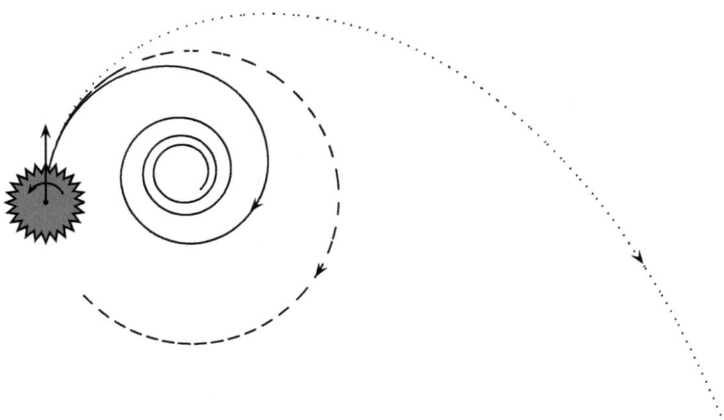

Fig. 7.9 Three kinds of asymptotic behavior of a rough disc with roughness formed by equilateral triangles and with $1.38 < \beta < 1.49$: (**I**) converging spiral (*solid line*); (**II**) circumference (*dashed line*); (**III**) curve approaching a straight line (*dotted line*)

intervals of monotonically decrease (as for the case of an equilateral triangle), then the trajectory may also approach a circumference. The length of the disc path is always proportional to the logarithm of time.

7.5 Conclusions and Comparison with Previous Works

In our opinion, the inverse Magnus effect in highly rarefied media is caused by two factors:

(i) Nonelastic interaction of particles with the body. Part of the tangential component of the particles' momentum is transmitted to the body, resulting in the creation of a transversal force.
(ii) Multiple collisions of particles with the body due to the fact that the body's surface is not convex but contains microscopic cavities.

In the papers [7, 27, 74, 75] the impact of factor (i) is studied. The body is supposed to be convex, and therefore factor (ii) is excluded from consideration. In these papers, the force acting on a spinning body moving through a rarefied gas is calculated, and in [27], additionally, the moment of this force slowing down the body's rotation is determined. The following shapes have been considered: a sphere, a cylinder [27, 75], convex bodies of revolution [27], and right parallelepipeds of regular polygon section [75]. The interaction of gas particles with a body is as follows: a fraction $1 - \alpha_\tau$ of the incident particles is elastically reflected according to the rule "the angle of incidence is equal to the angle of reflection," while the remaining fraction α_τ of the particles reaches thermal equilibrium with the body's surface and is reflected as a Maxwellian [7, 27, 74].

In the paper [75], a somewhat different model of interaction is considered, where the reflected particles acquire a fraction α_τ of the tangential momentum of a rotating body. The transversal force results from the tangential friction and acts on the body in the direction associated with the inverse Magnus effect.

It is remarkable that for different models and different shapes of the body, the formula for the transversal force is basically the same. If the rotation axis is perpendicular to the direction of the body's motion, then this force equals

$$\frac{1}{2}\,\alpha_\tau M_g \omega v, \tag{7.41}$$

where M_g is the mass of the gas displaced by the body, ω is the angular velocity of the body, and v is its translation velocity. (Note that in [7] this formula appears in the limit of infinite heat conductivity or zero gas temperature.) In [75] it was found that for parallelepipeds of regular n-gon section with n being odd, the transversal force depends on time, and the value of the time-dependent force was determined. It is easy to calculate, however, that the time-averaged force is equal to (7.41).

In this chapter, in contrast, we concentrate on the study of factor (ii). We suppose that all collisions of particles with a body are perfectly elastic (that is, $\alpha_\tau = 0$), and therefore there is no tangential friction. We restrict ourselves to the two-dimensional case and suppose that the body is a disc with small hollows on its boundary, or a *rough* disc. The Magnus effect is due to multiple reflections of particles in the hollows. We study all *geometrically possible* cases of hollows. According to (7.11), the transversal force equals

$$\frac{1}{2}\,\alpha(\gamma)\, M_g \omega v,$$

where $\gamma = \omega r / v$, $M_g = \pi r^2 \rho$ is the total mass of gas particles displaced by the body,

$$\alpha(\gamma) = \alpha(\gamma, \eta) = \frac{16}{3\pi} \frac{R_T[\eta, \gamma]}{\gamma},$$

and η is the measure characterizing the scattering law in the hollows. The function α depends on both η and γ. In particular, α varies between -0.409 and 2 for $\gamma = 0.1$, between -0.378 and 2 for $\gamma = 0.3$, and between -0.248 and 2 for $\gamma = 1$. We conjecture that

$$\lim_{\gamma \to \infty} \inf_{\eta} \alpha(\gamma, \eta) = 0 \text{ and } \lim_{\gamma \to \infty} \sup_{\eta} \alpha(\gamma, \eta) = 2.$$

The graphs of the function $\alpha(\gamma)$ with η corresponding to triangular hollows with the angles (i) $30°$, $120°$, $30°$ and (ii) $60°$, $60°$, $60°$ are shown in Fig. 7.8b. We see that this function significantly depends on the velocity of rotation γ; in general, the variation of $\alpha(\gamma)$ with η fixed can be more than twofold.

We conclude that the impact of both factors (i) and (ii) is unidirectional, and so they strengthen each other. Moreover, the formulas for the transversal force are similar; one should simply substitute the function $\alpha(\gamma, \eta)$ for α_τ. We have seen that $\alpha(\gamma, \eta)$ can be significantly greater than 1, whereas $\alpha_\tau \leq 1$. Actually, this can be just an artifact of our model's being two-dimensional.

Apart from its physical meaning, studying the dynamics of a spinning rough disc (or, more generally, of a noncircular body) in a rarefied medium represents a nice mathematical problem that originates in classical mechanics and is closely connected to Newton's aerodynamic problem [40]. According to our numerical simulations, most possible roughnesses (93.6 % for $\gamma = 1$) correspond to the *inverse* Magnus effect, and only a small portion of them correspond to the proper one. We know that roughnesses corresponding to the proper Magnus effect do exist, but we have no idea what they might look like, and no such roughness has been found. Another interesting question concerns the description of admissible trajectories and is closely related to the associated problem of (vector-valued) Monge–Kantorovich optimal mass transfer. In particular, the existence, in the same body, of a roughness corresponding to the proper Magnus effect for some values of γ and to the inverse one for others would imply the existence of a rough disc with a strange zigzag trajectory.

Chapter 8
Invisible Bodies

In this and the next chapters, we consider billiards that have extremal properties (from the viewpoints of aerodynamics and optics). In this chapter, we describe bodies that have zero resistance or are invisible and, in the next one, bodies that maximize the resistance averaged over all directions.

Let us describe the content of this chapter in more detail. We define the notions of (1) a body that has zero resistance in one direction, (2) a body leaving no trace when moving in one direction, and (3) a body invisible in one direction. The most surprising fact is that bodies of each kind do exist. We provide examples of such bodies and show that the bodies of third kind form a proper subset of the set of bodies of second kind, which is in turn a proper subset of the set of bodies of first kind. Next we show that there exist *connected* bodies of each kind. The largest number of reflections from a body of zero resistance and a body leaving no trace in our constructions equals 2, and from a body invisible in one direction it equals 3. These values cannot be decreased.

Further, we introduce a parameter $\kappa(B)$ equal to the relative volume of a body B in its convex hull. This parameter characterizes, in a sense, the "degree of nonconvexity" of a body; it equals 1 for convex bodies and is less than 1 for nonconvex ones. We prove that κ can be made arbitrarily close to 1 for bodies of zero resistance and bodies invisible in one direction.

We construct bodies of zero resistance and bodies invisible with respect to two different directions and prove the nonexistence of bodies that have these properties relative to all possible directions. Finally, we construct a body invisible from one point. The question on the maximum number of directions or points of invisibility or zero resistance remains open.

Most of the results presented in this chapter were published in [1, 57].

A. Plakhov, *Exterior Billiards: Systems with Impacts Outside Bounded Domains*, DOI 10.1007/978-1-4614-4481-7_8,

8.1 Main Constructions

8.1.1 Definitions and Statement of Main Result

Definition 8.1. We say that a body B has *zero resistance in a direction* $v_0 \in S^{d-1}$ if $v_B^+(\xi, v_0) = v_0$ for almost all $\xi \in \mathbb{R}^d$.

The following proposition provides two additional equivalent definitions of a body of zero resistance. Its proof is left to the reader.

Proposition 8.1. *(a) A body B has zero resistance in a direction v_0 if and only if for any function c satisfying (1.1) we have $R_{\delta_{v_0}}(B) = 0$.*

(b) A body B has zero resistance in a direction v_0 if and only if for some scalar function c satisfying $c(v, v^+) = 0$ for $v = v^+$ and $c(v, v^+) > 0$ for $v \neq v^+$ we have $R_{\delta_{v_0}}(B) = 0$.

Take a bounded convex body C containing B.

Definition 8.2. We say that the body B *leaves no trace in the direction* v_0 if it has zero resistance in this direction and, additionally, the map $\xi \mapsto \xi_{B,C}^+(\xi, v_0)$ between full-measure subsets of $\partial C \cap \{\xi : n(\xi) \cdot v_0 \leq 0\}$ and $\partial C \cap \{\xi : n(\xi) \cdot v_0 \geq 0\}$ preserves the measure $|n(\xi) \cdot v_0| \, d\xi$.

Definition 8.3. We say that B is *invisible in the direction* v_0 if it has zero resistance in this direction and, additionally, $\xi_{B,C}^+(\xi, v_0) - \xi$ is parallel to v_0.

Definitions 8.2 and 8.3 do not depend on the choice of the ambient body C.

Obviously, if a body is invisible in a direction, then it leaves no trace in this direction.

The following proposition can serve as an equivalent definition of an invisible body.

Proposition 8.2. *B is invisible in a direction v_0 if and only if every straight line with the directing vector v_0 contains (the trajectory of a billiard particle with initial velocity v_0) $\setminus$Conv B.*

These definitions admit the following (mechanical and optical) interpretations. Imagine a spaceship moving in open space through a rarefied cloud of solid particles; when colliding with the ship surface the particles are reflected elastically. A body of zero resistance experiences no drag force and can travel through the cloud infinitely without reducing its velocity. A body leaving no trace leaves the cloud behind it the same as it is before it; the particles are resting again, and the cloud density remains unchanged. An invisible body is indeed invisible from an infinitely distant point if its surface is specular: a parallel flow of light rays impinging on the body in the direction v_0, after several reflections, transforms again into a parallel flow in the same direction, and each particle of light after passing the body moves along the same line as before.

Note, however, that the bodies of zero resistance constructed below are such only with respect to a uniform motion. As it begins to accelerate or decelerate, the spaceship will experience a force impeding the change of velocity. In addition, when flying into a cloud the spaceship will experience a braking force, and when going away it will experience a compensatory force pushing it out of the cloud. A body leaving no trace, when coming into a cloud, will make a "dent" on its boundary and a "protrusion" where it goes away.

Let $\mathcal{B}^d_{zeroresist}(v_0)$, $\mathcal{B}^d_{notrace}(v_0)$, and $\mathcal{B}^d_{invisible}(v_0)$ be the classes of bodies of zero resistance, bodies leaving no trace, and invisible bodies, respectively. We have the inclusions

$$\mathcal{B}^d_{invisible}(v_0) \subset \mathcal{B}^d_{notrace}(v_0) \subset \mathcal{B}^d_{zeroresist}(v_0).$$

The following theorem states that these classes are nonempty and the inclusions are proper.

Theorem 8.1. *There exist (a) a body that has zero resistance in the direction v_0 but leaves a trace in this direction, (b) a body that leaves no trace in the direction v_0 but is not invisible in this direction, and (c) a body invisible in the direction v_0. If $d \geq 3$, then the bodies in (a)–(c) are connected.*

This theorem will be proved in the next section.

Remark 8.1. One easily sees that if B is invisible/leaves no trace in the direction v_0, then the same is true in the direction $-v_0$. In contrast, there exist bodies that have zero resistance in the direction v_0 but not in the direction $-v_0$. Thus, we have

$$\mathcal{B}_{invisible}(v_0) = \mathcal{B}_{invisible}(-v_0), \quad \mathcal{B}_{notrace}(v_0) = \mathcal{B}_{notrace}(-v_0),$$

but

$$\mathcal{B}_{zeroresist}(v_0) \neq \mathcal{B}_{zeroresist}(-v_0).$$

8.1.2 Proof of Theorem 8.1

First we consider the two-dimensional case. Construct two identical equilateral triangles ABC and A′B′C′, with C being the midpoint of the segment A′B′ and C′ the midpoint of AB. The vertical line CC′ is parallel to v_0. Let A″ (B″) be the point of intersection of segments AC and A′C′ (resp. BC and B′C′); see Fig. 8.1. The body B is the union of triangles AA′A″ and BB′B″; in Fig. 8.1 it is shown shaded. As seen from this figure, it has zero resistance in the direction v_0.

Additionally, it leaves no trace in this direction but is not invisible.

By slightly modifying this construction one can obtain a body that has zero resistance but leaves a trace (Fig. 8.2). The body is the union of curvilinear quadrangles ABCD and A′B′C′D′ symmetric to each other with respect to an axis parallel to v_0 (in the figure, this axis is vertical). The sides BC and C′D′ are arcs

Fig. 8.1 A body of zero resistance: basic construction

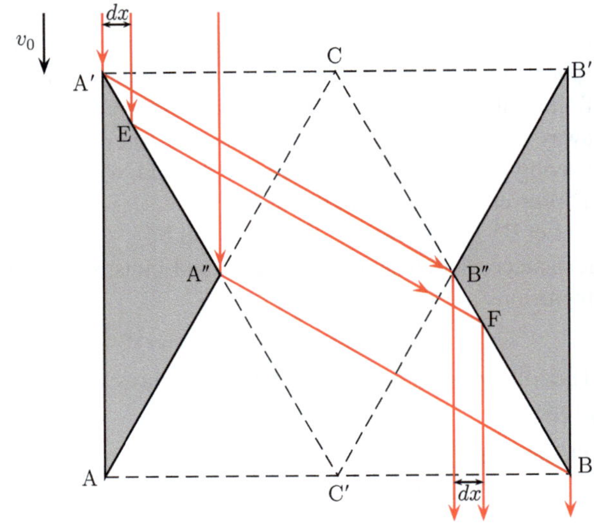

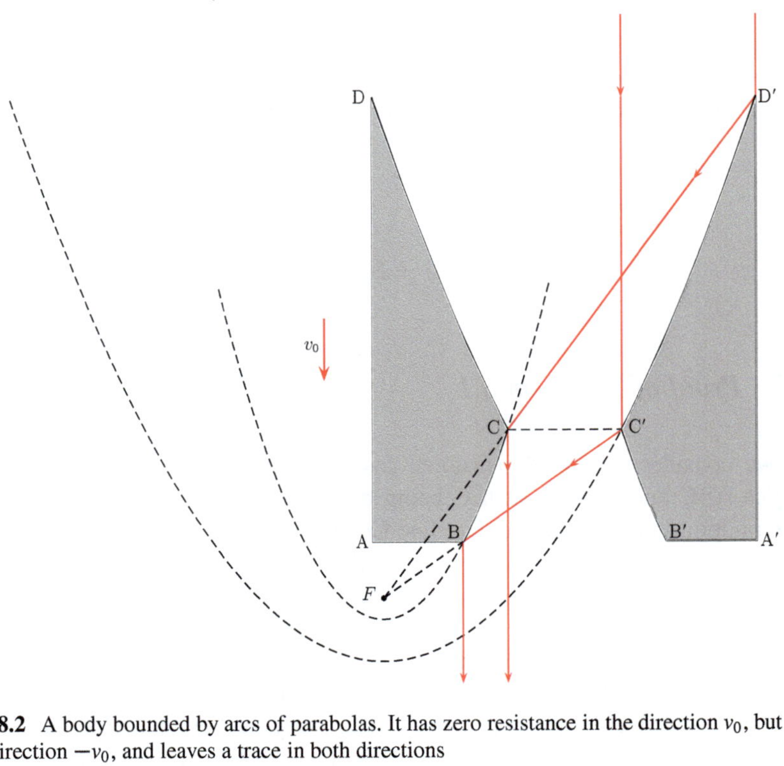

Fig. 8.2 A body bounded by arcs of parabolas. It has zero resistance in the direction v_0, but not in the direction $-v_0$, and leaves a trace in both directions

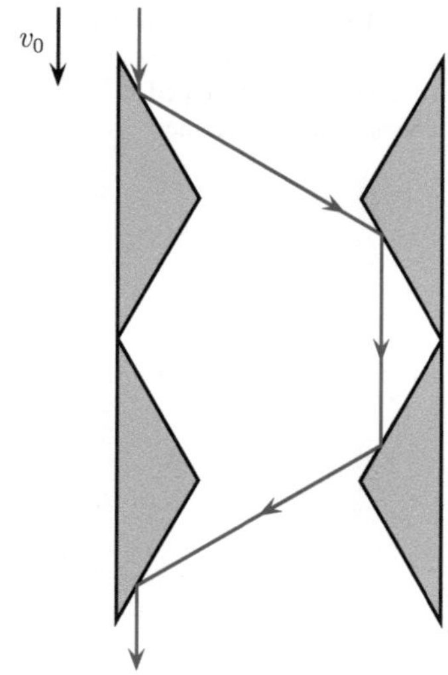

Fig. 8.3 A body invisible in the direction v_0

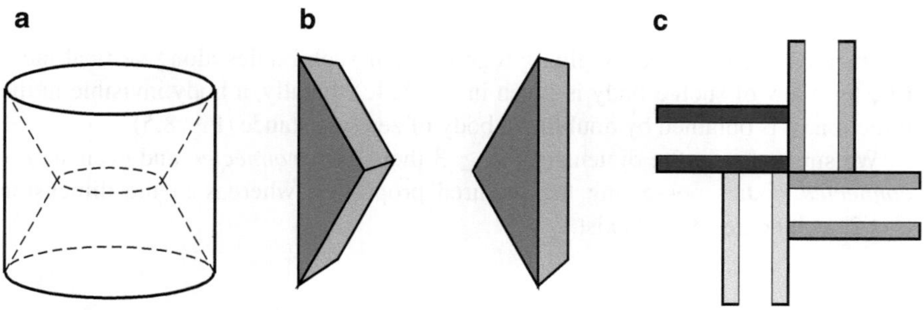

Fig. 8.4 (**a**) A body of zero resistance leaving a trace. (**b**) A disconnected body that is not invisible leaving no trace. (**c**) A simply connected body obtained by gluing together four bodies leaving no trace (top view)

of parabolas with vertical axis and common focus at F; the sides B′C′ and CD are obtained from them by a symmetry.

Finally, a body invisible in the direction v_0 is obtained by doubling the body in Fig. 8.1; see Fig. 8.3.

The corresponding bodies in higher dimensions $d \geq 3$ can be obtained from the two-dimensional construction. We restrict ourselves to three-dimensional bodies; bodies in higher dimensions are obtained by translation of three-dimensional ones in directions orthogonal to v_0.

The three-dimensional body obtained by rotation of the triangle AA′A″ (or BB′B″) in Fig. 8.1 about the straight line CC′ has zero resistance in the direction v_0 but leaves a trace (Fig. 8.4a). A body that leaves no trace but is not invisible is obtained by translation of a two-dimensional body in the direction orthogonal to the plane of Fig. 8.1 (Fig. 8.4b). A connected (and even simply connected) body leaving

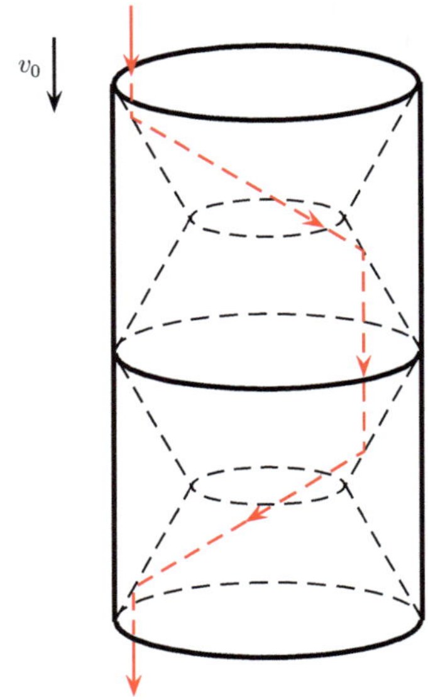

Fig. 8.5 A body invisible in the direction v_0, obtained by removing four truncated cones from the cylinder

no trace can be obtained by gluing together four such bodies along vertical faces (the top view of such a body is given in Fig. 8.4c). Finally, a body invisible in the direction v_0 is obtained by doubling a body of zero resistance (Fig. 8.5).

We stress that in the dimensions $d \geq 3$ there exist *connected* and even *simply connected* bodies possessing the required properties, whereas in the dimension $d = 2$ such bodies do not exist.

8.2 Other Constructions of Bodies of Zero Resistance

The following simple proposition allows one to produce invisible bodies by doubling bodies of zero resistance.

Proposition 8.3. *Assume that B has zero resistance in the direction v_0 and lies in the half-space $x \cdot v_0 \leq 0$. Let P_{v_0} be the symmetry relative to the hyperplane $\{v_0\}^\perp$, $P_{v_0}x = x - (x \cdot v_0)\, v_0$. Then the body $\tilde{B} = B \cup P_{v_0}B$ obtained by joining the original body and its image under the symmetry is invisible in the direction v_0.*

Proof. Indeed, almost every particle falling on a body with velocity v_0, after several reflections, will intersect $\{v_0\}^\perp$ with the same velocity v_0. Consider the part of its trajectory until the point of intersection and its image under the reflection from the hyperplane. The union of these lines is a trajectory of a billiard particle in $\mathbb{R}^d \setminus \tilde{B}$; moreover, the part of the trajectory outside Conv $\tilde{B}$ lies in a straight line with the directing vector v_0. The proof of the proposition is complete. □

Subsequently in this section we construct more examples of bodies of zero resistance, both in two- and in three-dimensional cases. By Proposition 8.3, each of these examples generates an invisible body.

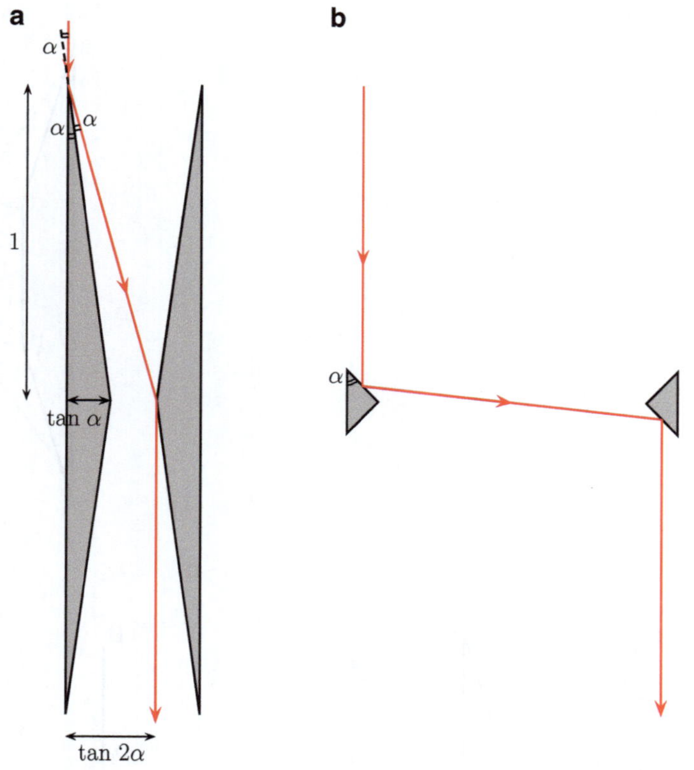

Fig. 8.6 The bodies B_α (**a**) with small α and (**b**) with α close to $\pi/4$ are displayed

Example 8.1. Consider a pair of isosceles triangles with angles α, α, and $\pi - 2\alpha$ ($0 < \alpha < \pi/4$). They are symmetric to each other relative to a point. This point lies on a symmetry axis of each triangle, at a distance $(\tan 2\alpha - \tan\alpha)/2$ from the obtuse angle [hence at a distance $(\tan 2\alpha + \tan\alpha)/2$ from its base]. The base of each triangle has a length of 2. The union of these triangles is a body of zero resistance B_α. Figure 8.6 displays two pairs of such triangles, with a small angle α and with an angle α close to $\pi/4$.

A three-dimensional body of zero resistance is obtained by rotating B_α about its vertical symmetry axis. Substituting $\alpha = \pi/6$ we obtain the main construction shown in Fig. 8.1.

Example 8.2. Consider the union of two isosceles trapezoids ABCD and A′B′C′D′ (Fig. 8.7). Later on we will define their parameters, ensuring that the body has zero resistance. Denote $r := |CC'|/|BB'|$ and $\alpha := \measuredangle ABC$ (and therefore $\alpha = \measuredangle BAD = \measuredangle A'B'C' = \measuredangle B'A'D'$); we assume that $\alpha < \pi/4$. Let us reflect the trapezoid BB′C′C relative to the side BC—the resulting trapezoid B_1BCC_1 is now reflected relative to the side B_1C_1—and continue the reflection process further (Fig. 8.8a). As a result of a series of these reflections, we obtain a sequence of

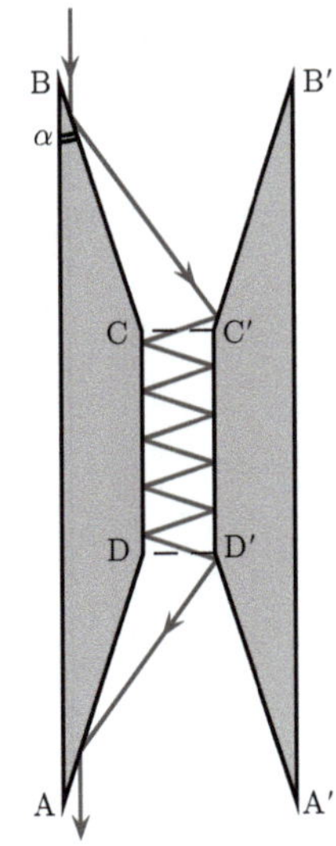

Fig. 8.7 Union of two trapezoids

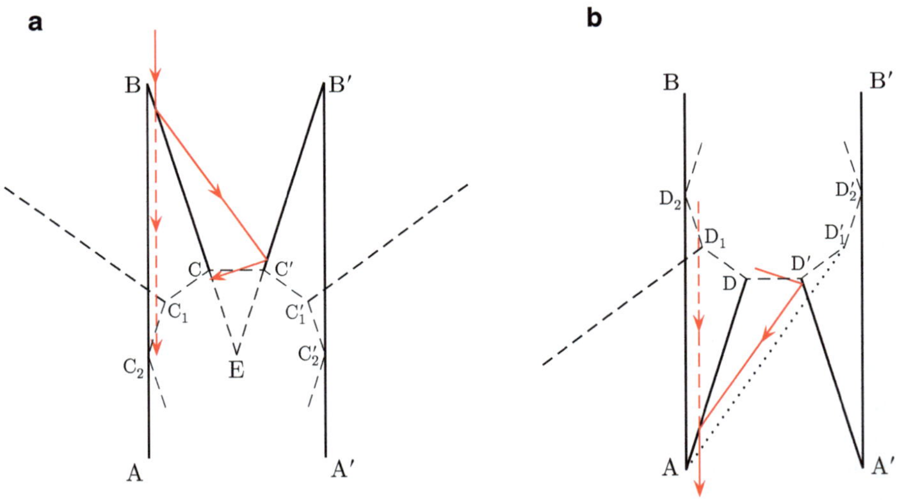

Fig. 8.8 Unfolding of a billiard trajectory

trapezoids B_1BCC_1, $B_2B_1C_1C_2, \ldots$, $B_kB_{k-1}C_{k-1}C_k, \ldots$. The process ends at the step $k_0 = \lfloor \pi/(2\alpha) + 1/2 \rfloor$; this is because at $k = k_0$ the segment $C_{k_0-1}C_{k_0}$ for the first time intersects the vertical symmetry axis of the figure. For each $\alpha \in (0, \pi/4)$ we choose a parameter $r = r(\alpha)$ such that the broken line $CC_1C_2 \ldots C_{k_0}$ touches the line AB, that is, intersects this line and is situated to the right of it (Fig. 8.8a). The corresponding value of r is $r(\alpha) = \sin\alpha/\sin(2\lfloor \pi/(4\alpha) \rfloor \alpha + \alpha)$. The function $r(\alpha)$ is continuous and monotonically increases from $r(0) = 0$ to $r(\pi/4) = 1$.

Proposition 8.4. *The union of the trapezoids* ABCD *and* A′B′C′D′ *is a two-dimensional body of zero resistance in the direction* $v_0 = (0, -1)$.

Proof. Consider a billiard particle with the initial velocity v_0 that has at least one reflection from the trapezoids. Without loss of generality we assume that it is first reflected from the side BC and apply the procedure of unfolding to the billiard trajectory. As a result, we obtain a vertical line situated between the line AB and the vertical symmetry axis of the figure; therefore, it inevitably intersects a segment $C_{k-1}C_k$ of the broken line $CC_1C_2 \ldots C_{k_0}$ (Fig. 8.8a). This means that the original particle after k successive reflections from the sides BC and B′C′ intersects the segment CC′ and enters the rectangle CC′D′D (Fig. 8.7).

After the first reflection the velocity of the particle forms the angle 2α with the vertical vector $v_0 = (0, -1)$ (we measure angles counterclockwise from the vertical); after the second reflection the angle becomes -4α, and so on. At the point of intersection with CC′ the angle is $(-1)^{k-1}2k\alpha$. Then the particle moves in the rectangle CC′D′D making successive reflections from the sides CD and C′D′. During that motion, the modulus of the velocity inclination angle remains equal to $2k\alpha$. When the particle leaves the rectangle and makes reflections from the sides AD and A′D′, the modulus of the inclination angle decreases, taking successively the values $2(k-1)\alpha$, $2(k-2)\alpha, \ldots$, and, finally, after the last reflection, the angle becomes $2k'\alpha$, where k' is a positive integer, $0 \le k' \le k$.

Let us show that $k' = 0$, and therefore the final velocity is vertical. To that end we apply the unfolding procedure again, this time to the part of the trajectory contained in the trapezium ADD′A′ (Fig. 8.8b). Assume without loss of generality that the point of last reflection lies in AD. Repeating the procedure as described above, one obtains a sequence of trapezoids ADD′A′, A_1D_1DA, $A_2D_2D_1A_1, \ldots$, $A_{k_0}D_{k_0}D_{k_0-1}A_{k_0-1}$. The unfolded trajectory is an interval with endpoints on the segment AA′ and on the broken line $D'DD_1D_2 \ldots$. By the choice of α and r, this broken line touches the line AB, whereas the indicated point on the broken line is situated between point D and the point of touch.

Let us extend the broken line symmetrically to the other side; as a result we obtain the line $\ldots D_2D_1DD'D_1'D_2' \ldots$ (Fig. 8.8b). Draw two tangent lines to this broken line from A (the lines AD_2 and AD_1' in Fig. 8.8b). We see that the angle of inclination of the segment DA is $-\alpha$, and of the tangent line D_2A is 0; by the symmetry of the broken line relative to D, we conclude that the angle of inclination of the tangent $D_1'A$ is -2α. Similarly, one concludes that the tangents D_1A' and $D_2'A'$ have inclination angles 2α and 0, respectively. This implies that the absolute values of the inclination angles of both tangents to the broken line drawn from each point on AA′ do not exceed 2α. Hence each line joining a point on AA′ with a point of a broken line (between points where this line touches the lines AB and A′B′) has an inclination angle of between -2α and 2α. Thus, we have $-2\alpha < 2k'\alpha < 2\alpha$, and since k' is an integer, we conclude that $k' = 0$. □

Remark 8.2. A body obtained by rotation of trapezoids about the vertical axis of symmetry is a three-dimensional body of zero resistance.

8.3 Properties of Bodies of Zero Resistance

Here we derive some properties of bodies of zero resistance and invisible bodies. For a body B denote by $\kappa(B)$ the relative volume of B in its convex hull, that is,

$$\kappa(B) = \frac{|B|}{|\mathrm{Conv}\, B|}.$$

The parameter κ may in a sense serve as a measure of "nonconvexity" of a body B; one obviously has $0 < \kappa(B) \le 1$ and $\kappa(B) = 1$ if and only if B is convex.

Let us calculate this value for the preceding examples of zero-resistance bodies. The convex hull of B_α in the three-dimensional case (Example 8.1) is a cylinder of radius $L_\alpha = (\tan 2\alpha + \tan\alpha)/2$ with height $H = 2$; thus, its relative height $h_\alpha = H/L_\alpha$ is $h_\alpha = 4/(\tan 2\alpha + \tan\alpha)$. One easily finds that $|B_\alpha| = \pi \tan\alpha(\tan 2\alpha + \tan\alpha/3)$ and $|\mathrm{Conv}\, B_\alpha| = \pi(\tan 2\alpha + \tan\alpha)^2$; hence

$$\kappa(B_\alpha) = \frac{2\tan\alpha(3\tan 2\alpha + \tan\alpha)}{3(\tan 2\alpha + \tan\alpha)^2}.$$

In the limit $\alpha \to 0$ we have $h_\alpha = \frac{4}{3\alpha}(1 + o(1)) \to \infty$ and $\kappa(B_\alpha) \to 14/27 \approx 0.52$. For the basic construction $\alpha = \pi/6$ (Fig. 8.1) we have $h_{\pi/6} = \sqrt{3}$ and $\kappa(B_\pi/6) = 5/12 \approx 0.42$. Taking $\alpha = (\pi - \varepsilon)/4$, $\varepsilon \to 0^+$, we obtain $h_{(\pi-\varepsilon)/4} = 2\varepsilon(1 + o(1))$ and $\kappa(B_{(\pi-\varepsilon)/4}) = \varepsilon(1 + o(1))$.

Now consider a three-dimensional body obtained by rotation of trapezoids (Example 8.2, Fig. 8.7). It can be described as a cylinder with a hole made inside it. The shape of this body $B_{\alpha\gamma}$ is uniquely defined by the parameters α and $\gamma = |\mathrm{CD}|/|\mathrm{BC}|$. As $\alpha \to 0$ and $\gamma \to \infty$, the relative height of the cylinder and the largest number of reflections tend to infinity and $\kappa(B_{\alpha\gamma})$ tends to 1.

By doubling this body we obtain a body invisible in the vertical direction inscribed in the doubled cylinder.

Thus we come to the following proposition.

Proposition 8.5. *We have*

(a) In the two-dimensional case $d = 2$, $\sup\{\kappa(B) : B$ *is invisible in one direction*$\} = 1$.

(b) In the case $d \ge 3$, $\sup\{\kappa(B) : B$ *is connected and invisible in one direction*$\} = 1$.

In other words, an invisible body can be obtained from a convex body by removing an arbitrarily small part of its volume.

In the three- and two-dimensional cases the proposition is proved by taking the body constructed previously and its central vertical cross section, respectively, and in higher dimensions the required family of bodies is produced by parallel translation of $B_{\alpha\gamma}$ in directions orthogonal to the subspace of the construction.

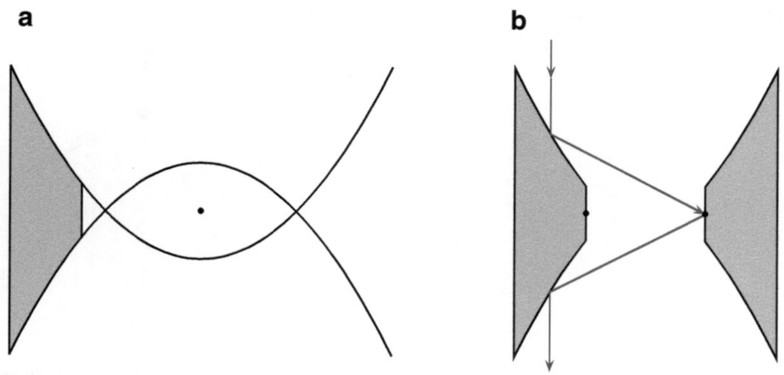

Fig. 8.9 An invisible body with three reflections

Let $\overline{m} = \overline{m}(B, v_0)$ and $\underline{m} = \underline{m}(B, v_0)$ be, respectively, the largest and smallest number of reflections a particle with initial velocity v_0 can make from B. From the basic construction and related constructions we see that the largest and smallest number of reflections for the constructed bodies of zero resistance and bodies leaving no trace are $\overline{m} = \underline{m} = 2$, and for invisible bodies they are $\overline{m} = \underline{m} = 4$. Below we present a construction of an invisible body with $\overline{m} = \underline{m} = 3$ proposed by Zilinskas.

Consider a curvilinear trapezoid formed by two arcs of parabolas and two parallel segments. The parabolas have a common focus and are centrally symmetric to each other relative to it. The common axis of the parabolas is parallel to the segments (Fig. 8.9a).

Now reflect the trapezoid with respect to the line parallel to the axis of the parabolas and equidistant from it and from the trapezoid. As a result we obtain two trapezoids, where the focus related to the first trapezoid lies on the rectilinear side of the second trapezoid and vice versa (Fig. 8.9b). A particle impinging in the direction of the axis of the parabolas first reflects from a curvilinear side of a trapezoid (this side is a parabolic arc), then from the corresponding focus that belongs to a side of the other trapezoid, and finally from the other curvilinear side of the original trapezoid. Then the particle moves uniformly in the direction of the axis, and, moreover, the parts of its trajectory before and after the reflections lie on the same straight line.

A three-dimensional invisible body with three reflections is obtained by rotating the body in Fig. 8.9b about the symmetry axis separating the two trapezoids.

One cannot make do with a smaller number of reflections in all three cases; see the following proposition.

Proposition 8.6. *(a) If B has zero resistance or leaves no trace in the direction v_0, then $\overline{m}(B, v_0) \geq \underline{m}(B, v_0) \geq 2$.*

(b) If B is invisible in the direction v_0, then $\overline{m}(B, v_0) \geq \underline{m}(B, v_0) \geq 3$.

Fig. 8.10 Two reflections are not enough for an invisible body

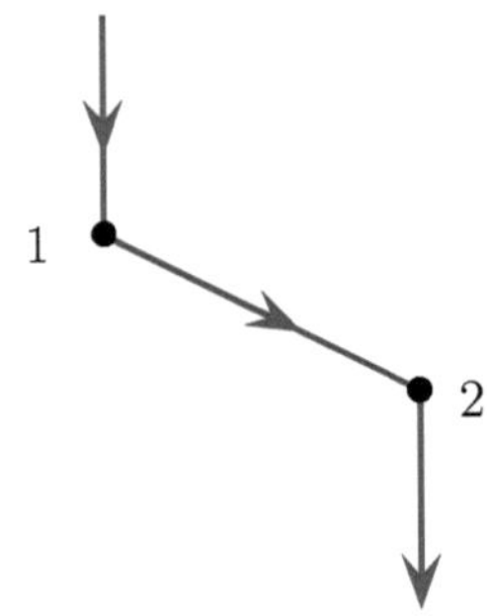

These inequalities are exact: there exists a body leaving no trace (and therefore having zero resistance) such that the number of reflections of each particle is exactly 2, and there exists an invisible body such that the number of reflections of each particle is exactly three.

Proof. Taking into account the preceding examples, it is sufficient to prove that the number of reflections of each particle is at least (a) two for bodies of zero resistance and (b) three for invisible bodies.

(a) If a particle makes a single reflection, then its final velocity does not coincide with the initial one, that is, $v_B^+(\xi, v_0) \neq v_0$ for some $\xi = \xi_0$, and therefore also for close values of ξ. Hence the resistance of B is nonzero.
(b) Note that if there are two reflections, then the initial and final parts of the trajectory do not lie on one straight line (Fig. 8.10).

Therefore, the number of reflections from an invisible body is at least three. □

Recall Definition 2.3 from Sect. 2.2. Consider a connected body $\Omega \subset \mathbb{R}^2$; then $\mathcal{S}(\Omega, h)$ denotes the class of connected bodies that lie in the cylinder $\Omega \times [-h, 0]$ and contain a section $\Omega \times \{c\}$, $-h \leq c \leq 0$ of this cylinder. Each body from this class is called a *body inscribed in a cylinder.*

Combining the results on zero-resistance bodies in this chapter and Theorem 2.1 from Chap. 2 we obtain the following statement in the spirit of Newton's aerodynamic problem.

Theorem 8.2. *For any $h > 0$ and for $v_0 = (0, 0, -1)$,*

$$\inf_{B \in \mathcal{S}(\Omega, h)} |R_{\delta_{v_0}}(B)| = 0 \tag{8.1}$$

holds. If Ω is convex, then the infimum is not attained. On the other hand, for some nonconvex (and even for some simply connected) bodies Ω and some values of h the infimum is attained, that is, there exists a body of zero resistance inscribed in the corresponding cylinder.

Theorem 8.2 means that one can select *"almost" perfectly streamlined* bodies from the class of bodies inscribed in a given cylinder. In addition, for some nonconvex (and never for convex) cylinders there exist *perfectly streamlined* bodies inscribed in them.

Formula (8.1) is proved in Sect. 2.4. The existence of bodies of zero resistance inscribed in a cylinder is demonstrated above in this chapter in the cases where Ω is a ring or a simply connected polygon of special shape with mutually orthogonal and parallel sides (Figs. 8.1c and 8.7). Thus, to finish the proof of Theorem 8.2 it suffices to prove the following proposition that claims the nonexistence of bodies of zero resistance inscribed in a convex cylinder.

Proposition 8.7. *Let Ω be a convex body and let B be a body inscribed in the cylinder $\Omega \times [-h, 0]$ such that the scattering on B in the direction $v_0 = (0, 0, -1)$ is regular. Then $R_{\delta_{v_0}}(B) \neq 0$.*

Proof. Since the scattering is regular, the function $v_B^+(\cdot, v_0)$ is defined for almost all $\xi \in \Omega$ and is measurable. B contains a horizontal section $\Omega \times \{c\}$, $-h \leq c \leq 0$ of the cylinder. If $c = 0$, then all particles are reflected from the upper end of the cylinder, and therefore $R_{\delta_{v_0}}(B) \neq 0$. Now assume that $c < 0$ and consider the smaller cylinder $\Omega \times [c, 0]$ bounded below by this section. The particle trajectory intersects this smaller cylinder but cannot intersect its lower end $\Omega \times \{c\}$ in the downward direction. Hence it leaves the cylinder through either its lateral surface $\partial\Omega \times (c, 0)$ or its upper end $\Omega \times \{0\}$ and never returns to the cylinder afterward since it is convex. In both cases, the final velocity of the particle is not v_0. This implies that $R_{\delta_{v_0}}(B) \neq 0$. □

8.4 Invisibility in Several Directions

In this section we construct three-dimensional bodies invisible in two mutually orthogonal directions and prove that there do not exist bodies invisible in all possible directions. These claims are also true for bodies of zero resistance.

8.4.1 Bodies Invisible in Two Directions

The following theorem and the underlying construction are due to Roshchina.

Theorem 8.3. *For any two orthogonal vectors v_1 and $v_2 \in S^2$*

(a) There exists a body that has zero resistance in these directions.
(b) There exists a body invisible in these directions.

Proof. Take a plane Π containing v_1 and orthogonal to v_2, and consider two parabolas in this plane that have common focus and a common axis parallel to v_1 and are centrally symmetric to each other with respect to the focus. Take two straight lines in this plane parallel to the axis and situated at the same distance on both

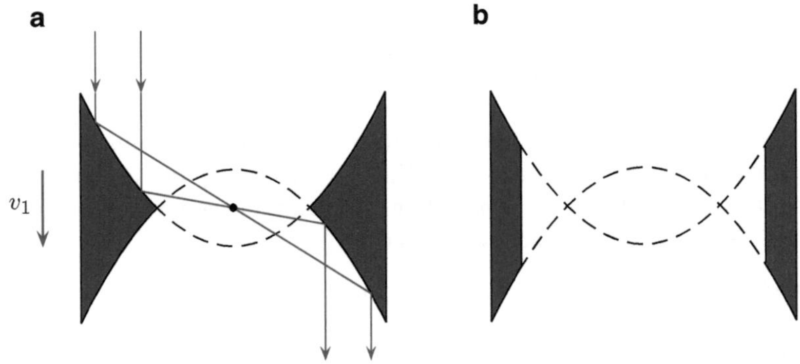

Fig. 8.11 Two-dimensional bodies that have zero resistance in one direction: (**a**) a union of two curvilinear triangles; (**b**) a union of two curvilinear trapezoids

sides of it. Next, consider two curvilinear triangles formed by segments of these straight lines and by arcs of the parabolas (Fig. 8.11a). The union of these triangles is a (disconnected) two-dimensional body having zero resistance in the direction v_1. Indeed, taking into account the focal property of a parabola we see that each incident particle in the direction v_1, after reflecting from a parabola, passes through the common focus, then reflects from the other parabola, and moves afterward freely with velocity v_1. That is, a parallel flow with velocity v_1 is transformed into a parallel flow with the same velocity.

Note also that the union of two trapezoids bounded by arcs of the parabolas and by two pairs of line segments (where two segments in each pair are parallel to the axis and symmetric to each other with respect to it) is also a body of zero resistance in the direction v_1 (Fig. 8.11b).

Then we obtain a three-dimensional body B_1 having zero resistance in the same direction v_1 by parallel translation of the two-dimensional body of Fig. 8.11a in the direction v_2 orthogonal to the plane of the body (Fig. 8.12a). The length h of this translation is equal to the height of the body (that is, to the length of the rectilinear side of a triangle). Then we rotate B_1 through the angle $\pi/2$ about its symmetry axis perpendicular to v_1 and v_2. The resulting body B_2 has zero resistance in the direction v_2 (Fig. 8.12b). Finally, we show that the body $B = B_1 \cap B_2$ has zero resistance in both directions v_1 and v_2 (Fig. 8.12c).

Indeed, the intersection of B with any plane parallel to Π is a union of two curvilinear trapezoids (which is a two-dimensional body of zero resistance); in addition, the outward normal to ∂B at each point of a curvilinear side of the trapezoids is parallel to Π. Therefore, each incident particle that initially moves in this plane with velocity v_1, after two reflections from curvilinear sides of the trapezoids, will eventually move in the same plane with the same velocity v_1. Therefore, B has zero resistance in the direction v_1. For v_2 the argument is the same.

To obtain a body invisible in the directions v_1 and v_2, it suffices to take a union of four identical bodies obtained from B by shifts by 0, hv_1, hv_2, and $hv_1 + hv_2$ (Fig. 8.13). □

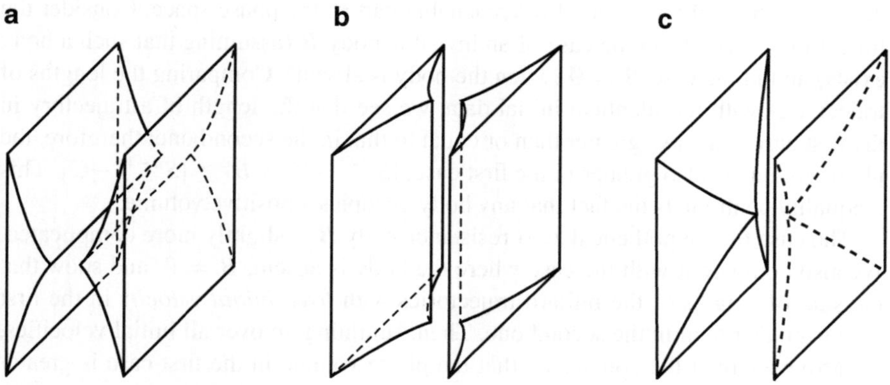

Fig. 8.12 Construction of a body of zero resistance in two directions

Fig. 8.13 A body invisible in two directions

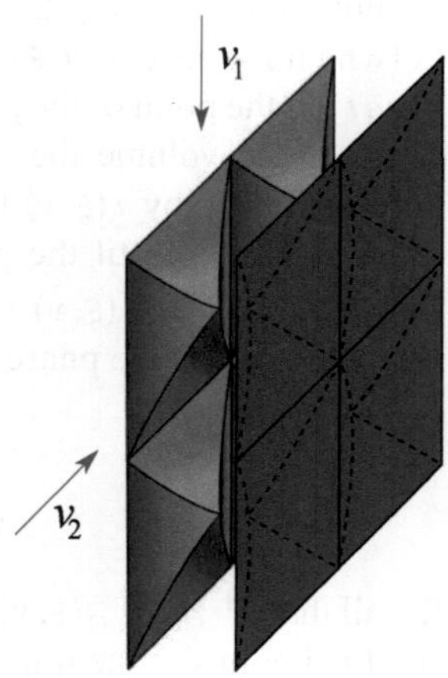

8.4.2 *Nonexistence of Bodies Invisible in All Directions*

Theorem 8.4. *There do not exist bodies that*

(a) Are invisible in all directions or
(b) Have zero resistance in all directions.

Proof. Let us first outline the idea of the proof. Note that statement (b) of the theorem implies statement (a), but for methodological reasons we first prove (a) and then (b).

The phase space of the billiard in $C \setminus B$ is $(C \setminus B) \times S^{d-1}$, with the coordinate (x, v) and the element of Liouville phase volume $dx\,dv$. The total volume of the phase space equals $|S^{d-1}| \cdot |C \setminus B|$.

The phase volume can be estimated in a different way. Summing up the lengths of all billiard trajectories (of course, summation amounts to integration over the initial data), we obtain the volume of the reachable part of the phase space. Consider the trajectories inside C in the case of an invisible body B (assuming that such a body exists) and in the case $B = \emptyset$ (when the body is absent). Comparing the lengths of trajectories with the identical initial data, we see that the length of a trajectory in the first case is always greater than or equal to that in the second one; therefore, the phase volume is also greater in the first case, $|S^{d-1}| \cdot |C \setminus B| \geq |S^{d-1}| \cdot |C|$. This inequality contradicts the fact that any body occupies a positive volume.

The case of a hypothetical zero-resistance body B is slightly more complicated. We also compare it with the case where the body is absent, $B = \emptyset$, and show that the sum of lengths of the billiard trajectories with *fixed initial velocity* in the first case is greater than in the second one. Then, summing up over all initial velocities, we arrive again at the conclusion that the phase volume in the first case is greater than or equal to that in the second one, which is a contradiction.

Let us pass to a more precise exposition. Suppose that a billiard particle starts its motion at a point $\xi \in \partial C$ and with the initial velocity $v \in S^{d-1}$ turned inside C [which means that $n(\xi) \cdot v \leq 0$], and let $t \geq 0$; then assign the new coordinate (ξ, v, t) to the point of the phase space reached by the particle in time t. The element of the phase volume then takes the form $(-n(\xi) \cdot v)\, d\xi\, dv\, dt = d\mu_-(\xi, v)\, dt$. Further, denote by $\tau(\xi, v)$ the length of the particle's trajectory inside C, from the starting point ξ until the point $\xi^+ = \xi^+_{B,C}(\xi, v)$, where it leaves C. Recall that $(\partial C \times S^{d-1})_\pm = \{(\xi, v) \in \partial C \times S^{d-1} : \pm n(\xi) \cdot v \geq 0\}$. Then the volume of the reachable part of the phase space equals

$$\int_{(\partial C \times S^{d-1})_-} \tau(\xi, v)\, d\mu_-(\xi, v). \tag{8.2}$$

Recall that $\xi^+ = \xi^+_{B,C}(\xi, v)$. Taking into account that the distance between the initial and final points of the trajectory does not exceed its length,

$$|\xi^+ - \xi| \leq \tau(\xi, v), \tag{8.3}$$

and at some point (ξ, v) (and therefore in its neighborhood) the inequality in (8.3) is strict, and we obtain

$$\int_{(\partial C \times S^{d-1})_-} |\xi^+ - \xi|\, d\mu_-(\xi, v) < |S^{d-1}||C \setminus B|. \tag{8.4}$$

Now let $\xi_0^+ = \xi_0^+(\xi, v)$ be the point where the particle leaves C in the case $B = \emptyset$. In other words, ξ_0^+ is the point of intersection of the ray $\xi + vt$, $t > 0$ with ∂C. In this case the entire phase space is reachable; in addition, one has equality in (8.3), and therefore in place of (8.4) one has the equality

$$\int_{(\partial C \times S^{d-1})_-} |\xi_0^+ - \xi|\, d\mu_-(\xi, v) = |S^{d-1}||C|. \tag{8.5}$$

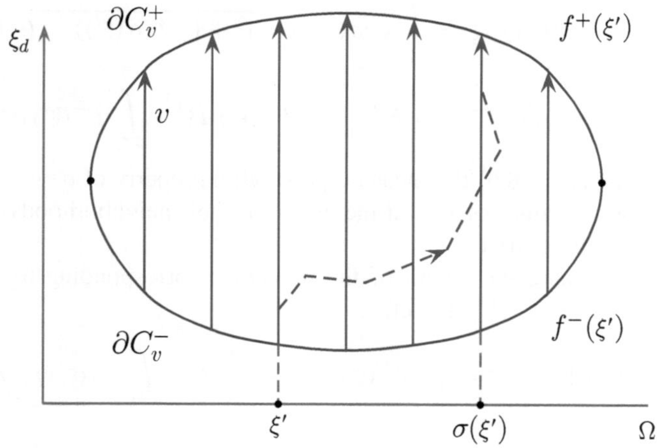

Fig. 8.14 Restriction of phase space to subspace $v = \text{const}$

If B is invisible in all directions, then $\xi_0^+ = \xi^+$; therefore, from (8.4) and (8.5) one obtains

$$|C| < |C \setminus B|,$$

which is a contradiction.

Remark 8.3. Formulas (8.2) and (8.5) are known in integral geometry, geometric probability, and billiards [36, 66, 76]; see also a brief review in [14].

Now let B have zero resistance in all directions, that is, $v_{B,C}^+(\xi, v) = v$ for all ξ and v. Denote by $\partial C_v^\pm$ the set of points ξ such that $\pm n(\xi) \cdot v \geq 0$ with the induced measure $\pm n(\xi) \cdot v\, d\xi$. Since the mapping $(\xi, v) \mapsto (\xi_{B,C}^+(\xi, v), v)$ from $((\partial C \times S^{d-1})_-, \mu_-)$ to $((\partial C \times S^{d-1})_+, \mu_+)$ preserves the measure, we conclude that the induced mapping $\xi \mapsto \xi_{B,C}^+(\xi, v)$ from ∂C_v^- to ∂C_v^+ preserves the induced measure for almost every v. Fix v and introduce an orthogonal coordinate system $\xi_1, \ldots, \xi_d$ in $\mathbb{R}^d$ such that $v = (0, \ldots, 0, 1)$. Denote $\xi' = (\xi_1, \ldots, \xi_{d-1})$; then the subsets $\partial C_v^\pm$ take the form

$$\partial C_v^\pm = \{(\xi', \xi_d) : \xi' \in \Omega,\ \xi_d = f^\pm(\xi')\},$$

where Ω is a convex domain in $\mathbb{R}^{d-1}$, f^- is a convex function on Ω, and f^+ is a concave function on Ω; in addition, $f^- \leq f^+$. Then both measures $\pm n(\xi) \cdot v\, d\xi$ on $\partial C_v^\pm$ take the form $d\xi_1 \ldots d\xi_{d-1}$ and the mapping $\xi \mapsto \xi_{B,C}^+(\xi, v)$ takes the form $(\xi', f^-(\xi')) \mapsto (\sigma(\xi'), f^+(\sigma(\xi')))$, where σ is a transformation of Ω preserving the Lebesgue measure $d\xi' = d\xi_1 \ldots d\xi_{d-1}$ (Fig. 8.14).

The length $\tau(\xi, v)$ of the billiard trajectory starting at $\xi = (\xi', f^-(\xi'))$ does not exceed the distance between the initial and final points of the trajectory $(\xi', f^-(\xi'))$ and $(\sigma(\xi'), f^+(\sigma(\xi'))$; therefore, we obtain the estimate

$$\int_{\partial C_v^-} \tau(\xi, v)\,(-n(\xi)\cdot v)\,d\xi \geq \int_{\Omega} \sqrt{|\sigma(\xi') - \xi'|^2 + (f^+(\sigma(\xi')) - f^-(\xi'))^2}\,d\xi'$$

$$\geq \int_{\Omega} (f^+(\sigma(\xi')) - f^-(\xi'))\,d\xi' = \int_{\Omega} f^+(\xi')\,d\xi' - \int_{\Omega} f^-(\xi')\,d\xi'. \tag{8.6}$$

In the last equality in (8.6) the measure-preserving property of σ was used. Note also that for some values of v (and therefore for their neighborhoods) one of the inequalities in (8.6) is strict.

On the other hand, the length of the trajectory corresponding to $B = \emptyset$ is $\tau_0(\xi, v) = f^+(\xi') - f^-(\xi')$; therefore,

$$\int_{\partial C_v^-} \tau_0(\xi, v)\,(-n(\xi)\cdot v)\,d\xi = \int_{\Omega} (f^+(\xi') - f^-(\xi'))\,d\xi' \leq \int_{\partial C_v^-} \tau(\xi, v)\,(-n(\xi)\cdot v) d\xi. \tag{8.7}$$

Here, again, for an open set of values of v the inequality is strict. Integrating both parts in (8.7) over v, we obtain the phase volume $|S^{d-1}|\cdot|C|$ on the left-hand side and the reachable phase volume (which is less or equal than $|S^{d-1}|\cdot|C\setminus B|$) on the right-hand side. Thus, we obtain

$$|S^{d-1}||C| < |S^{d-1}||C \setminus B|,$$

which is a contradiction. □

Remark 8.4. Literally repeating this proof for piecewise smooth surfaces (which, in contrast to bodies, have zero volume), we come to an analogous conclusion: there are no surfaces invisible (or having zero resistance) in all directions.

8.5 Bodies Invisible from One Point

A body invisible in one direction (and with specular surface) is indeed invisible when observed from a sufficiently large distance (as compared with the body size) in that direction. In this case, light rays passing through the observation point and hitting the body can be considered approximately parallel. In practice, however, the distance to the body is too small to ensure its invisibility. So it is natural to give the following definition of a body invisible from a point.

Definition 8.4. We say that B is *invisible from a point* $O \notin B$ if each billiard trajectory emanating from O, except for the part of the trajectory situated between the first and last points of reflection, belongs to a half-line with vertex at O.

Theorem 8.5. *For each d there exists a body $B \subset \mathbb{R}^d$ invisible from a point. If $d \geq 3$, then the body is connected.*

The proof of the theorem is based on a direct construction. In the proof we use the following lemma.

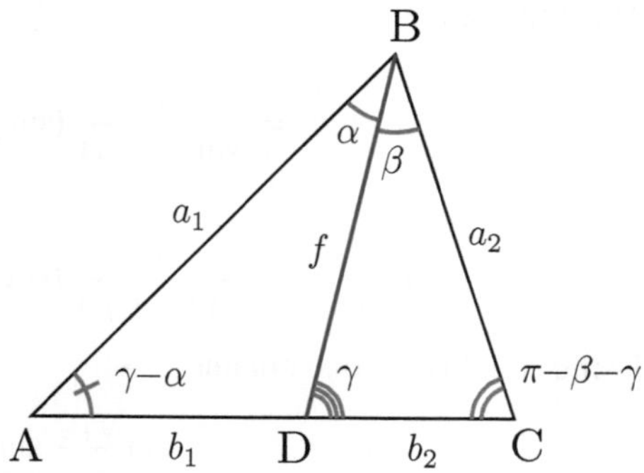

Fig. 8.15 Characteristic property of a bisector in a *triangle*

Consider a triangle ABC and a point D lying on the side AC. Let AB $= a_1$, BC $= a_2$, AD $= b_1$, DC $= b_2$, and BD $= f$ (Fig. 8.15).

Lemma 8.1 (A Characteristic Property of a Bisector in a Triangle). *The segment* BD *is the bisector of the angle* $\measuredangle$ABC *if and only if*

$$(a_1 + b_1)(a_2 - b_2) = f^2.$$

Proof. Consider the following relations on the values a_1, a_2, b_1, b_2, and f:

(a) $a_1/a_2 = b_1/b_2$;
(b) $a_1a_2 - b_1b_2 = f^2$;
(c) $(a_1 + b_1)(a_2 - b_2) = f^2$.

Equalities (a) and (b) are well known; each is a characteristic property of a triangle bisector. It is interesting to note that each of these algebraic relations is a direct consequence of the two others.

Assume that BD is the bisector of the angle $\measuredangle$ABC. Then equalities (a) and (b) are true; therefore, (c) is also true. The direct statement of the lemma is thus proved.

To derive the inverse statement, we need to apply the sine rule and some trigonometry. Denote $\alpha = \measuredangle$ABD, $\beta = \measuredangle$CBD, and $\gamma = \measuredangle$BDC (Fig. 8.15). Applying the sine rule to $\triangle$ABD we have

$$\frac{a_1}{\sin\gamma} = \frac{b_1}{\sin\alpha} = \frac{f}{\sin(\gamma - \alpha)},$$

and applying the sine rule to $\triangle$BDC we have

$$\frac{a_2}{\sin\gamma} = \frac{b_2}{\sin\beta} = \frac{f}{\sin(\gamma + \beta)}.$$

This implies that

$$a_1 + b_1 = \frac{f}{\sin(\gamma - \alpha)}\,(\sin\gamma + \sin\alpha) = f\,\frac{\sin\frac{\gamma+\alpha}{2}}{\sin\frac{\gamma-\alpha}{2}},$$

$$a_2 - b_2 = \frac{f}{\sin(\gamma + \beta)}\,(\sin\gamma - \sin\beta) = f\,\frac{\sin\frac{\gamma-\beta}{2}}{\sin\frac{\gamma+\beta}{2}}.$$

Using equality (c) one obtains

$$f^2\,\frac{\sin\frac{\gamma+\alpha}{2}\,\sin\frac{\gamma-\beta}{2}}{\sin\frac{\gamma-\alpha}{2}\,\sin\frac{\gamma+\beta}{2}} = f^2,$$

whence

$$\sin\frac{\gamma+\alpha}{2}\,\sin\frac{\gamma-\beta}{2} = \sin\frac{\gamma-\alpha}{2}\,\sin\frac{\gamma+\beta}{2}.$$

After some algebra, as a result we have

$$\cos\left(\gamma + \frac{\alpha-\beta}{2}\right) = \cos\left(\gamma - \frac{\alpha-\beta}{2}\right).$$

The last equation and the conditions $0 < \alpha,\ \beta,\ \gamma < \pi$ imply that $\alpha = \beta$. The inverse statement of the lemma is also proved. □

Proof. Draw confocal ellipse and hyperbola on a plane such that a segment joining a point of their intersection with a focus is orthogonal to the segment joining the foci. In Fig. 8.16 the ellipse is indicated by $\mathcal{E}$, the right branch of the hyperbola by $\mathcal{H}$ (the other branch is not considered), and the foci by F_1 and F_2. The points of intersection of $\mathcal{E}$ with $\mathcal{H}$ are H and H', and the segments HF_1 and $H'F_1$ are orthogonal to F_1F_2. In a convenient coordinate system, the ellipse is given by the equation

$$\frac{x^2}{a^2} + \frac{y^2}{b^2} = 1$$

and the hyperbola by the equation

$$\frac{x^2}{\alpha^2} - \frac{y^2}{\beta^2} = 1,$$

and the equation

$$a^2 - b^2 = \alpha^2 + \beta^2 \tag{8.8}$$

means that they are confocal. Let $c = \sqrt{a^2 - b^2}$ be one half of the focal distance; then the condition that HF_1 and F_1F_2 are orthogonal takes the form

$$\frac{1}{\beta^2} - \frac{1}{b^2} = \frac{1}{c^2}. \tag{8.9}$$

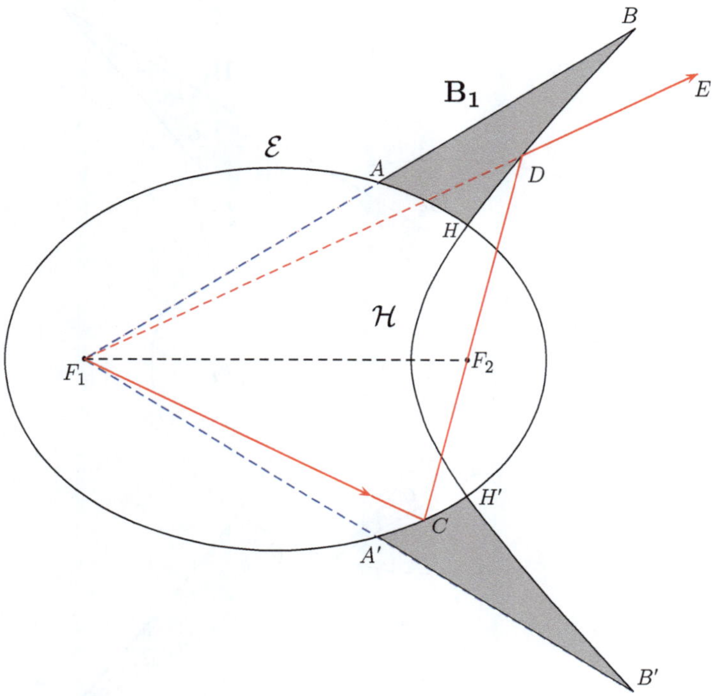

Fig. 8.16 A body having zero resistance to a flow of particles emanating from a point

Draw a ray with vertex at F_1 that is situated above H and intersects $\mathcal{H}$. Let A and B be the points of its intersection with $\mathcal{E}$ and $\mathcal{H}$. Reflect the ray with respect to the straight line F_1F_2 and denote the points of intersection of the resulting ray with $\mathcal{E}$ and $\mathcal{H}$ by A' and B'; they are symmetric to A and B with respect to F_1F_2. Thus we have

$$\measuredangle AF_1F_2 = \measuredangle A'F_1F_2. \tag{8.10}$$

Denote by $\mathbf{B_1}$ the union of the triangles ABH and $A'B'H'$. The body $\mathbf{B_1}$ is shown in gray in Fig. 8.16. Consider a particle emanating from F_1 and making a reflection from $\mathbf{B_1}$. The first reflection is from one of the arcs AH and $A'H'$. Suppose without loss of generality that the point of first reflection lies in $C \in A'H'$. We have

$$\measuredangle CF_1F_2 < \measuredangle A'F_1F_2. \tag{8.11}$$

After the reflection, the particle passes through the focus F_2 and then intersects $\mathcal{H}$ at a point D.

By the focal property of an ellipse, we have

$$|F_1C| + |F_2C| = |F_1H| + |F_2H|, \tag{8.12}$$

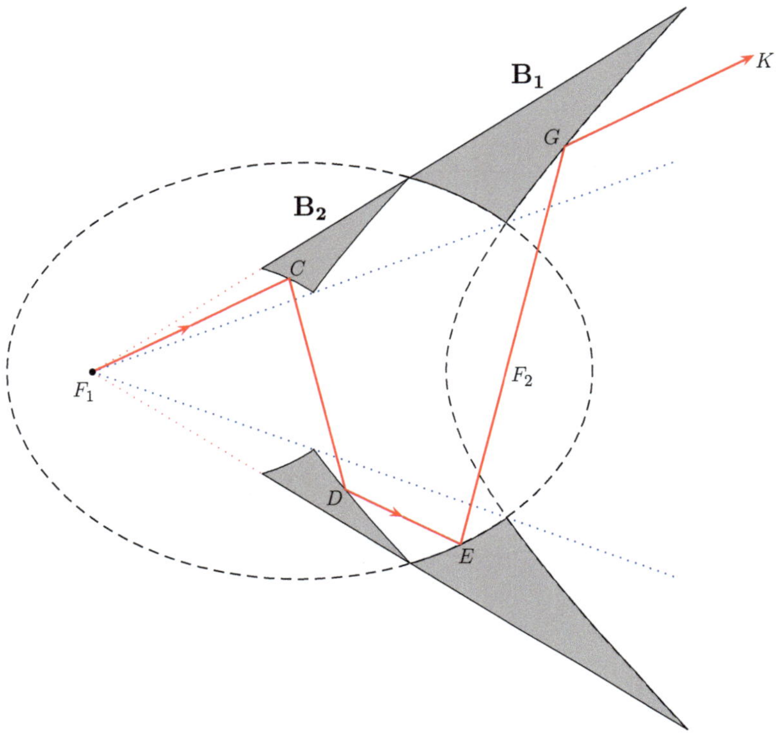

Fig. 8.17 A body invisible from one point

and by the focal property of a hyperbola,

$$|F_1D| - |F_2D| = |F_1H| - |F_2H|. \tag{8.13}$$

Multiplying both parts of (8.12) and (8.13) and bearing in mind that F_1F_2 is orthogonal to F_2H, we obtain

$$(|F_1C| + |F_2C|)(|F_1D| - |F_2D|) = |F_1H|^2 - |F_2H|^2 = |F_1F_2|^2. \tag{8.14}$$

Applying Lemma 8.1 to the triangle CF_1D and using (8.14) we conclude that F_1F_2 is a bisector of this triangle, that is,

$$\measuredangle CF_1F_2 = \measuredangle DF_1F_2. \tag{8.15}$$

Using (8.10), (8.11), and (8.14), we obtain that $\measuredangle DF_1F_2 < \measuredangle AF_1F_2$; therefore, D lies on the arc HB.

After reflecting at D the particle moves along the line DE containing F_1.

Now consider the body $\mathbf{B_2}$ obtained from $\mathbf{B_1}$ by dilation with center at F_1 and such that $\mathbf{B_1}$ and $\mathbf{B_2}$ have exactly two points in common (in Fig. 8.17 the dilation coefficient is smaller than 1). A particle emanating from F_1 and reflected from $\mathbf{B_2}$ at C and D further moves along the line DE containing F_1; in addition, the equality (8.15) takes place.

Then the particle makes two reflections from $\mathbf{B_1}$ at E and G and moves freely afterwards along a line containing F_1, besides the equality

$$\measuredangle EF_1F_2 = \measuredangle GF_1F_2. \tag{8.16}$$

takes place. Using (8.15) and (8.16), as well as the (trivial) equality $\measuredangle DF_1F_2 = \measuredangle EF_1F_2$, we find that

$$\measuredangle CF_1F_2 = \measuredangle GF_1F_2.$$

This means that the initial segment F_1C of the trajectory and its final ray GK lie in the same ray F_1K. The rest of the trajectory, the broken line CDEG, belongs to the convex hull of the set $\mathbf{B_1} \cup \mathbf{B_2}$. Thus we have proved that $\mathbf{B_1} \cup \mathbf{B_2}$ is a two-dimensional body invisible from point F_1.

In the case of a higher dimension d, the (connected) body invisible from F_1 is obtained by rotation of $\mathbf{B_1} \cup \mathbf{B_2}$ about the axis F_1F_2. □

Remark 8.5. From the proof of the theorem we see that the invisible body is determined by four parameters: a, b, α, β, with two conditions imposed by (8.8) and (8.9). Thus, the construction is defined by two parameters. One of them is the scale, and the second one can be taken to be $|F_2H|/|F_1F_2|$, the angular degree of the cone that touches the outer surface of the invisible body.

8.6 Possible Applications of Invisible Bodies and Open Questions

We believe that the models proposed in this chapter can find applications in optics and in the aerodynamics of space flights.

A body of zero resistance with a specular surface can be used, for instance, as a constituent element of a structure (curtain) that lets light through only in a fixed direction. By slightly modifying the construction, a surface can be designed that, like a lens, focuses sunlight onto one point. Bodies with a mirror surface invisible in one direction may also be of interest.

Above 150 km, the atmosphere is so rarefied that the effect of intermolecular collisions can be neglected when calculating the resistance of a body of reasonable size (several meters) [26]. As regards the body (union of two prisms) in Fig. 8.4a, the flow density in some zones between the prisms duplicates and triplicates as compared with the density outside the body, that is, still remains small. The bodies depicted in Figs. 8.4a and 8.7 create an infinite density along their symmetry axis; however, this effect may be of little importance in practice because of the thermal motion of flow particles and not completely specular reflection from the body surface.

Our model is robust with respect to small changes of physical parameters. This means that in the case of relatively slow thermal motion of gas molecules and nearly specular gas–surface interaction, the resistance is still small. Notice in this respect that the velocity of artificial satellites on low Earth orbits is much greater than the mean thermal motion of atmospheric particles [20].

The gas–surface interaction is currently being intensively studied. It is very sensitive to many factors, including spacecraft material, atmospheric composition (which in turn depends on the altitude), angle of incidence, velocity of the satellite. It is commonly accepted now that the interaction of atmospheric particles with the surface of existing spacecraft at altitudes of between 150 and 300 km is mostly diffuse [26, 39]; however, it has been argued [20] that carefully manufactured clean smooth metallic surfaces would favor specular reflections.

Therefore, we believe that spacecraft of the shapes indicated in Fig. 8.4a,b with a suitably manufactured surface may experience reduced air resistance and, consequently, have increased lifetime and decreased deflection from the predicted trajectory.

Many questions remain open. Do there exist bodies invisible from two or more points? Do there exist bodies that are invisible/have zero resistance in three or more directions, or even in a set of directions of positive measure? We suppose that the answer to the last part of the question is negative, but we cannot prove it.

Chapter 9
Retroreflectors

In everyday life, optical devices that reverse the direction of all (or a significant part of) incident beams of light are called *retroreflectors*. They are widely used in technology, for example, in road safety. Some artificial satellites in Earth's orbit also carry retroreflectors. We are mostly interested in *perfect retroreflectors* that reverse the direction of *any* incident beam of light to the *exact opposite* direction. An example of perfect retroreflector based on *light refraction* is the Eaton lens, a transparent ball with varying radially symmetric refractive index going to infinity at the center [19, 71]. Each incident light ray, after making a curve in the lens around the center, leaves in the direction exactly opposite to the direction of incidence. On the other hand, no *perfect* retroreflector based solely on *light reflection* (rather than refraction) is known.

In what follows, only retroreflectors using light reflection (or billiard retroreflectors) will be considered. First we define the notion of an asymptotically perfect retroreflector: a sequence of bodies such that the property of light reflection from them tends, in a sense, to the property of perfect retroreflection, and then we present several asymptotically perfect retroreflectors and discuss and compare their properties.

The results of this chapter were published mainly in [3, 23, 24, 49, 54, 56].

9.1 Preliminaries

The most commonly used retroreflector based solely on *light reflection* is the so-called cube corner (its two-dimensional analog, square corner, is shown in Fig. 9.1). Both cube and square corners are not perfect, however: part of the incoming light is reflected in the wrong direction. This is clearly seen in Fig. 9.1 for the square corner: one of the rays depicted in the figure changes to the opposite direction, while the other one does not.

Our aim here is to bring together known billiard retroreflectors. They form a small collection of four objects; the first three, called *mushroom*, *tube*, and *notched*

A. Plakhov, *Exterior Billiards: Systems with Impacts Outside Bounded Domains*, DOI 10.1007/978-1-4614-4481-7_9,

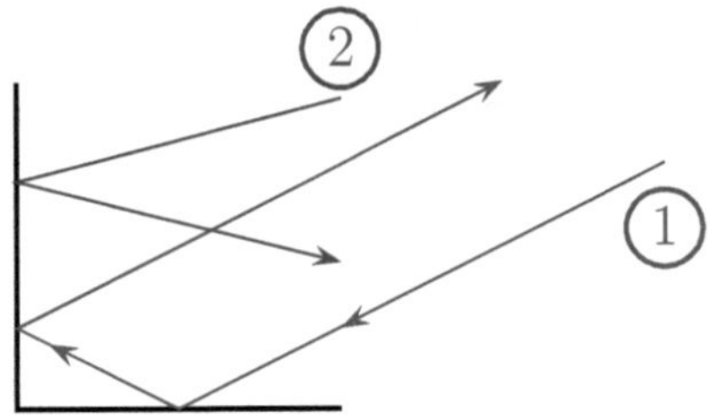

Fig. 9.1 *Square corner*: a retroreflector based on light reflection. Two incident light rays are shown: ray 1 is retroreflected, whereas ray 2 is not

angle, [54, 56] are asymptotically perfect retroreflectors, whereas the fourth one, called *helmet*, is a retroreflector that is very close to perfect [23, 24]. Note that the proof of retroreflectivity for the tube reduces to a quite nontrivial ergodic problem considered in [3], while helmet is examined numerically in [23, 24].

While bodies are generally supposed to be bounded in this book, in Sect. 9.1.1 *unbounded bodies* (regions with piecewise smooth boundary) will be briefly considered. Several definitions adapted to unbounded bodies will be given there.

As before, the coordinate and velocity of a billiard particle in $\mathbb{R}^d \setminus B$ at moment t are denoted by $x(t)$ and $v(t) = x'(t)$. Asymptotic velocities of the particle (if they exist) are denoted by $v = \lim_{t\to-\infty} v(t)$ and $v^+ = \lim_{t\to+\infty} v(t)$. We shall consider without notice only particles whose motion is defined for all $t \in \mathbb{R}$, that is, particles that do not hit singular points of ∂B and do not make infinitely many reflections in a finite time. Such particles form a full-measure set in the phase space.

We say that a particle is incident on B if it makes at least one reflection from B and there exists the asymptotic velocity $v = \lim_{t\to-\infty} v(t)$.

Definition 9.1. A body B is called a *perfect retroreflector* if for almost all incident particles the asymptotic velocity $v^+ = \lim_{t\to+\infty} v(t)$ exists and is opposite to the asymptotic velocity $v = \lim_{t\to-\infty} v(t)$, that is, $v^+ = -v$.

9.1.1 Unbounded Bodies

Unbounded perfect retroreflectors are of little interest for both mathematics and applications. However, we provide here several examples just to make the picture complete. Recall that the question of the existence of *bounded* perfect retroreflectors remains open.

Example 1. $B = B_P$ is the exterior of a parabola in $\mathbb{R}^2$. There exists a unique velocity of incidence that is parallel to the parabola axis. The initial and final velocities of any incident particle are mutually opposite, and the segment of the trajectory between the two consecutive reflections passes through the focus, as shown in Fig. 9.2.

Remark 9.1. If B is the exterior of a parabola perturbed within a bounded set (that is, $B = B_P \triangle K$, with K bounded), then B is again a perfect retroreflector.

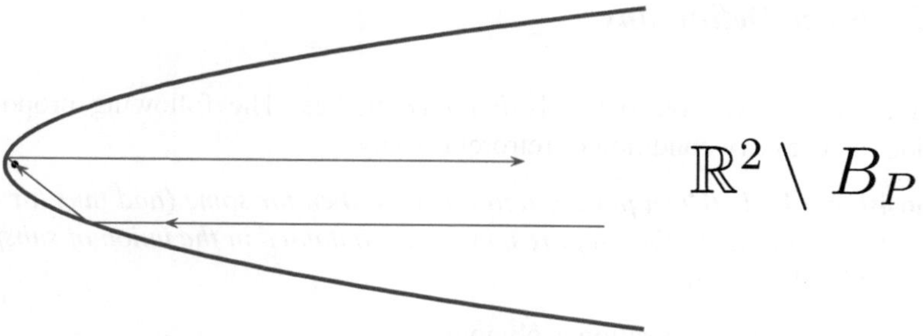

Fig. 9.2 The exterior of a parabola is an unbounded perfect retroreflector with a unique velocity of incidence

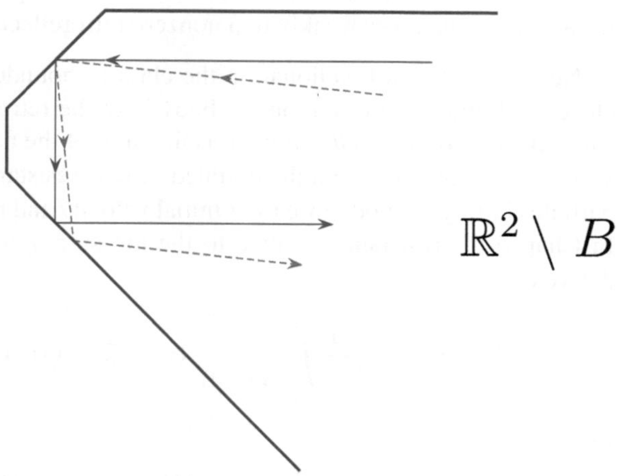

Fig. 9.3 The two-dimensional unbounded retroreflector shown here is a convex polygon contained in an angle of size $\pi/4$, with all angles at its vertices being multiples of $\pi/4$. Two billiard trajectories in $\mathbb{R}^2 \setminus B$ are shown

Indeed, any segment (or the extension of a segment) of a billiard trajectory within the parabola touches a confocal parabola with the same axis. The branches of this confocal parabola are codirectional or counterdirectional with respect to the original parabola. This implies that the segments of an incident trajectory, when going away to infinity, are becoming "straightened," that is, increasingly parallel to the parabola axis, and therefore $v^+ = -v$.

There also exist unbounded retroreflectors that admit a continuum of incidence velocities.

Example 2. Let $\mathbb{R}^d \setminus B$ be determined by the relations $x_1 > 0, \dots, x_d > 0$ in an orthonormal reference system $x_1, \dots, x_d$; then B is a perfect retroreflector.

Example 3. Let the set $\mathbb{C} \setminus B$ in the complex plane $\mathbb{C} \sim \mathbb{R}^2$ be given by the relations $\mathrm{Re}(e^{\frac{i\pi k}{2m}} z) > a_k$, $k = 0, 1, \dots, 2m - 1$, with $m \in \mathbb{N}$ and arbitrary constants a_k; then B is a perfect retroreflector; see Fig. 9.3 for the case $m = 2$.

9.1.2 Basic Definitions

In what follows we consider only *bounded* bodies. The following proposition provides a necessary condition of retroreflectivity.

Proposition 9.1. *If B is a perfect retroreflector, then for some (and thus for each) ambient convex body C the support spt $\nu_{B,C}$ is contained in the union of subspaces* $\{v^+ = -v\} \cup \{v^+ = v\}$.

The proof of this proposition is obvious.

Now we introduce the notion of an *asymptotically perfect retroreflector.*

Definition 9.2. We say that ν is a *retroreflector measure* if spt $\nu \subset \{v^+ = -v\}$. A family of bounded bodies B_ε, $\varepsilon > 0$ is called an *asymptotically perfect retroreflector* if $\nu_{B_\varepsilon, \text{Conv} B_\varepsilon}$ converges weakly to a nonzero retroreflector measure as $\varepsilon \to 0$.

Next we define a functional on the class of bounded bodies that indicates how close the billiard scattering on the body is to the retroreflector scattering. We will call it the *rate of retroreflectivity*; it is the ratio of the mean resistance (under elastic scattering) of the body and the doubled diffuse resistance related to the case where particles hitting the body lose their initial velocity and remain near the body forever. In Chap. 6 the resistance relative to the measure χ under the elastic reflection is derived:

$$R_\chi(B) = \frac{d+1}{4} \int_{(\partial C \times S^{d-1})_-} (v - v_B^+(\xi, v)) \cdot v \, |n(\xi) \cdot v| \, d\xi \, d\chi(v).$$

Let

$$I_B(\xi, v) = \begin{cases} 1, & \text{if the corresponding particle hits } B \text{ at most once;} \\ 0 & \text{elsewhere.} \end{cases}$$

Then the mean diffuse resistance is

$$D_\chi(B) = \frac{d+1}{4} \int_{(\partial C \times S^{d-1})_-} I_B(\xi, v) \, |n(\xi) \cdot v| \, d\xi \, d\chi(v).$$

In other words, the mean diffuse resistance is proportional to the total number of particles hitting the body per unit time.

Remark 9.2. The notion of diffuse scattering has a strong physical motivation originating, in particular, from space aerodynamics. The interaction of artificial satellites in low Earth orbits with the rarefied atmosphere is considered to be mainly diffuse by some researches (see, e.g., [39]); then the resistance is $D(B)$. Others [7, 27, 74] prefer to use a Maxwellian representation of interaction as a linear combination of elastic scattering and a diffuse one. In the latter case, the resistance equals $\alpha D(B) + (1-\alpha) R(B)$, where α is the so-called accommodation coefficient.

One obviously has

$$R_\chi(B) \le 2D_\chi(B), \tag{9.1}$$

and we have equality if B is a hypothetical perfect retroreflector. Conversely, if χ is supported on S^{d-1} and we have equality in (9.1), then B is a perfect retroreflector. Actually, if $A \subset S^{d-1}$ is the support of χ, then the ratio $R_\chi(B)/(2D_\chi(B))$ is a measure of retroreflectivity of B with respect to the set of directions A. It takes values between 0 and 1 and equals 1 if the body B is retroreflective with respect to this set of directions. We are interested in retroreflectivity over all directions; therefore, we choose the uniform measure $\chi = u$ on S^{d-1}. In this case we have

$$R_u(B) = \frac{d+1}{4} \int_{(S^{d-1})^3} \frac{1}{2} |v - v^+|^2 \, d\nu_{B,C}(v, v^+, n).$$

It was found in Chap. 6 that the elastic resistance of a convex body equals the area of its surface, and the diffuse resistance equals $(d+1)/4$ times the surface area; hence if B is convex, then

$$R_u(B) = |\partial B| \quad \text{and} \quad D_u(B) = \frac{d+1}{4} |\partial B|.$$

Definition 9.3. The quantity $r(B) = \frac{R_u(B)}{2D_u(B)}$ is called the *rate of retroreflectivity.*

For each B we have $0 \le r(B) \le 1$. As was proved in the previous chapter, bodies of zero resistance do not exist, therefore $r(B) > 0$. If B is a (hypothetical) retroreflector, then $r(B) = 1$.

Let us bring together the properties of r.

1. $0 < r(B) \le 1$.
2. If B is convex, then $r(B) = 2/(d+1)$; in particular, $r(B) = 2/3$ for $d = 2$ and $r(B) = 1/2$ for $d = 3$.
3. $\sup_B r(B) = 1$ in any dimension.
4. The infimum of r depends on the dimension d.
   If $d = 2$, then the infimum over connected bodies is $\inf_B r(B) = 0.6585\ldots$ [50].
   If $d \ge 3$, then only estimates are known. In particular, for $d = 3$ we have $\inf_B r(B) < 0.4848$ [52].
5. If B_ε is an asymptotically perfect retroreflector, then $\lim_{\varepsilon \to 0} r(B_\varepsilon) = 1$.

Property 3 is a consequence of the existence, in any dimension, of asymptotically perfect retroreflectors, which will be proved in Sect. 9.2.

Remark 9.3. The quantity $r(B)$ is proportional to the (elastic) resistance of B divided by the number of particles hitting B per unit time interval. It can also be interpreted as the mathematical expectation of the longitudinal component of the momentum transmitted to the body by a randomly chosen incident particle of mass $1/2$, that is, $r(B) = \frac{1}{2}\mathbb{E}\langle v - v^+, v\rangle$.

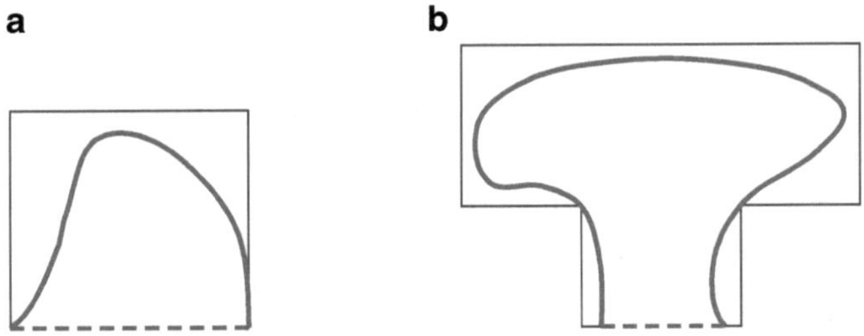

Fig. 9.4 (**a**) A convenient hollow (**b**) An inconvenient hollow

In the two-dimensional case, the rate of retroreflectivity admits a comfortable representation through the retroreflectivity of hollows.

Definition 9.4. A hollow (Ω, I) on a plane is called *convenient* if the orthogonal projection of Ω on a straight line containing I coincides with I. Otherwise, it is called *inconvenient*. See Fig. 9.4a, b for examples of convenient and inconvenient hollows.

Recall that the measure η_{retr} on $\square = [-\pi/2, \pi/2] \times [-\pi/2, \pi/2]$ with the density $\frac{1}{2} \cos\varphi \, \delta(\varphi - \varphi^+)$ is also called a *retroreflector measure.*

Definition 9.5. A family of hollows $(\Omega_\varepsilon, I_\varepsilon)$ is called *asymptotically retroreflecting* if $\eta_{\Omega_\varepsilon, I_\varepsilon}$ weakly converges to η_{retr}.

The following proposition is obvious.

Proposition 9.2. *A family of hollows* $(\Omega_\varepsilon, I_\varepsilon)$ *is asymptotically retroreflecting if and only if for all* $\varepsilon > 0$ *we have* $\lim_{\varepsilon\to 0} \mu(\{(\xi, \varphi) : |\varphi - \varphi^+_{\Omega_\varepsilon, I_\varepsilon}(\xi, \varphi)| \geq \varepsilon\}) = 0$.

Introduce the functional

$$\mathcal{F}(\eta) = \frac{1}{2} \iint_\square (1 + \cos(\varphi - \varphi^+)) \, d\eta(\varphi, \varphi^+). \tag{9.2}$$

We have, in particular, $\mathcal{F}(\eta_0) = 2/3$ and $\mathcal{F}(\eta_{\text{retr}}) = 1$.

Let B be a connected body in $\mathbb{R}^2$, and let (Ω_i, I_i) be the hollows on its boundary and I_0 the convex part of ∂B. Denote $c_i = |I_i|/|\partial(\text{Conv}B)|$; we have $\sum c_i = 1$. Then the rate of retroreflectivity $r(B)$ can be written as

$$r(B) = \frac{2}{3} c_0 + \sum_{i\neq 0} c_i \mathcal{F}(\eta_{\Omega_i, I_i}). \tag{9.3}$$

Formula (9.3) suggests a strategy of constructing two-dimensional asymptotically perfect retroreflectors. First, find an asymptotically retroreflecting family of hollows $(\Omega_\varepsilon, I_\varepsilon)$; that is, $\lim_{\varepsilon\to 0} \mathcal{F}(\eta_{\Omega_\varepsilon, I_\varepsilon}) = 1$. Then find a family of bodies B_ε with all hollows on their boundary similar to $(\Omega_\varepsilon, I_\varepsilon)$ and such that the relative

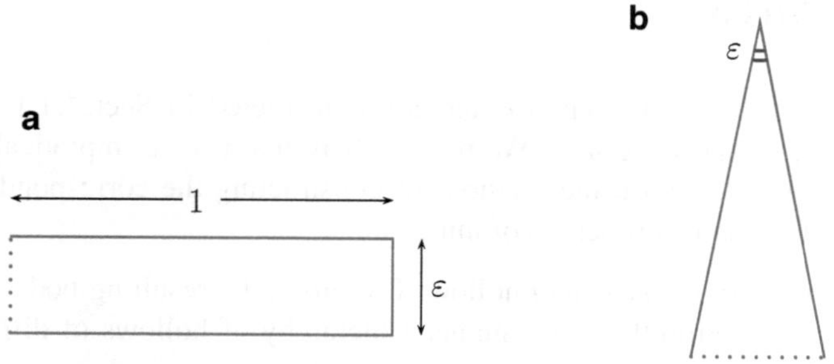

Fig. 9.5 A rectangular hollow (**a**) and a triangular hollow (**b**)

length of the convex part of ∂B_ε vanishes, $\lim_{\varepsilon\to 0} c_0^\varepsilon = 0$, and the sequence of convex hulls Conv B_ε converges to a fixed convex body as $\varepsilon \to 0$. In this case one has

$$\lim_{\varepsilon\to 0} r(B_\varepsilon) = \lim_{\varepsilon\to 0}\left(\frac{2}{3}c_0^\varepsilon + (1-c_0^\varepsilon)\mathcal{F}(\eta_{\Omega_\varepsilon,I_\varepsilon})\right) = 1,$$

and therefore the family B_ε is an asymptotically perfect retroreflector.

If all the hollows are convenient (Fig. 9.4a), then one can find bodies B_ε with identical hollows. On the other hand, if the hollows are not convenient (Fig. 9.4b), then each body B_ε must contain on its boundary a hierarchy of hollows of different sizes.

Let us finally define semiretroreflecting hollows.

Definition 9.6. We say that (Ω, I) is a *semiretroreflecting* hollow if $\eta_{\Omega,I} = \frac{1}{2}(\eta_0 + \eta_{\text{retr}})$. A family of hollows $(\Omega_\varepsilon, I_\varepsilon)$, $\varepsilon > 0$ is called *asymptotically semiretroreflecting* if the family of measures $\eta_{\Omega_\varepsilon,I_\varepsilon}$ converges weakly to $\frac{1}{2}(\eta_0 + \eta_{\text{retr}})$.

According to this definition, half of incident particles are reflected elastically $\varphi^+ = -\varphi$ in the hollow, and the other half is retroreflected, $\varphi^+ = \varphi$.

We provide two such families, a rectangle and a triangle, as shown in Fig. 9.5. The ratio of the width to the height of the rectangle equals ε. The triangle is isosceles, and the angle at the apex equals ε. Denote by $\eta_\sqcup^\varepsilon$ and $\eta_\vee^\varepsilon$ the measures generated by the rectangle and the triangle, respectively.

Proposition 9.3. *Both $\eta_\sqcup^\varepsilon$ and $\eta_\vee^\varepsilon$ converge weakly to $\frac{1}{2}(\eta_0 + \eta_{retr})$ as $\varepsilon \to 0$.*

The proof of this proposition is not difficult, but it is a bit lengthy, and therefore put off to Sect. 9.7. The proposition implies that both functionals $\mathcal{F}(\eta_\sqcup^\varepsilon)$ and $\mathcal{F}(\eta_\vee^\varepsilon)$ converge to $5/6$. Note also that the measures $\eta_\sqcup^\varepsilon$ and $\eta_\vee^\varepsilon$ do not converge in the norm. These shapes serve as starting points for developing *true* retroreflectors: tube (Sect. 9.3) and notched angle (Sect. 9.4).

In the following sections we describe several asymptotically perfect retroreflectors in two dimensions.

9.2 Mushroom

The hollow called mushroom was already constructed in Sect. 4.1.1 (step 1 of Theorem 4.1; see also Fig. 4.7). We proved there that it is asymptotically retroreflecting, and we described the method for constructing the corresponding bodies B_ε. We point out some properties of mushroom.

1. Mushroom is an *inconvenient* hollow. Therefore, the resulting body (asymptotically perfect retroreflector) contains a hierarchy of hollows of different sizes (Fig. 9.14a).
2. The difference $\varphi - \varphi^+$ is always nonzero; this means that the measures associated with mushroom converge to η_{retr} *weakly* but *not in the norm.*
3. If the semiellipse is replaced by a semicircle, then the resulting hollow (which is also called mushroom) will also be asymptotically retroreflecting. This modified construction can be generalized to any dimension; that is, there exist multidimensional asymptotically perfect retroreflectors with mushroom-shaped hollows (for a more detailed description, see [49]).
4. Most incident particles make *exactly one reflection.* This means that the portion of incident particles making one reflection tends to 1 as $\varepsilon \to 0$.

Note, however, that this shape is not suitable for practical implementation. An estimate placed in Sect. 9.7 shows that the size of the smallest hollows in the corresponding retroreflecting body B with $r(B) = 0.99$ of total size 1 m should be much smaller than the size of atoms. In practice, square or cube corners are more suitable, and helmet is even better suited for purposes of retroreflection.

9.3 Tube

Tube is a rectangle of length a and height 1 with two rows of rectangles of smaller size $\delta \times \varepsilon$ removed (Fig. 9.6). The lower and upper rows of rectangles are adjacent to the lower and upper sides of the tube, respectively. The distance between neighbor

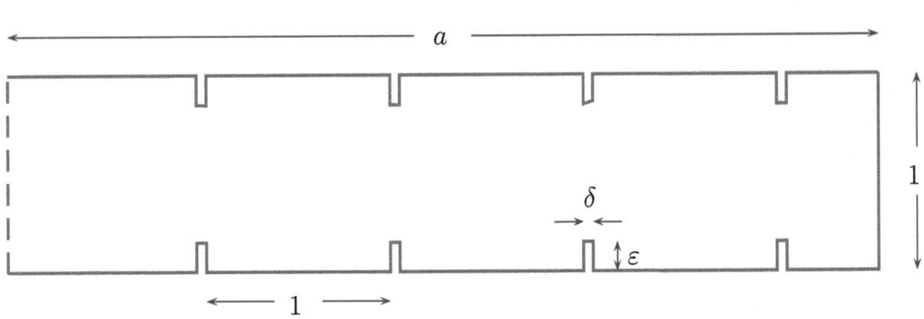

Fig. 9.6 Tube

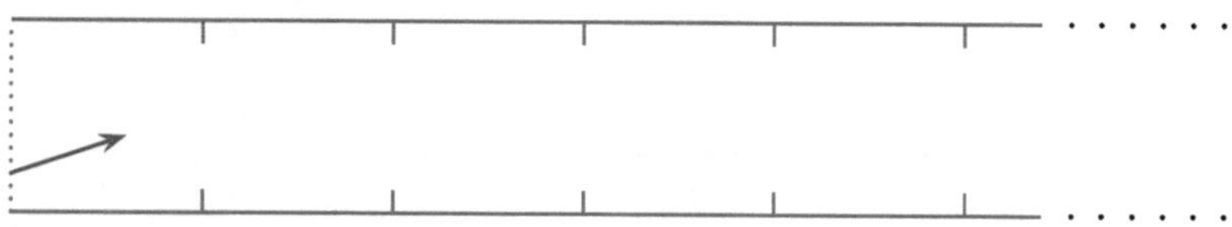

Fig. 9.7 Semi-infinite tube

rectangles of each row equals 1. The inlet of the tube is the left vertical side of the large rectangle. Denote by $\eta_{\varepsilon,\delta,a}$ the measure associated with the tube.

For each particle incident in the tube, with φ and φ^+ being the angles of entering and leaving, only two cases may happen: $\varphi^+ = \varphi$ or $\varphi^+ = -\varphi$. Letting $a \to \infty$ and $\delta \to 0$ (with ε fixed), we get a semi-infinite tube where small rectangles are replaced with vertical segments of length ε (Fig. 9.7). Studying the dynamics in this tube leads to the following ergodic theorems.

Theorem 9.1. *Almost all particles incident in a semi-infinite tube eventually escape from it.*

Thanks to this theorem, the measure η_ε associated with a semi-infinite tube is well defined.

Take the circle $\mathbb{R}/\mathbb{Z}$ with the coordinate $\xi \bmod 1$ and consider the iterated rotation of the circle by a fixed angle $\alpha \in \mathbb{R}/\mathbb{Z}$, $\xi_n = \xi + \alpha n \bmod 1$, $n = 1, 2, \ldots$. Mark the successive moments $n = n_1,\ n_1 + n_2,\ n_1 + n_2 + n_3, \ldots$, when $\xi_n \in [-\varepsilon, \varepsilon] \bmod 1$. Denote by $l = l_\varepsilon(\xi, \alpha)$ the smallest value such that $n_1 - n_2 + \ldots + n_{2l-1} - n_{2l} \le 0$. Let $\mathbb{P}$ be a probability measure on $\mathbb{R}/\mathbb{Z} \times \mathbb{R}/\mathbb{Z}$ absolutely continuous with respect to a Lebesgue measure. Then we have the following theorem.

Theorem 9.2. *There exists the limiting distribution* $p_k = \lim_{\varepsilon\to 0} \mathbb{P}(\{(\xi, \alpha) : l_\varepsilon(\xi, \alpha) = k\})$, *with* $\sum_{k=1}^{\infty} p_k = 1$.

Proofs of Theorems 9.1 and 9.2 can be found in [3]. It is additionally proved in [3] that the limiting distribution p_k, $k = 1, 2, \ldots$ does not depend on $\mathbb{P}$. Now we use Theorem 9.2 to show that the semi-infinite tube is an asymptotically retroreflecting "hollow" (it is not a true hollow since it is unbounded and its boundary is not piecewise smooth). That is, the following theorem holds true.

Let $D = \{\varphi = \varphi^+\}$ be the diagonal of the square $\square$.

Theorem 9.3. $\lim_{\varepsilon\to 0} \eta_\varepsilon(D) = 1$.

Proof. Let $\varphi_\varepsilon^+ = \varphi_\varepsilon^+(\xi, \varphi)$ be the final inclination angle of an incident particle with the initial data $\xi \in [0, 1]$, $\varphi \in [-\pi/2, \pi/2]$. Recall that the measure $\mu = \mu_I$ in $[0, 1] \times [-\pi/2, \pi/2]$ was defined in Sect. 4.1.1.

We have only two possibilities: $\varphi_\varepsilon^+ = \varphi$ or $\varphi_\varepsilon^+ = -\varphi$; therefore, it suffices to prove that

$$\lim_{\varepsilon\to 0} \mu(\varphi_\varepsilon^+ = -\varphi) = 0. \tag{9.4}$$

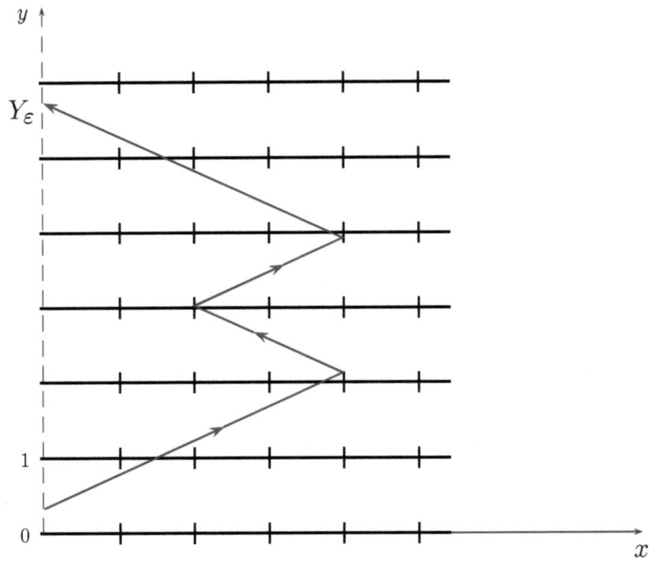

Fig. 9.8 An unfolded trajectory. In this example $l_\varepsilon = 2$, $\lfloor Y_\varepsilon \rfloor = 5$ and $n_1 = 4$, $n_2 = n_3 = 2$, $n_4 \geq 4$

The equality $\varphi_\varepsilon^+ = \varphi$ takes place when the number of reflections from the horizontal sides of the tube is odd and $\varphi_\varepsilon^+ = -\varphi$ when this number is even.

Without loss of generality we assume that $\varphi > 0$ and denote $\alpha = \tan\varphi \mod 1$. The mapping $T : (\xi, \varphi) \mapsto (\xi, \alpha)$ induces the probability measure $\mathbb{P} = 2T^\# \mu$ on $\mathbb{R}/\mathbb{Z} \times \mathbb{R}/\mathbb{Z}$.

Consider the (partial) unfolding of a billiard trajectory in a tube induced by successive reflections from the horizontal sides of the tube (Fig. 9.8). We see that reflections from the vertical sides correspond to the inclusions

$$(\xi + \alpha n) \mod 1 \in [-\varepsilon, \varepsilon] \mod 1.$$

The corresponding values of n are n_1, $n_1 + n_2$, $n_1 + n_2 + n_3, \ldots, n_1 + n_2 + \ldots + n_{2l-1}$, with $l = l_\varepsilon(\xi, \alpha)$. The unfolded trajectory is a zigzag line. The horizontal projections of its segments have the lengths n_1, $n_2, \ldots, n_{2l-1}$, n'_{2l}, where n'_{2l} is less than or equal to n_{2l} and is determined by the equality

$$n_1 - n_2 + \ldots + n_{2l-1} - n'_{2l} = 0. \tag{9.5}$$

The vertical projections of the segments have the lengths αn_1, $\alpha n_2, \ldots, \alpha n_{2l-1}$, $\alpha n'_{2l}$.

The point of escape of the unfolded trajectory lies on the y-axis and has the y-coordinate $Y = Y_\varepsilon(\xi, \alpha)$, where

$$Y = \xi + \alpha \left(n_1 + n_2 + \ldots + n_{2l-1} + n'_{2l}\right).$$

Using (9.5) one obtains

$$Y = 2(\xi + \alpha n_1 + \alpha n_3 + \ldots + \alpha n_{2l-1}) - \xi. \tag{9.6}$$

The number of reflections of the original trajectory from the horizontal sides of the tube equals $\lfloor Y_\varepsilon \rfloor$; therefore, (9.4) is equivalent to

$$\lim_{\varepsilon\to 0} \mathbb{P}(\lfloor Y_\varepsilon \rfloor \text{ is even}) = 0. \tag{9.7}$$

Let $\|\cdot\|$ denote the distance to the nearest integer. We have $\|\xi + \alpha n_1\| < \varepsilon$. On the other hand, for $i \ge 2$ we have $\|\xi + \alpha n_1 + \ldots + \alpha n_{i-1}\| < \varepsilon$ and $\|\xi + \alpha n_1 + \ldots + \alpha n_{i-1} + \alpha n_i\| < \varepsilon$; therefore, $\|\alpha n_i\| < 2\varepsilon$. Thus,

$$\|\xi + \alpha n_1 + \alpha n_3 + \ldots + \alpha n_{2l-1}\| \le \|\xi + \alpha n_1\| + \|\alpha n_3\| + \ldots \|\alpha n_{2l-1}\| < 2l\varepsilon. \tag{9.8}$$

It follows from (9.6) and (9.8) that for some integer m,

$$2m - 4l\varepsilon - \xi < Y < 2m + 4l\varepsilon - \xi. \tag{9.9}$$

By (9.9), if $\lfloor Y_\varepsilon \rfloor$ is even, then either $4l\varepsilon \ge \xi$ or $4l\varepsilon \ge 1 - \xi$. Therefore, it remains to prove that

$$\lim_{\varepsilon\to 0} \mathbb{P}(4\varepsilon l_\varepsilon \ge \min\{\xi, 1-\xi\}) = 0; \tag{9.10}$$

this relation will imply (9.7).

For any $0 < a < 1/2$ we have

$$\mathbb{P}(4\varepsilon l_\varepsilon \ge \min\{\xi, 1-\xi\}) < \mathbb{P}(\xi < a/2 \text{ or } \xi > 1 - a/2) + \mathbb{P}(4\varepsilon l_\varepsilon \ge a/2). \tag{9.11}$$

Fix arbitrary $\delta > 0$ and choose a such that the first term on the right-hand side of (9.11) is smaller than $\delta/2$. Further, using Theorem 9.2, choose ε sufficiently small so that the second term on the right-hand side of (9.11) is smaller than $\delta/2$. Thus, we obtain that for ε sufficiently small $\mathbb{P}(4\varepsilon l_\varepsilon \ge \min\{\xi, 1-\xi\}) < \delta$, and so (9.10) is proved. □

Let us show that there exists a family of *true* tube-shaped hollows that is asymptotically retroreflecting. To this end, define the function $H(\xi, \varphi, \varepsilon, \delta, a)$, which is equal to 0 if a billiard particle with initial data (ξ, φ) satisfies the equality $\varphi^+ = \varphi$ and equals 1 if $\varphi^+ = -\varphi$ (there are no other possibilities). For a semi-infinite tube this function takes the form $H(\xi, \varphi, \varepsilon, 0, +\infty) =: H(\xi, \varphi, \varepsilon)$. The asymptotic retroreflectivity of the semi-infinite tube means that

$$\lim_{\varepsilon\to 0} \iint_{[0,1]\times[-\pi/2,\pi/2]} H(\xi, \varphi, \varepsilon)\, d\xi\, d\varphi = 0.$$

Note that for fixed ξ, φ and for $1/a$ and δ small enough the corresponding particle makes the same sequence of reflections (and therefore has the same velocity of exit) as in the limiting case $\delta = 0$, $a = +\infty$. This implies that $H(\xi, \varphi, \varepsilon, \delta, a)$ converges pointwise (stabilizes) to $H(\xi, \varphi, \varepsilon)$ as $\delta \to 0$, $a \to +\infty$, and therefore,

$$\lim_{\delta\to 0, a\to+\infty} \iint_{[0,1]\times[-\pi/2,\pi/2]} H(\xi, \varphi, \varepsilon, \delta, a)\, d\xi\, d\varphi = \iint_{[0,1]\times[-\pi/2,\pi/2]} H(\xi, \varphi, \varepsilon)\, d\xi\, d\varphi.$$

Then, using the diagonal method, one selects $\delta = \delta(\varepsilon)$ and $a = a(\varepsilon)$ such that $\lim_{\varepsilon\to 0} a(\varepsilon) = \infty$, $\lim_{\varepsilon\to 0} \delta(\varepsilon) = 0$, and

$$\lim_{\varepsilon\to 0} \iint_{[0,1]\times[-\pi/2,\pi/2]} H(\xi, \varphi, \varepsilon, \delta(\varepsilon), a(\varepsilon))\, d\xi\, d\varphi = 0.$$

Thus, the corresponding family of tubes is asymptotically retroreflecting.

The obtained result can be formulated as follows.

Theorem 9.4. *η_{retr} is a limit point of the set of measures associated with tubes $\{\eta_{\varepsilon,\delta,a}\}$ equipped with the norm topology.*

A tube has the following properties.

1. A tube is a *convenient* hollow. This property makes it possible to construct an asymptotically perfect retroreflector with identical tube-shaped hollows (Fig. 9.14b).
2. The measure associated with the tube [with $\delta = \delta(\varepsilon)$ and $a = a(\varepsilon)$ properly chosen] converges *in the norm* to the retroreflector measure. In other words, the portion of retroreflected particles (that is, particles reflected in the *exactly* opposite direction) tends to 1 as $\varepsilon \to 0$.
3. We believe this construction admits a generalization to higher dimensions, but we cannot yet prove it.
4. The average number of reflections in a tube is of order $1/\varepsilon$ and therefore goes to infinity as $\varepsilon \to 0$.

9.4 Notched Angle

Consider two isosceles triangles $\bar{A}\bar{O}\bar{B}$ and $\bar{A}'\bar{O}\bar{B}'$ with common vertex at $\bar{O}$ and require that the base of one of them is contained in the base of the other one, $\bar{A}\bar{B} \subset \bar{A}'\bar{B}'$. The segment $\bar{A}\bar{B}$ is horizontal in Fig. 9.9. Denote $\measuredangle\bar{A}\bar{O}\bar{B} = \alpha$ and $\measuredangle\bar{A}'\bar{O}\bar{B}' = \alpha + \beta$. Draw two broken lines with horizontal and vertical segments with the origin at $\bar{A}$ and $\bar{B}$, respectively, and require that the vertices of the first line belong to the segments $\bar{O}\bar{A}$ and $\bar{O}\bar{A}'$ and that the vertices of the second line belong to the segments $\bar{O}\bar{B}$ and $\bar{O}\bar{B}'$ (Fig. 9.9). The endpoint of both broken lines is $\bar{O}$; both lines have infinitely many segments and finite length. We will consider the

Fig. 9.9 Notched angle

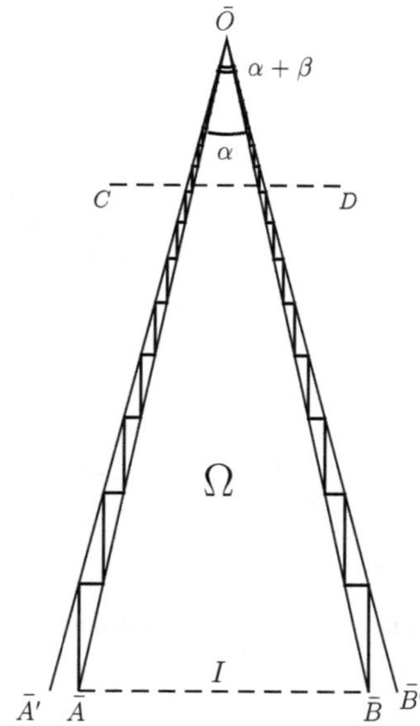

"hollow" (Ω, I) with the inlet $I = I_\alpha = \bar{A}\bar{B}$ and with the set $\Omega = \Omega_{\alpha,\beta}$ bounded by $\bar{A}\bar{B}$ and the two broken lines. This "hollow" is called a *notched angle* of size (α, β), or just an (α, β)-*angle*. The boundary $\partial\Omega$ is not piecewise smooth ($\bar{O}$ is a limit point of singular points of $\partial\Omega$); therefore, the word *hollow* is put in quotes; however, the measure associated with this "hollow" is defined in the standard way. This measure depends only on α and β and is denoted by $\eta_{\alpha,\beta}$.

Theorem 9.5. *There exists a function* $\beta = \beta(\alpha)$, $\lim_{\alpha\to 0}(\beta/\alpha) = 0$ *such that* $\eta_{\alpha,\beta}$ *converges in the norm to the retroreflector measure* η_{retr} *as* $\alpha \to 0$.

Remark 9.4. Using this theorem, one easily constructs a family of *true* hollows for which convergence in the norm to η_{retr} takes place. That is, draw a straight line CD parallel to $\bar{A}\bar{B}$ at a small distance δ from $\bar{O}$; the *true hollow* is the part of the original "hollow" situated between $\bar{A}\bar{B}$ and CD, with the same inlet (Fig. 9.9). The measure associated with this hollow converges to η_{retr} as $\alpha \to 0$, with properly chosen $\beta = \beta(\alpha)$ and vanishing $\delta = \delta(\alpha)$ as $\alpha \to 0$.

Proof. For arbitrary initial data ξ, φ the angle of going away $\varphi^+ = \varphi^+_{\alpha,\beta}(\xi, \varphi)$ satisfies either $\varphi^+ = \varphi$ or $\varphi^+ = -\varphi$. To prove the theorem, it suffices to check that the measure μ of the set of initial data ξ, φ satisfying $\varphi^+_{\alpha,\beta}(\xi, \varphi) = -\varphi$ and $|\varphi| > \alpha$ tends to 0 as $\alpha \to 0$, $\beta = \beta(\alpha)$.

Make a uniform extension along the horizontal axis in such a way that the resulting angle $\bar{A}\bar{O}\bar{B}$ becomes right. Then the angle $\bar{A}'\bar{O}\bar{B}'$ becomes equal to $\pi/2 + \gamma$, $\gamma = \gamma(\alpha, \beta)$ (Fig. 9.10); in addition, the conditions $\alpha \to 0$, $\beta/\alpha \to 0$ imply that $\gamma \to 0$. This extension takes the (α, β)-angle to a $(\pi/2, \delta)$-angle,

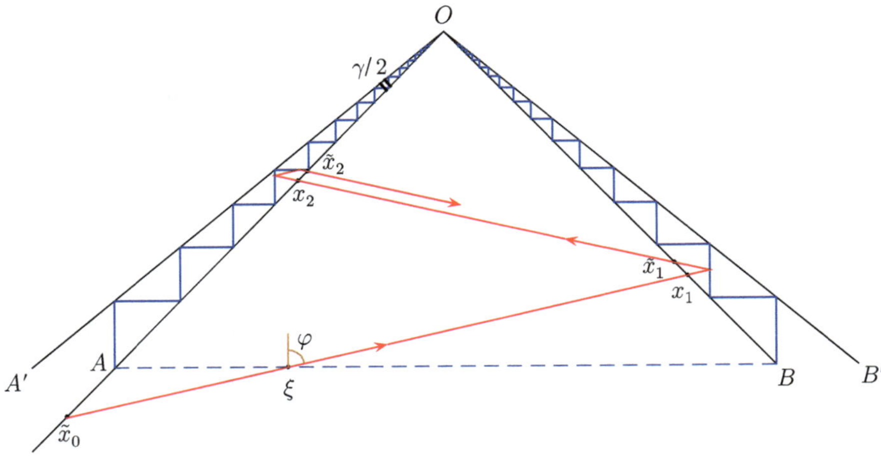

Fig. 9.10 Reduced notched angle

takes each billiard trajectory to another billiard trajectory, and takes the measure $\frac{1}{2}\cos\varphi\,d\varphi\,d\xi$ to a measure absolutely continuous with respect to it.

The vertices of the resulting notched angle will be denoted by $O,\ A,\ B,\ A',\ B'$ (without the overline) to distinguish them from the previous notation.

Without loss of generality we assume that $|OA| = |OB| = 1$. Introduce the uniform parameter ξ on the segment AB, where A corresponds to the value $\xi = 0$ and B to the value $\xi = 1$. Extend the trajectory of an incident particle with initial data $\xi,\ \varphi < -\pi/4$[1] back to the intersection with an extension of OA. Denote by $\tilde{x}_0$ the distance from O to the point of intersection (Fig. 9.10). (In what follows, a point on ray OA or OB will be identified with the distance from vertex O to this point.) In the new representation, the particle starts its motion at a point $\tilde{x}_0$ and intersects the segment AB at a point ξ and at an angle φ. Continuing the straight-line motion, it intersects the side OB at a point x_1 $(0 < x_1 < 1)$, then makes one or two reflections from the broken line and intersects OB again at a point $\tilde{x}_1$. Denote $x_1/\tilde{x}_0 = \lambda$; obviously one has $0 < \lambda < 1$. The value λ is the tangent of the angle of trajectory inclination relative to OA; thus, one has $\varphi = -\pi/4 - \arctan\lambda$. It is convenient to change the variables in the space of particles entering the hollow at an angle $\varphi < -\pi/4$. That is, we pass from the parameters $\xi \in [0,\ 1],\ \varphi \in [-\pi/2,\ -\pi/4]$ to the parameters $\lambda \in [0,\ 1],\ \tilde{x}_0 \in [1,\ 1/\lambda]$. This change of variables can be written as $\xi = \frac{\lambda}{1-\lambda}(\tilde{x}_0 - 1),\ \varphi = \pi/4 + \arctan\lambda$; it transforms the measure $\frac{1}{2}\cos\varphi\,d\varphi\,d\xi$ into the measure $\frac{\lambda}{2\sqrt{2}(1+\lambda^2)^{3/2}}\,d\lambda\,d\tilde{x}_0$.

By considering successive alternating reflections of the particle from the broken lines resting on the sides OB and OA, we define the sequence of values $x_1,\ \tilde{x}_1, \ldots,$

[1] Recall that the angle φ is measured counterclockwise from the vertical vector (0, 1) to the velocity of the incident particle, so one has $\varphi < 0$ in Fig. 9.10.

x_{m-1}, $\tilde{x}_{m-1}$. Obviously, all these values are smaller than 1. Then the particle leaves the hollow and intersects the extension of the side OA or OB at a point $x_m > 1$. If m is even, then intersection with OA takes place, and $\varphi^+ = \varphi$. If m is odd, then intersection with OB takes place, with $\varphi^+ = -\varphi$. Clearly, m depends on the initial data $\tilde{x}_0$, λ and on the parameter γ, $m = m_\gamma(\tilde{x}_0, \lambda)$.

Proposition 9.4. *For any λ the measure of the set of values $\tilde{x}_0$ such that $m_\gamma(\tilde{x}_0, \lambda)$ is odd goes to 0 as $\gamma \to 0$.*

Let us derive the theorem from this proposition. Indeed, let $f_\gamma(\lambda)$ be the measure of the set indicated in the proposition, $f_\gamma(\lambda) = |\{\tilde{x}_0 : m_\gamma(\tilde{x}_0, \lambda) \text{ is odd}\}|$. Introduce the measure η on the segment $[0, 1]$ according to $d\eta(\lambda) = \frac{\lambda\, d\lambda}{2\sqrt{2}(1+\lambda^2)^{3/2}}$; then $\int_0^1 f_\gamma(\lambda)\, d\eta(\lambda)$ is the measure of the set of initial values $(\lambda, \tilde{x}_0)$ such that $m_\gamma(\tilde{x}_0, \lambda)$ is odd. The value $f_\gamma(\lambda)$ does not exceed the full Lebesgue measure of the segment $[1, \lambda^{-1}]$,

$$f_\gamma(\lambda) \le \lambda^{-1} - 1, \tag{9.12}$$

and the function $\lambda^{-1} - 1$ is integrable relative to η, $\int_0^1 (\lambda^{-1} - 1)\, d\eta(\lambda) = \frac{\sqrt{2}-1}{2\sqrt{2}}$. According to Proposition 9.4, for any λ

$$\lim_{\gamma \to 0} f_\gamma(\lambda) = 0 \tag{9.13}$$

holds. Taking into account (9.12) and (9.13) and applying Lebesgue's dominated convergence theorem, one obtains

$$\lim_{\gamma \to 0} \int_0^1 f_\gamma(\lambda)\, d\eta(\lambda) = 0.$$

This means that the measure of the set of values (ξ, φ), $\varphi \le -\pi/4$ for which the equality $\varphi^+_{\pi/2,\gamma}(\xi, \varphi) = -\varphi$ is true tends to 0 as $\gamma \to 0$. The same statement, due to the axial symmetry of the billiard, is also valid for $\varphi \ge \pi/4$.

Now make a uniform contraction along the abscissa axis transforming the $(\pi/2, \gamma)$-angle into an (α, β)-angle (where β depends on γ and α). Taking into account that the measures associated with these angles are mutually absolutely continuous, we get that the measure $\mu(\{(\xi, \varphi) : |\varphi| \ge \alpha \text{ and } \varphi^+_{\alpha,\beta}(\xi, \varphi) = -\varphi\})$ goes to 0 at fixed α and $\beta \to 0$.

Finally, choose a diagonal family of parameters α, $\beta(\alpha)$, $\lim_{\alpha \to 0}(\beta(\alpha)/\alpha) = 0$ such that the measure

$$\mu(\{(\xi, \varphi) : |\varphi| \ge \alpha \text{ and } \varphi^+_{\alpha,\beta(\alpha)}(\xi, \varphi) = -\varphi\}) \to 0 \quad \text{as } \alpha \to 0.$$

It remains to note that $\mu(\varphi^+_{\alpha,\beta(\alpha)} = -\varphi) \le \mu(|\varphi| \ge \alpha \text{ and } \varphi^+_{\alpha,\beta(\alpha)} = -\varphi) + \mu(|\varphi| < \alpha)$ and $\mu(|\varphi| < \alpha) \to 0$ as $\alpha \to 0$. This completes the proof of Theorem 9.5. □

Proof of Proposition 9.4. Note that the broken lines intersect the sides OA and OB at the points $x = e^{-n\delta}$, $n = 0, 1, 2, \ldots$, where δ is defined by the relation

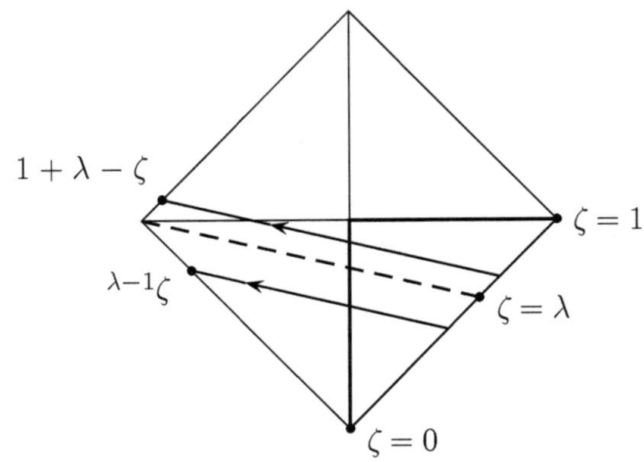

Fig. 9.11 Dynamics in a *small right triangle*

$\tanh\delta = \sin\gamma$. Consider an arbitrary pair of values $x_k,\ \tilde{x}_k$; they belong to a segment bounded by a pair of points $x = e^{-n\delta}$ and $e^{-(n+1)\delta}$. Consider also a right triangle with the hypotenuse being this segment and with the legs being segments of the broken line.

Two cases may emerge: either (I) $x_k/\tilde{x}_{k-1} = \lambda$ or (II) $x_k/\tilde{x}_{k-1} = \lambda^{-1}$, the first case corresponding to the "forward" motion in the direction of point O, and the second, to the "backward" motion. Introduce the local variable ζ on the hypotenuse according to $x = e^{-n\delta}[1+\zeta(e^{-\delta}-1)]$ (Fig. 9.11). Thus, the value $\zeta = 0$ corresponds to the point $x = e^{-n\delta}$, and $\zeta = 1$ to the point $x = e^{-(n+1)\delta}$. The sequences $x_k,\ \tilde{x}_k$ generate two sequences $\zeta_k,\ \tilde{\zeta}_k \in (0,\ 1)$ and an integer-valued sequence n_k. Consider the two cases separately.

(I) $x_k/\tilde{x}_{k-1} = \lambda$.

(a) If $0 < \zeta_k < \lambda$, then $\tilde{\zeta}_k = \lambda^{-1}\zeta_k$ and the particle, after leaving the triangle, continues the forward motion, that is, $x_{k+1}/\tilde{x}_k = \lambda$.
(b) If $\lambda < \zeta_k < 1$, then $\tilde{\zeta}_k = 1 + \lambda - \zeta_k$ and the particle, after leaving the triangle, proceeds to the backward motion, $x_{k+1}/\tilde{x}_k = \lambda^{-1}$.

(II) $x_k/\tilde{x}_{k-1} = \lambda^{-1}$. In this case, one has $\tilde{\zeta}_k = \lambda\zeta_k$ and the backward motion continues, $x_{k+1}/\tilde{x}_k = \lambda^{-1}$.

Introduce the logarithmic scale $z = -\frac{1}{\delta}\ln x$; then we have a sequence of values $\tilde{z}_0,\ z_1,\ \tilde{z}_1, \ldots, z_{m-1},\ \tilde{z}_{m-1},\ z_m$. The first and last terms in this sequence are negative, and the remaining terms are positive. One has $-\frac{1}{\delta}\ln\frac{1}{\lambda} < \tilde{z}_0 < 0$. The following equations establish the connection between $z_k,\ \tilde{z}_k$ and $\zeta_k,\ \tilde{\zeta}_k$.

$$z_k = n_k - \frac{1}{\delta}\ln[1 + \zeta_k(e^{-\delta} - 1)], \tag{9.14}$$

$$\tilde{z}_k = n_k - \frac{1}{\delta}\ln[1 + \tilde{\zeta}_k(e^{-\delta} - 1)]. \tag{9.15}$$

One has $z_k = n_k + \zeta_k + O(\delta)$ and $\tilde{z}_k = n_k + \tilde{\zeta}_k + O(\delta)$ as $\delta \to 0$, where the estimates $O(\delta)$ are uniform over all k and all initial data; thus, ζ_k and $\tilde{\zeta}_k$ are approximately equal to the fractional parts of z_k and $\tilde{z}_k$, respectively.

For several initial values $k = 1, 2, \ldots, k_\delta - 1$ corresponding to the forward motion of the particle, according to (Ia) one has

$$z_k = \tilde{z}_{k-1} + \frac{1}{\delta}\ln\frac{1}{\lambda}; \qquad 0 < \zeta_k < \lambda, \quad \tilde{\zeta}_k = \lambda^{-1}\zeta_k; \qquad z_{k+1} = \tilde{z}_k + \frac{1}{\delta}\ln\frac{1}{\lambda}. \tag{9.16}$$

Here and in the following formulas (9.17), (9.18) ζ_k is determined by z_k and $\tilde{z}_k$ is determined by $\tilde{\zeta}_k$, according to (9.14) and (9.15). For the value $k = k_\delta$ corresponding to the transition from the forward motion to the backward one, according to (Ib) one has

$$z_{k_\delta} = \tilde{z}_{k_\delta - 1} + \frac{1}{\delta}\ln\frac{1}{\lambda}; \quad \lambda < \zeta_{k_\delta} < 1, \quad \tilde{\zeta}_{k_\delta} = 1 + \lambda - \zeta_{k_\delta}; \quad z_{k_\delta+1} = \tilde{z}_{k_\delta} - \frac{1}{\delta}\ln\frac{1}{\lambda}. \tag{9.17}$$

Finally, for the values $k = k_\delta + 1, \ldots, m-1$ corresponding to the backward motion, according to (II) one has

$$z_k = \tilde{z}_{k-1} - \frac{1}{\delta}\ln\frac{1}{\lambda}; \qquad \tilde{\zeta}_k = \lambda\zeta_k; \qquad z_{k+1} = \tilde{z}_k - \frac{1}{\delta}\ln\frac{1}{\lambda}. \tag{9.18}$$

Notice that $k_\delta = 2$ in Fig. 9.10.

Formulas (9.14)–(9.18) define iterations of the pairs of mappings

$$\tilde{z}_{k-1} \mapsto z_k \mapsto \tilde{z}_k \tag{9.19}$$

with positive integer time k. These mappings commute with the shift $z \mapsto z + 1$. The initial value $\tilde{z}_0$ satisfies $\tilde{z}_0 \in (-\frac{1}{\delta}\ln\frac{1}{\lambda}, 0)$, and the relation $z_m \in (-\frac{1}{\delta}\ln\frac{1}{\lambda}, 0)$ defines the time m when the corresponding value leaves the positive semiaxis $z \geq 0$ and the process stops.[2]

During the forward motion, the first mapping in (9.19) increases the value of z by $\frac{1}{\delta}\ln\frac{1}{\lambda}$, and the second one changes it by a value smaller than 1. During the backward motion, the first mapping decreases z by $\frac{1}{\delta}\ln\frac{1}{\lambda}$, and the second mapping changes it again by a value smaller than 1. Therefore, if the initial value satisfies $\tilde{z}_0 \in (-\frac{1}{\delta}\ln\frac{1}{\lambda} + 2k, -2k)$ with $k > k_\delta$, then $z_{2k_\delta} \in (-\frac{1}{\delta}\ln\frac{1}{\lambda}, 0)$, and so $m = 2k_\delta$. This means that m is always even, except for a small portion $4k/(\frac{1}{\delta}\ln\frac{1}{\lambda})$ of the initial values. Thus, to complete the proof of Proposition 9.4, we only need a result stating that the transition time k_δ remains bounded when $\delta \to 0$.

---

[2]Notice that m depends on δ and $\tilde{z}_0$; thus, strictly speaking, one should write $m = m_\delta(\tilde{z}_0)$. Then the equality holds $m_\delta(\tilde{z}_0) = m_\gamma(\tilde{x}_0, \lambda)$, where $\sin\gamma = \tanh\delta$ and $\tilde{x}_0 = e^{-\delta\tilde{z}_0}$; recall that the parameter λ is fixed.

Due to invariance with respect to integer shifts, formulas (9.14)–(9.18) determine iterated maps on the unit circumference with the coordinate z mod 1. The value $k_\delta = k_\delta(\tilde{z}_0 \bmod 1)$ is a Borel-measurable function; it can be interpreted as a random variable, where the random event is represented by the variable $\tilde{z}_0$ mod 1 on the circumference with a Lebesgue measure.

Proposition 9.5. *The limiting distribution of k_δ as $\delta \to 0$ equals $P_\lambda(k) = \lambda^{k-1}(1-\lambda)$, $k = 1, 2, \dots$.*

Let us derive Proposition 9.4 using Proposition 9.5. Indeed, one has $1 - P_\lambda(1) - \dots - P_\lambda(k) = \lambda^k$. Take an arbitrary $\varepsilon > 0$ and choose k such that $\lambda^k < \varepsilon$. Then, using Proposition 9.5, choose $\delta_0 > 0$ such that $\mathbb{P}(k_\delta > k) < \varepsilon$ for any $\delta < \delta_0$. This implies that the inequality $|\tilde{z}_0 - z_{2k_\delta}| < 2k$ holds with a probability of at least $1 - \varepsilon$. Therefore, if δ satisfies $\delta < \delta_0$ and $4k/(\frac{1}{\delta}\ln\frac{1}{\lambda}) < \varepsilon$, the relative Lebesgue measure of the set of points $\tilde{z}_0 \in (-\frac{1}{\delta}\ln\frac{1}{\lambda}, 0)$ producing the value $m = 2k_\delta$ is greater than $1 - 2\varepsilon$. Passing from the variable $\tilde{z}_0$ to the variable $\tilde{x}_0 = e^{-\delta\tilde{z}_0}$, one concludes that the Lebesgue measure of the set of values of $\tilde{x}_0$ corresponding to odd m tends to 0 as $\delta \to 0$. This completes the proof of Proposition 9.4. □

Proof of Proposition 9.5. For convenience write iterations of a pair of mappings until the transition time k_δ in the form

$$z_k = \tilde{z}_{k-1} + \frac{1}{\delta}\ln\frac{1}{\lambda} \mod 1, \quad \tilde{z}_k = f_\delta^{-1}(z_k) \quad (1 \le k < k_\delta), \tag{9.20}$$

where the function f_δ is given by relations (9.14)–(9.16); one easily derives that $f_\delta(\tilde{z}) = \zeta^{-1}(\lambda\,\zeta(\tilde{z}))$, with $\zeta(z) = (1-e^{-\delta z})/(1-e^{-\delta})$. The function f_δ is monotone and injectively maps the circumference $\mathbb{R}/\mathbb{Z}$ with the coordinate z mod 1 into itself and is discontinuous at 0 mod 1. In the limit $\delta \to 0$, $f_\delta(\tilde{z})$ uniformly converges to $\lambda\tilde{z}$ and the derivative f_δ' uniformly converges to λ; the last means that

$$\liminf_{\delta\to 0} f_\delta' = \limsup_{\delta\to 0} f_\delta' = \lambda. \tag{9.21}$$

Iterations (9.20) are defined while $z_k \in \text{Range}(f_\delta)$; the first moment when $z_k \notin \text{Range}(f_\delta)$ is $k = k_\delta$.

Denote by $\mathcal{A}_\delta(k) = \{\tilde{z}_0 \bmod 1 : k_\delta(\tilde{z}_0 \bmod 1) > k\}$ the set of initial values $\tilde{z}_0$ mod 1 $\in \mathbb{R}/\mathbb{Z}$ for which the inequality $k_\delta > k$ holds. Then one has $\mathbb{P}(k_\delta > k) = |\mathcal{A}_\delta(k)|$, where $|\cdot|$ means the Lebesgue measure on $\mathbb{R}/\mathbb{Z}$. The following inductive formulas are valid: $\mathcal{A}_\delta(0) = \mathbb{R}/\mathbb{Z}$ and $\mathcal{A}_\delta(k+1) = f_\delta(\mathcal{A}_\delta(k)) - \frac{1}{\delta}\ln\frac{1}{\lambda} \bmod 1$. They imply that $|\mathcal{A}_\delta(0)| = 1$ and

$$\inf_z f_\delta'(z) \le \frac{|\mathcal{A}_\delta(k+1)|}{|\mathcal{A}_\delta(k)|} \le \sup_z f_\delta'(z). \tag{9.22}$$

Formulas (9.21) and (9.22) imply that $\lim_{\delta\to 0}|\mathcal{A}_\delta(k)| = \lambda^k$; therefore, $\lim_{\delta\to 0}\mathbb{P}(k_\delta = k) = \lim_{\delta\to 0}(\mathcal{A}_\delta(k-1)) - \mathcal{A}_\delta(k))) = \lambda^{k-1}(1-\lambda)$. Proposition 9.5 is proved. □

From Theorem 9.5 we conclude that a family of hollows $(\Omega_{\alpha,\beta(\alpha)}, I_\alpha)$ with suitably chosen $\beta(\alpha)$ is asymptotically retroreflecting. This family is called a notched angle, and the corresponding body is shown in Fig. 9.14d. Here we point out its properties.

1. A notched angle is a *convenient* hollow.
2. The corresponding measure converges *in the norm* to the retroreflector measure η_{retr}.
3. We are unaware of multidimensional generalizations of this shape.
4. The mean number of reflections in a notched angle goes to infinity as α tends to zero.
5. The boundary of a notched angle is the graph of a function.

9.5 Helmet

Another remarkable hollow called *helmet* was discovered and studied by Gouveia in [23] (see also [24]). It is a curvilinear triangle, with the inlet being the base of the triangle. Its lateral sides are arcs of parabolas, where the vertex of each parabola coincides with the focus of the other one (and also coincides with a vertex of the triangle at its base). The base is a segment contained in the common axis of the parabolas; see Fig. 9.12.

A helmet can also be defined by simple formulae. It is the domain $0 < y < \sqrt{2 - 4|x|}$ in a plane with cartesian coordinates x, y, and the inlet is the segment $|x| \le 1/2$, $y = 0$.

Helmet is a nearly perfect retroreflector; the measure η associated with this hollow satisfies $\mathcal{F}(\eta) = 0.9977$ [recall that $\mathcal{F}$ is given by (9.2)]; this value is only 0.23 % smaller than the maximum value of $\mathcal{F}$. A body bounded by helmets is shown in Fig. 9.14c.

Helmet has the following properties.

1. It is a *convenient* hollow.
2. There always exists a small discrepancy between the initial and final directions, which is maximal for perpendicular incidence and vanishes for nearly tangent incidence. See Fig. 9.13, where the support of η is shown. The figure is obtained

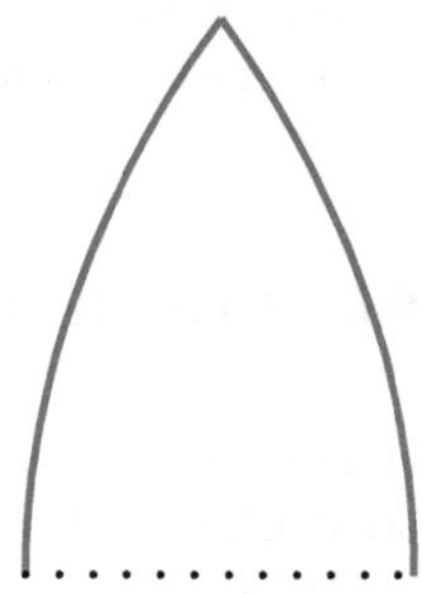

Fig. 9.12 Helmet

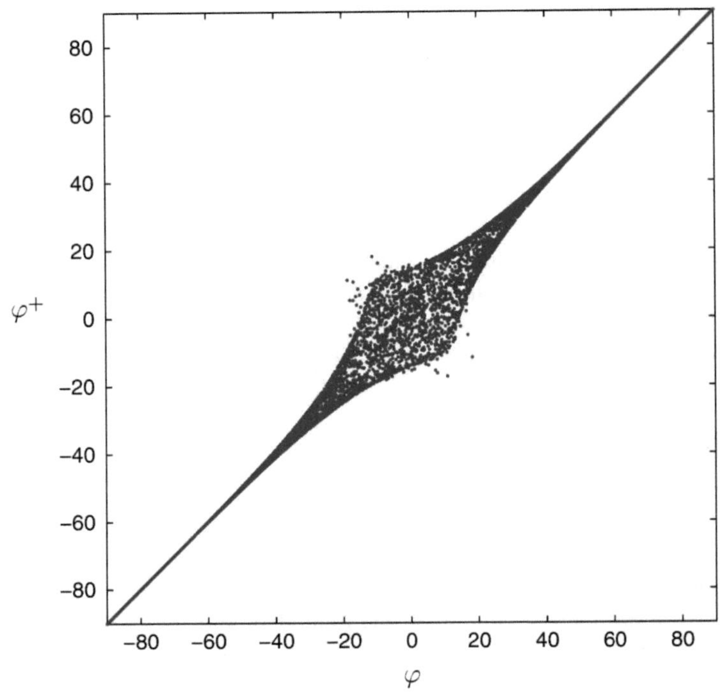

Fig. 9.13 The support of the measure associated with helmet is obtained numerically by calculating 10,000 randomly chosen pairs (φ, φ^+)

numerically by calculating the pairs (φ, φ^+) for 10,000 values of φ chosen at random. This means that, when illuminated, the contour of the retroreflector is seen best of all, which is useful for visual reconstruction of its shape.

3. We do not know if there exist multidimensional generalizations of this shape. By now, the greatest value of the parameter $\mathcal{F}$ attained by numerical simulation in three dimensions equals only 0.9.
4. For most particles, the number of successive reflections equals 3, although 4, 5, etc. (up to infinity) reflections are also possible. When the number of reflection increases, the number of corresponding particles rapidly decreases.
5. The boundary of helmet is the graph of a function. This means that this shape may be easy to manufacture.

9.6 Collection of Retroreflectors

Here we put together the billiard retroreflectors. For convenience, their properties are tabulated below. The limiting values of r are equal to 1 in all shapes, except for the Helmet.

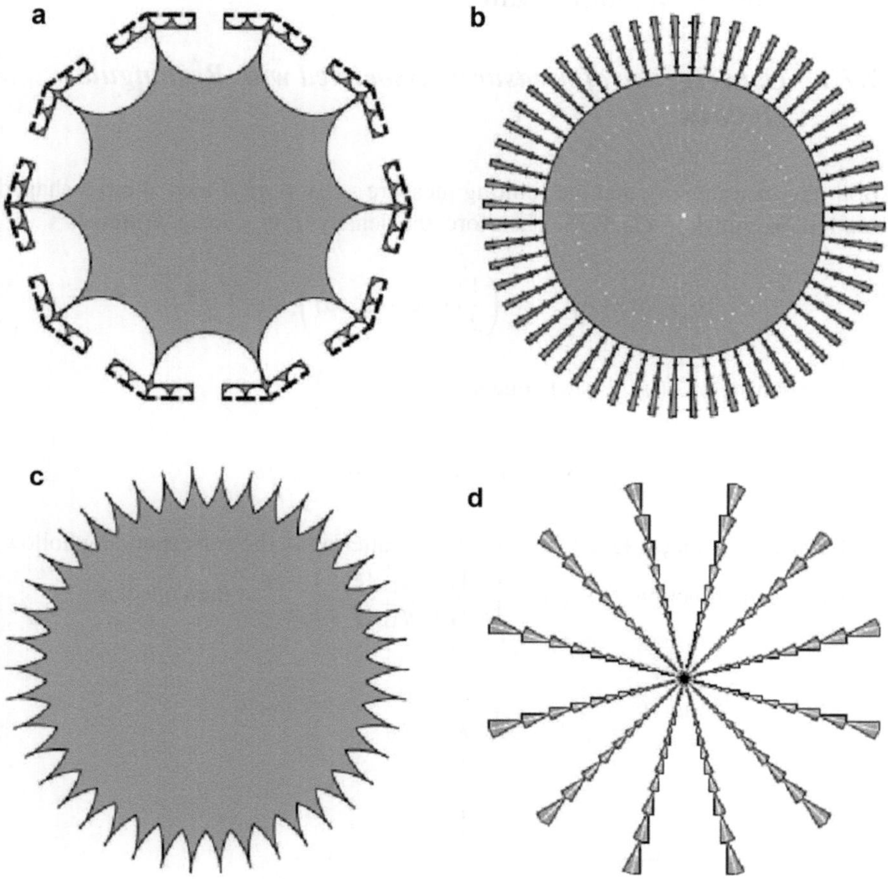

Fig. 9.14 Bodies with boundaries formed by retroreflecting hollows. (**a**) Mushroom. (**b**) Tube. (**c**) Helmet. (**d**) Notched angle

Figure 9.14 shows four bodies with boundaries formed by corresponding retroreflecting hollows.

As for possible applications of these shapes, each of them seems to have some advantages and disadvantages. The tube and notched angle ensure exact direction reversal, while in the mushroom and helmet a small discrepancy between initial and final directions is always present, which can make them inefficient at very large distances. On the other hand, the number of reflections for most incident particles in a mushroom and helmet equals 1 and 3, respectively, whereas the mean number of reflections from bodies representing a tube and notched angle goes to infinity, which may mean very high-quality requirements for the reflecting boundary.

9.7 Proofs of Auxiliary Statements

9.7.1 Convergence of Measures Associated with Rectangular Hollows

Both the measures $\eta_{\sqcup}^{\varepsilon}$ and the limiting measure $\frac{1}{2}(\eta_0 + \eta_{\text{retr}})$ have a cross-shaped support, as shown in Fig. 9.15. Therefore, the density of $\eta_{\sqcup}^{\varepsilon}$ can be written as

$$\rho_\varepsilon(\varphi)\,\delta(\varphi - \varphi^+) + \left(\frac{1}{2}\cos\varphi - \rho_\varepsilon(\varphi)\right)\delta(\varphi + \varphi^+),$$

and the density of $\frac{1}{2}(\eta_0 + \eta_{\text{retr}})$ equals

$$\frac{1}{4}\cos\varphi\,(\delta(\varphi - \varphi^+) + \delta(\varphi + \varphi^+)).$$

Let the function $\varphi_{\sqcup\varepsilon}^{+}(\xi, \varphi)$ determine the scattering in the corresponding hollow, and define the function $f_\varepsilon(\xi, \varphi) = \begin{cases} 1, \text{ if } \varphi_{\sqcup\varepsilon}^{+}(\xi, \varphi) = \varphi \\ -1, \text{ if } \varphi_{\sqcup\varepsilon}^{+}(\xi, \varphi) = -\varphi \end{cases}$ then one has

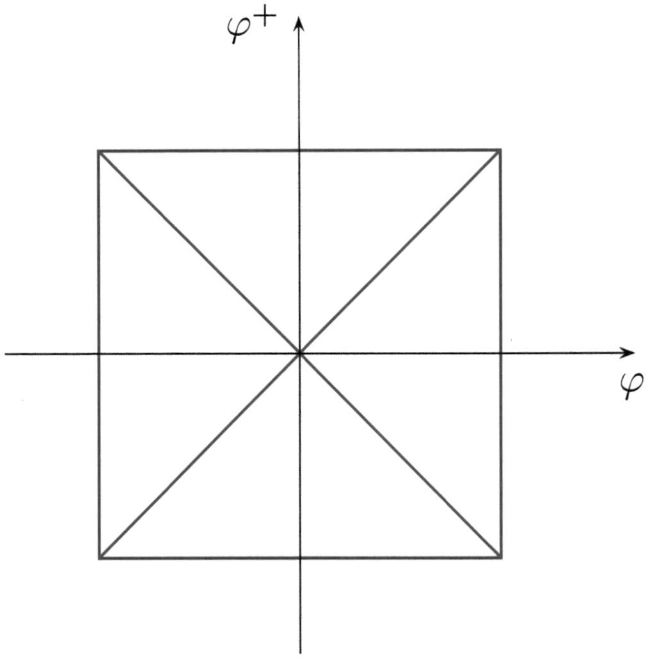

Fig. 9.15 Support of semiretroreflecting measure

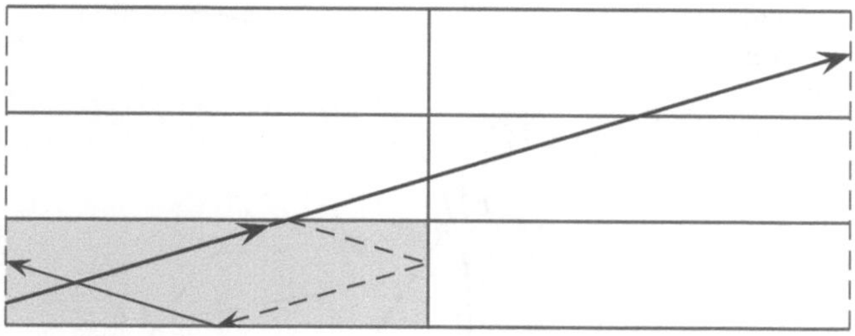

Fig. 9.16 An unfolded billiard trajectory in a tube

$$\rho_\varepsilon(\varphi) - \left(\frac{1}{2}\cos\varphi - \rho_\varepsilon(\varphi)\right) = \cos\varphi \cdot \int_0^1 f_\varepsilon(\xi,\varphi)\,d\xi.$$

The value of f_ε is determined by the parity of the number of reflections in the tube and can be easily found by unfolding the billiard trajectory (Fig. 9.16). One easily sees that $f(\xi,\varphi) = 1$ if $\lfloor \xi + \frac{2}{\varepsilon}\tan\varphi \rfloor$ is odd, and $f(\xi,\varphi) = -1$ if $\lfloor \xi + \frac{2}{\varepsilon}\tan\varphi \rfloor$ is even, where $\lfloor \ldots \rfloor$ denotes the integer part of a real number.

To prove the weak convergence, it suffices to check that for any $-\pi/2 < \Phi_1 < \Phi_2 < \pi/2$,

$$\lim_{\varepsilon\to 0} \int_0^1 \int_{\Phi_1}^{\Phi_2} f_\varepsilon(\xi,\varphi)\cos\varphi\,d\varphi\,d\xi = 0. \tag{9.23}$$

Fix ξ and denote $\varphi_m = \arctan(\frac{\varepsilon}{2}(m-\xi))$. One has $f_\varepsilon(\xi,\varphi) = 1$ if $\varphi_{2n-1} < \varphi < \varphi_{2n}$ and $f_\varepsilon(\xi,\varphi) = -1$ if $\varphi_{2n} < \varphi < \varphi_{2n+1}$. One easily deduces from this that the integral $\int_{\Phi_1}^{\Phi_2} f_\varepsilon(\xi,\varphi)\cos\varphi\,d\varphi$ converges to zero as $\varepsilon \to 0$ [and is obviously bounded, $|\int_{\Phi_1}^{\Phi_2} f_\varepsilon(\xi,\varphi)\cos\varphi\,d\varphi| < 2$], and therefore, the convergence in (9.23) takes place.

9.7.2 Convergence of Measures Associated with Triangular Hollows

The images of the triangular hollow AOB obtained by the unfolding procedure form a polygon inscribed in a circle (Fig. 9.17). Introduce the angular coordinate $x \bmod 2\pi$ (measured clockwise from point B) on the circumference. Given an incident particle, denote by x and x^+ the two points of intersection of the unfolded trajectory with the circumference. We are given $\measuredangle AOB = \varepsilon$; therefore $x \in [0, \varepsilon]$.

Let ϕ be the angle between the direction vector of the unfolded trajectory and the radius at the first point of intersection; then the angle at the second point of intersection will be $-\phi$. Both angles are measured counterclockwise from the corresponding radius to the velocity; thus, we have $\phi > 0$ in Fig. 9.17.

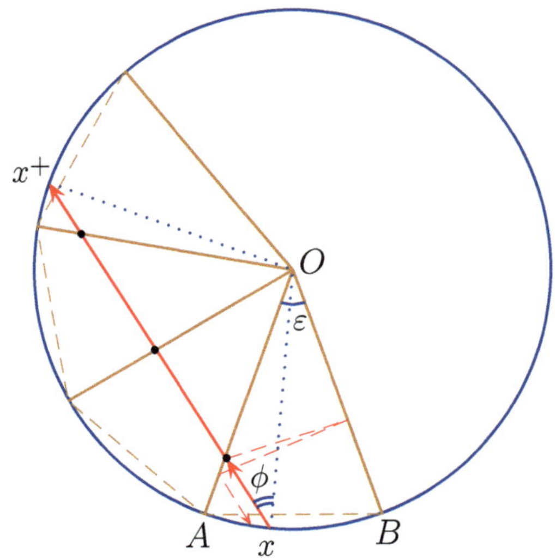

Fig. 9.17 An unfolded billiard trajectory in an isosceles triangle

One has $x^+ = x + \pi - 2\phi$. The number of intersections of the unfolded trajectory with the images of the radii OA and OB coincides with the number of reflections of the *true* billiard trajectory and is equal to $n = n_\varepsilon(x, \phi) = \lfloor \frac{x+\pi-2\phi}{\varepsilon} \rfloor$. In Fig. 9.17, $n = 3$.

Denote by φ and φ^+, respectively, the angles formed by the velocity of the true billiard trajectory with the outer normal to AB at the moments of the first and second intersections with the opening AB. One easily sees that

$$|\varphi - \phi| \le \varepsilon/2 \quad \text{and} \quad |\varphi^+ - (-1)^{n+1}\phi| \le \varepsilon/2. \tag{9.24}$$

The mapping $(x, \phi) \mapsto (\varphi, \varphi^+)$ defines a measure preserving one-to-one correspondence between a subspace of the space $[0, \varepsilon] \times [-\pi/2, \pi/2]$ with the measure $\frac{1}{2\sin(\varepsilon/2)} dx \cdot \frac{1}{2} \cos\phi \, d\phi$ and the space $\square = [-\pi/2, \pi/2]^2$ with the measure $\eta^\varepsilon_\vee$. Consider also the mapping

$$(x, \phi) \mapsto (\phi, (-1)^{n_\varepsilon(x,\phi)+1}\phi)$$

and the measure $\tilde\eta^\varepsilon_\vee$ induced on $\square$ by this mapping. One easily deduces from inequalities (9.24) that the difference $\eta^\varepsilon_\vee - \tilde\eta^\varepsilon_\vee$ converges weakly to zero as $\varepsilon \to 0$; therefore, it is sufficient to prove the weak convergence

$$\tilde\eta^\varepsilon_\vee \to \frac{1}{2}(\eta_0 + \eta_{\text{retr}}) \quad \text{as } \varepsilon \to 0. \tag{9.25}$$

Introduce the function

$$g_\varepsilon(x, \phi) = \begin{cases} 1, & \text{if } n_\varepsilon(x,\phi) \text{ is odd,} \\ -1, & \text{if } n_\varepsilon(x,\phi) \text{ is even.} \end{cases}$$

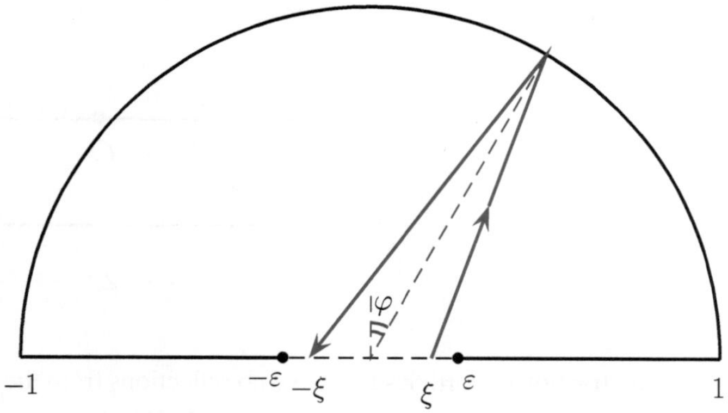

Fig. 9.18 Trajectory of a particle in a mushroom with a single reflection

Similarly to the previous Sect. 9.7.1, it suffices to prove that for any $-\pi/2 < \Phi_1 < \Phi_2 < \pi/2$,

$$\lim_{\varepsilon\to 0}\frac{1}{\varepsilon}\int_0^{\varepsilon}\int_{\Phi_1}^{\Phi_2} g_\varepsilon(x,\phi)\cos\phi\,d\phi\,dx = 0. \tag{9.26}$$

Fix $x \in [0, \varepsilon]$ and set $\phi_m = \frac{1}{2}(x + \pi - m\varepsilon)$. One has $g_\varepsilon(x,\phi) = 1$ if $\phi_{2n-1} < \phi < \phi_{2n}$ and $g_\varepsilon(x,\phi) = -1$ if $\phi_{2n} < \phi < \phi_{2n+1}$. It is easy to check that the integral $\int_{\Phi_1}^{\Phi_2} g_\varepsilon(x,\phi)\cos\phi\,d\phi$ uniformly converges to zero as $\varepsilon \to 0$ (actually, it is less than 2ε), and therefore, the convergence in (9.26) also takes place.

9.7.3 Size of the Smallest Hollows in a Mushroom Body

Here we derive a heuristical estimate of the size of the smallest retroreflector in a "mushroom seedling." All mushrooms are similar to a mushroom pattern. Their cap is a semiellipse with major semiaxis 1 and focal distance 2ε, and the stem is a rectangle of size $2\varepsilon \times \delta$, with $\delta/\varepsilon \to 0$ as $\varepsilon \to 0$. The seedling is formed by a hierarchy of mushrooms of n levels, with each level being δ times the previous one. Thus, the mushrooms of the first level have size 1, the mushrooms of the second level size δ, and the mushrooms of the nth level size δ^{n-1}. The cup height of the smallest mushrooms is δ^n. This quantity is the smallest size of construction details.

The point of entry is parameterized by $\xi \in [-\varepsilon, \varepsilon]$. Assume that the particle makes only one reflection from the cap and does not make reflections from the stem; then the point of exit is $-\xi$ (terms of higher order are neglected here). Let the direction of entry make an angle φ with the normal to the inlet of the hollow (Fig. 9.18); then the angle between the direction of entry and the direction of exit is $2\xi\cos\varphi$.

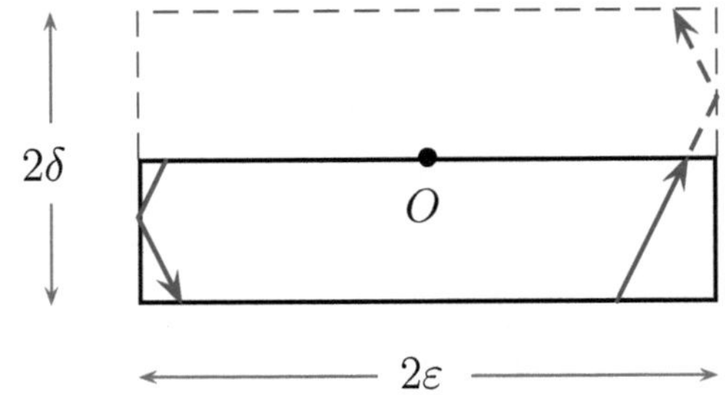

Fig. 9.19 Trajectory in a mushroom stem

Let $1 - \kappa$ be the fraction of particles that have no reflections from the stem; then the resistance of the mushroom (up to a normalizing constant) is

$$(1-\kappa)\,\frac{1}{2\varepsilon}\int_{-\varepsilon}^{\varepsilon} d\xi \cdot \frac{1}{2}\int_0^{\pi/2} (1+\cos(2\xi\cos\varphi))\,\cos\varphi\, d\varphi =$$

$$= (1-\kappa)\,\frac{1}{2\varepsilon}\int_{-\varepsilon}^{\varepsilon} d\xi \cdot \int_0^{\pi/2} (1-\xi^2\cos^2\varphi)\,\cos\varphi\, d\varphi = (1-\kappa)\left(1-\frac{2\varepsilon^2}{9}\right).$$

Let us now estimate the fraction κ of particles that make at least one reflection from the mushroom stem. The intersection of a particle trajectory with the stem is partitioned into two parts: the initial and final ones, and the directions of entering the cap and exiting it (that is, the directions at the points of intersection of the trajectory with the large semiaxis of the ellipse) coincide up to $O(\varepsilon)$. Let us reflect the final part of the trajectory relative to the center of the ellipse (point O in Fig. 9.19); then the union of the initial part and the image of the final part is a trajectory in the doubled rectangle of size $2\varepsilon \times 2\delta$ (Fig. 9.19).

If the angle of incidence of the particle satisfies the inequalities $0 < \varphi < \arctan(\varepsilon/\delta)$, then two cases are possible: (i) for $\xi \in [-\varepsilon,\ \varepsilon - 2\delta\tan\varphi]$ there are no reflections from the stem; (ii) for $\xi \in [\varepsilon - 2\delta\tan\varphi,\ \varepsilon]$ there is only one such reflection. The portion of particles satisfying (i) is

$$\frac{1}{2\varepsilon}\int_0^{\arctan\frac{\varepsilon}{\delta}} 2(\varepsilon - \delta\tan\varphi)\,\cos\varphi\, d\varphi = 1 - \delta/\varepsilon + O(\delta^2/\varepsilon^2)$$

(recall that $\delta/\varepsilon \to 0$ as $\varepsilon \to 0$). The portion of particles satisfying (ii) is $\delta/\varepsilon + O(\delta^2/\varepsilon^2)$, and the contribution of these particles to the resistance is

$$\frac{1}{2\varepsilon}\int_0^{\arctan\frac{\varepsilon}{\delta}} 2\delta\tan\varphi\,\frac{1+\cos(2\varphi)}{2}\,\cos\varphi\, d\varphi = \delta/(3\varepsilon) + O(\delta^2/\varepsilon^2).$$

On the other hand, the contribution of the particles satisfying $\arctan(\varepsilon/\delta) < \varphi < \pi/2$ is $O(\delta^2/\varepsilon^2)$.

Thus, we obtain at a first approximation that $\kappa = \delta/\varepsilon$ and the (normalized) resistance of the particles coming to the mushroom is

$$\left(1-\frac{\delta}{\varepsilon}\right)\left(1-\frac{2\varepsilon^2}{9}\right)+\frac{\delta}{3\varepsilon}=1-\frac{2\varepsilon^2}{9}-\frac{2\delta}{3\varepsilon}.$$

Finally, the relative length of the convex part of the boundary that remains unoccupied after placing the mushrooms of the nth level is $(1-\varepsilon)^n = e^{-n\varepsilon}$, whereas its specific resistance equals $2/3$. Then the rate of retroreflectivity—the resistance of the body as a whole—is defined by

$$\left(1-\frac{2\varepsilon^2}{9}-\frac{2\delta}{3\varepsilon}\right)(1-e^{-n\varepsilon})+\frac{2}{3}\,e^{-n\varepsilon}=1-\frac{2\varepsilon^2}{9}-\frac{2\delta}{3\varepsilon}-\frac{1}{3}\,e^{-n\varepsilon}.$$

Assume that the rate of retroreflectivity equals $1-c$, with c small; then we have $\varepsilon < 3\sqrt{c/2}$, $\delta < 9\sqrt{c^3/8}$, $n > -\ln(3c)\cdot\sqrt{2/(9c)}$, and we obtain a very optimistic estimate for the size of the smallest hollows

$$d(c) = e^{n\ln\delta} < \exp\left\{-\frac{1}{\sqrt{2c}}\ln(3c)\left[\ln c+\frac{1}{3}\ln(81/8)\right]\right\}. \tag{9.27}$$

Let us calculate this size in the case where the rate of retroreflectivity is the same as in the right isosceles triangle, $r = 2\sqrt{2}/3 \approx 1-0.057$. Substituting $c = 0.057$ into (9.27), we obtain $d < 0.0002$. If the size of the largest hollow is 1 m, the smallest hollow should be less than 0.02 mm.

Suppose now that we want to obtain the rate of retroreflectivity 0.99. It is much smaller than the retroreflectivity rate 0.9977 of a helmet. Substituting $c = 0.01$ into (9.27) we obtain an unrealistic quantity $d < 10^{-40}$.

List of Open Problems

1. Summarizing some results of Chaps. 2 and 6, we see that the infimum of resistance in one direction (for bodies in reasonable classes) is zero, while the infimum of the resistance averaged over all possible directions is positive. In the first case, the corresponding measure χ in (1.3) is concentrated at a point, $\chi = \delta_{v_0}$, while in the second case it is uniformly distributed over a sphere, $\chi = u$. We believe that the following more general result holds for an arbitrary measure χ.

Consider a function $c : S^{d-1} \times S^{d-1} \to \mathbb{R}$ satisfying $c(v_1, v_2) > 0$ for $v_1 \neq v_2$ and $c(v, v) = 0$. Let C_1 and C_2 be bounded convex bodies such that $C_1 \subset C_2 \subset \mathbb{R}^d$ and $\partial C_1 \cap \partial C_2 = \emptyset$, and consider the class of (not necessarily connected) bodies B such that $C_1 \subset B \subset C_2$. Recall that the resistance functional R_χ is defined by (1.3).

Conjecture 9.1. (a) If spt χ has a zero Lebesgue measure in S^{d-1}, then $\inf_{C_1 \subset B \subset C_2} R_\chi(B) = 0$.
(b) Otherwise we have $\inf_{C_1 \subset B \subset C_2} R_\chi(B) > 0$.

In the case where $c(v_1, v_2) = v_1 \cdot (v_1 - v_2)$, the conjecture can be interpreted as follows. A body moves through a medium in a random direction; we are going to minimize the mathematical expectation of the body resistance along this direction.

We claim that the mathematical expectation of the resistance can be made arbitrarily small by small (in the C^0-norm) variation of the body boundary if and only if the set of admissible directions has measure zero.

2. The class of measures associated with hollows is characterized in Chap. 4, and the infimum and supremum of the resistance of a hollow is found in Chap. 6. However, little is known about measures associated with *special* hollows and their resistances. The calculations have only been made in two dimensions, where the hollow Ω is an isosceles triangle with an angle $\alpha \in [\pi/2, \pi)$ at the apex and the inlet I is its base, and in several limiting cases. Recall that the measure associated with a hollow (Ω, I) is denoted by $\eta_{\Omega,I}$ in the two-dimensional case, and by $\nu_{\Omega,I}$ in three (and higher) dimensions. The resistance of a two-dimensional hollow equals $R_{[2]}(\eta_{\Omega,I})$, where

$$R_{[2]}(\eta) = \frac{3}{4} \iint_{\square} (1 + \cos(\varphi - \varphi^+))\, d\eta(\varphi, \varphi^+).$$

The resistance of a three-dimensional hollow equals $R_{[3]}(\nu_{\Omega,I})$, where

$$R_{[3]}(\nu) = \iint_{(S^2)^2} \frac{1}{2} |v - v^+|^2\, d\nu(v, v^+).$$

The resistance of a hollow can also be given by the following formulae. Recall that the function $\varphi^+ = \varphi^+_{\Omega,I}(\xi, \varphi)$ determines the relation between the initial position $\xi \in I$ and angle $\varphi \in [-\pi/2, \pi/2]$ of a particle incident in a two-dimensional hollow (Ω, I) and its final angle φ^+ (Sect. 4.1.1). Assume that the length of segment I equals 1 and denote by $d\xi$ the Lebesgue measure in I; then the resistance equals

$$R_{[2]}(\Omega, I) = \frac{3}{4} \int_I \int_{-\pi/2}^{\pi/2} (1 + \cos(\varphi - \varphi^+_{\Omega,I}(\xi, \varphi)))\, \frac{1}{2} \cos\varphi\, d\varphi\, d\xi.$$

Similarly, $v' = v'_{\Omega,I}(\xi, v)$ determines the relation between the initial position $\xi \in I$ and velocity v of a particle incident in a three-dimensional hollow and its final velocity v' (Sect. 4.2). Let n be the outward normal to $\partial\Omega$ at a point of I; assume that the plane domain I has area 1, and introduce the cartesian coordinates $\xi = (\xi_1, \xi_2)$ in I. The unit vectors v and v' can be expressed in polar coordinates as φ, θ and φ', θ', where φ (φ') indicates the angle between pole $-n$ and v (between n and v', respectively), and θ (θ') measures the angle along the equator. The resistance is

$$R_{[3]}(\Omega, I) = \frac{1}{\pi} \iint_I d\xi_1 \, d\xi_2 \int_0^{\pi/2} \sin\varphi \cos\varphi \, d\varphi$$

$$\int_0^{2\pi} (1 + \sin\varphi \sin\varphi' \cos(\theta - \theta') + \cos\varphi \cos\varphi') \, d\theta;$$

here we use the shorthand notation $\varphi' = \varphi'_{\Omega,I}(\xi_1, \xi_2, \varphi, \theta)$, $\theta' = \theta'_{\Omega,I}(\xi_1, \xi_2, \varphi, \theta)$.

In Problems 2.1 and 2.2 it is required to calculate measures and resistances for several special hollows.

Problem 2.1. Find the measures and resistances corresponding to the following two-dimensional hollows.

(a) The hollow Ω is an isosceles triangle with an angle $\pi/3 \le \alpha < \pi/2$ at the apex. The inlet I is the base of the triangle.
(b) The hollow is a circle segment, and the inlet is the chord bounding the segment.

Certainly the resistance in Problem 2.1(a) should be given by a unique analytic function $f(\alpha)$. The maximum resistance of a triangular hollow is attained when the triangle is isosceles with an angle $\alpha^* \in (\pi/3, \pi/2)$ at the apex [56]; therefore, the solution of Problem 2.1(a) will provide exact (implicit) formulae for that maximal value. That is, we will have $R_{\max} = f(\alpha)$, with $f'(\alpha) = 0$.

Problem 2.2. Find the measures and resistances corresponding to the following three-dimensional hollows.

(a) The hollow is a triangular pyramid. Its lateral sides are identical right isosceles triangles lying in mutually perpendicular planes. The inlet is the base of the pyramid.
(b) The hollow is a hemisphere. The inlet is the circular section bounding the hemisphere.

3. It is known (Chap. 6) that in any dimension there exists a hollow with resistance smaller than 1. That is, for a two-dimensional hollow (Ω, I) we have $R_{[2]}(\eta_{\Omega,I}) < 1$, and for a three-dimensional hollow (Ω, I) we have $R_{[3]}(\eta_{\Omega,I}) < 1$. This result, together with the fact that the mean resistance of a body with its boundary formed by hollows equals a weighted sum of resistances of the hollows, implies that in the two-dimensional case there exist a circle and a connected body containing it such that the mean resistance of the body is smaller than that of the circle. In the three-dimensional case the same is true, with a circle replaced with a ball.

Notice, however, that the hollows constructed in Chap. 4 (and in particular hollows with resistances smaller than 1) have extremely complicated shapes.

Problem 3.1. (a) Find a two-dimensional hollow with a simple shape and resistance smaller than 1.
(b) The same problem for a three-dimensional hollow.

We state a related conjecture.

Conjecture 3.2. (a) There exist a circle and a connected two-dimensional body containing it such that the resistance of the body in each direction is smaller than the resistance of the circle in this (and therefore in any) direction.

(b) The same claim in three dimensions, with the circle replaced by a ball. [Recall that the resistance of a body B in a direction v is $R_{\delta_v}(B)$.]

4. It is known (Chap. 6) that the maximum mean resistance of a two-dimensional hollow equals 1.5. A hollow called the helmet (Fig. 9.12) has a resistance of 1.4965, which is very close to the maximum value. In the three-dimensional case, the maximum resistance of a hollow is 2; however, no hollow with a resistance greater than 1.8 has been found.

Problem 4. Find a three-dimensional hollow with a resistance greater than 1.99.

5. The problem stated by Comte and Lachand-Robert in [18] (see also Sect. 2.5 of this book) is, in slightly modified form, as follows. Consider a bounded open domain $\Omega \subset \mathbb{R}^2$ and a piecewise smooth function $u : \bar{\Omega} \to \mathbb{R}$ such that $u(x) < 0$ for all $x \in \Omega$ and $u\rfloor_{\partial\Omega} = 0$. We impose the *single-impact condition* (s.i.c.) on u, which means that each incident particle falling vertically downward, after an elastic reflection from the graph of u, will no longer intersect it. This condition is expressed analytically in (2.22). The specific resistance of u equals

$$F(u) = \frac{1}{|\Omega|} \iint_\Omega \frac{d^2x}{1 + |\nabla u(x)|^2}. \tag{9.28}$$

Problem 5.1. Minimize $F(u)$ in (9.28) over all functions u satisfying the s.i.c.

One easily sees that $F(u) \geq 0.5$ for any u. Note also that $\sup_u F(u) = 1$ is attained at any sequence u_n of functions with uniformly vanishing gradient. It was shown in [18] that $\inf_u F(u) \leq 0.593$; later on this value was reduced to 0.581 in [53]. It is even unknown whether the infimum equals or is greater than 0.5.

As far as we know, the following two-dimensional analog of the problem also remains open. [Note that the analytic formula for the s.i.c. in two dimensions is obtained by replacing ∇u with u' in (2.22).]

Problem 5.2. Minimize $F_{[2]}(u) = \int_0^1 (1 + u'^2(x))^{-1} dx$ over all piecewise smooth functions $u : [0, 1] \to \mathbb{R}$ such that $f(0) = f(1) = 0$ and $f(x) < 0$ for $0 < x < 1$ and satisfying the s.i.c.

We conjecture that the minimum value, $F_{[2]}(u_*) = \pi/2 - 2 \arctan \frac{1}{2} \approx 0.644$, is attained at the function $u_*(x) = \begin{cases} \frac{(x-1)^2-1}{2} & \text{if } 0 \leq x \leq 1/2 \\ \frac{x^2-1}{2} & \text{if } 1/2 \leq x \leq 1 \end{cases}$ (Fig. 9.20).

6. Perfect billiard retroreflectors are defined in this chapter. It is known that there exist perfect *unbounded* retroreflectors, but no perfect *bounded* retroreflectors have

Fig. 9.20 Hypothetical minimizer of Problem 5.2

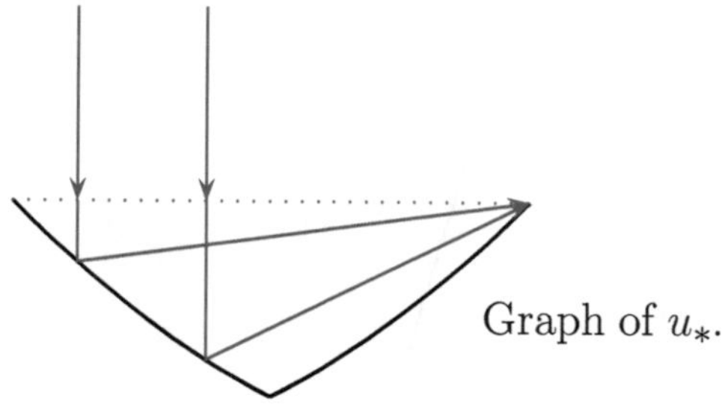

been found. A possible way of constructing a perfect retroreflector would be, first, to find a hollow with maximum resistance, and second, to take a body with its boundary formed by copies of this hollow.

Problem 6. (a) Do there exist perfect bounded billiard retroreflectors? The question remains open in any dimension.
(b) Do there exist hollows that have maximum resistance? (Notice that the maximum resistance equals 1.5 in the two-dimensional case and 2 in the three-dimensional case.)

7. It was shown in Chap. 7 that for a spinning rough disc on a plane both inverse and proper Magnus effects can happen. Recall that the transversal component $R_T[\eta, \gamma]$ of resistance force is defined in Theorem 7.1, where the measure η is associated with the disc roughness, and γ is the relative angular velocity of the disc. The inverse Magnus effect occurs when $R_T[\eta, \gamma] > 0$ and the proper one when this value is negative. However, for all simple shapes of roughness and values of γ examined by us only the inverse effect has been realized.

Problem 7.1. (a) Find a hollow with a simple shape and a value γ for which the proper Magnus effect takes place.
(b) Do there exist a single hollow and two different values γ_1 and γ_2 corresponding to the inverse and proper Magnus effect, respectively?

It was found in Chap. 7 that at least three possible kinds of trajectory of the disc center may be realized: a converging spiral, a curve approaching a circle, and a curve approaching a straight line. In addition, a special rough disc was found (bounded by a broken line with small segments and the angle nearly $\pi/3$ between consecutive segments) that exhibits all three kinds of trajectory: different trajectories correspond to different initial values of γ.

The question is: what other kinds of trajectory of a rough disc can occur? This question is closely related to the study of the set of functions $\{\gamma \mapsto R_T[\eta, \gamma] : \eta \in \mathcal{M}\}$. If for each segment $[a, b] \subset \mathbb{R}_+$ and each smooth function $f : [a, b] \to \mathbb{R}$ there exist $c > 0$ and $\eta \in \mathcal{M}$ such that $R_T[\eta, \gamma] = cf(\gamma)$, then this would probably imply that any smooth curve of finite length is the trajectory of a rough disc.

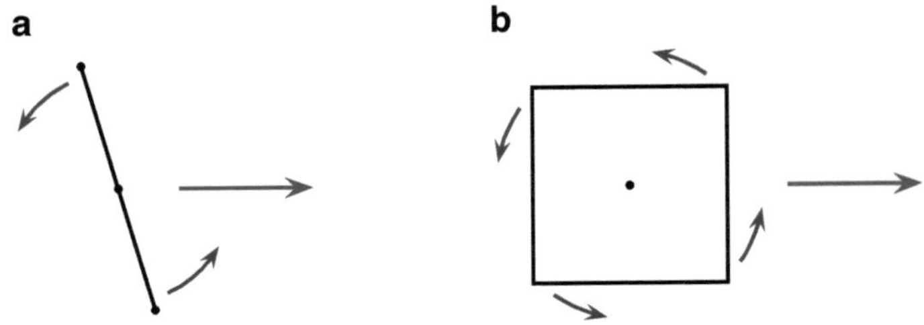

Fig. 9.21 Rotating figures: (**a**) rod; (**b**) square

Conjecture 7.2. (a) For each sufficiently smooth function $f : [a, b] \to \mathbb{R}$, $0 < a < b$, there exist $\eta \in \mathcal{M}$ and a constant $c > 0$ such that $R_T[\eta, \gamma] = cf(\gamma)$, $\gamma \in [a, b]$.

(b) Each sufficiently smooth curve of finite length in $\mathbb{R}^2$ is the trajectory of the center of a spinning rough disc.

Problem 7.3. Describe the dynamics of rotating bodies on a plane; take, for example, a rod or a square with uniform distribution of mass (Fig. 9.21) or a figure with a special shape.

8. There exist connected bodies with mirror surface in three dimensions invisible in one direction and invisible from one point. There exist disconnected bodies in three dimensions that are invisible in two directions and disconnected bodies in two dimensions that are invisible in one direction and invisible from one point. Bodies invisible in *all* directions do not exist. These results are proved in Chap. 8.

Problem 8.1. *(a) Do there exist three-dimensional bodies invisible in* 3, 4, ... *directions and three-dimensional bodies invisible from* 2, 3, ... *points?*

(b) The same question for connected bodies.

(c) The same question in other dimensions.

A body B is called *invisible for a billiard trajectory* L in $\mathbb{R}^d \setminus B$ if there exists a straight line l that intersects B and coincides with L everywhere except for the set $\text{Conv}\, B$; that is, $l \cap B \neq \emptyset$ and $L \setminus \text{Conv}\, B = l \setminus \text{Conv}\, B$.

Conjecture 8.2. Take a set $A \subset \mathbb{R}^d \times S^{d-1}$ with a positive Lebesgue measure. There do not exist bodies in $\mathbb{R}^d$ that are invisible for all trajectories with initial data $(x, v) \in A$.

We managed to prove the statement of nonexistence in this conjecture only for *polyhedral* bodies.

The following statement is a weaker version of Conjecture 8.2 corresponding to a particular case of two dimensions and four reflections. Consider a system $\Gamma = \{\gamma_1, \gamma_2, \gamma_3, \gamma_4\}$ of four smooth curves of finite length. Consider a ball $\mathcal{B} = B_\varepsilon(O)$

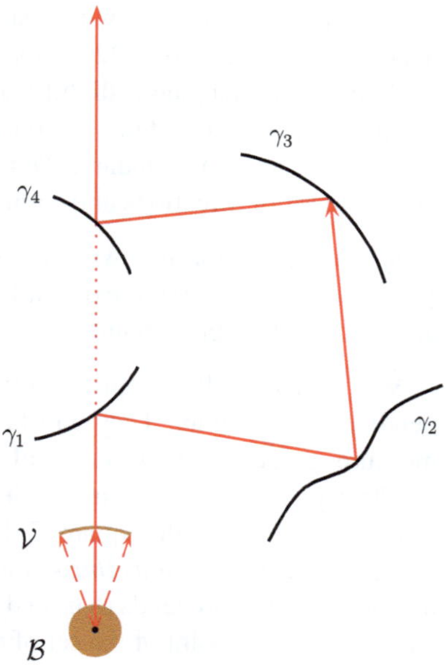

Fig. 9.22 Conjecture on four invisible curves

in $\mathbb{R}^2$ and an arc $\mathcal{V} \subset S^1$, and take the set S of billiard trajectories with initial data (ξ, v) from $\mathcal{B} \times \mathcal{V}$ (Fig. 9.22). Each trajectory from S makes four successive reflections from γ_1, γ_2, γ_3, and γ_4.

Conjecture 8.3. There does not exist a system of curves Γ invisible for all trajectories from S.

If one of the sets $\mathcal{B}$ or $\mathcal{V}$ reduces to a point, the conjecture is not true. In other words, there does exist a system of curves Γ invisible for all trajectories with initial data from a set $\mathcal{B} \times \{v\}$ or from $\{O\} \times \mathcal{V}$. See the constructions of bodies invisible in one direction and from one point (Figs. 8.3 and 8.17).

Actually, Conjecture 8.2 (and therefore also Conjecture 8.3) is closely connected with a long-standing conjecture on periodic trajectories. This conjecture, proposed by Ivrii in 1978, states that the set of periodic orbits in a billiard has measure zero. It sprang from studies on spectral asymptotics of the Laplace operator in bounded domains [28]. For a detailed review of the history of Ivrii's conjecture see [25] or [22]. The conjecture was only proved in some special cases, in particular in the case of triangular orbits [65,67,73,77] and, quite recently, in the much more complicated case of rectangular orbits in planar billiards [21,22].

The point is that a trajectory invisible for a body B can be treated as a "periodic" trajectory in $\mathbb{R}^d \setminus B$ in the following "projective" sense. The particle goes to infinity, passes through an "infinitely distant" point, and then returns along the same line, so the trajectory becomes closed. Using the same technique as in the periodic case, one can prove that the set of trajectories with three reflections for which a given two-dimensional body is invisible has measure zero. We hope that the analogous statement for trajectories with four reflections can be proved using the (suitably adapted) technique from [22]. This will prove Conjecture 8.3.

Abusing the language slightly, a bounded set is called a *fractal body* if it has infinitely many connected components, and any such component is a domain with a piecewise smooth boundary. We believe that there exist fractal bodies invisible in any finite set of directions and from any finite set of points.

Conjecture 8.4. For any two finite sets of vectors $\{v_1, \dots, v_n\} \subset S^{d-1}$ and points $\{x_1, \dots, x_m\} \subset \mathbb{R}^d$ there exists a fractal body invisible in each of these directions and from each of these points.

9. Consider hollows formed by broken lines with segments parallel and perpendicular to an inlet. They will be referred to as *R-hollows*. We are interested in measures associated with such hollows.

Clearly, the support of any such measure lies in the union of diagonals $\{\varphi^+ = -\varphi\} \cup \{\varphi^+ = \varphi\}$ of the square $\square$. Let the hollow be a rectangle and the inlet be its lower side. If the *(height)/(base)* ratio of the rectangle goes to 0 or to infinity, then the induced measure tends to η_0 and to $\frac{1}{2}(\eta_0 + \eta_{\text{retr}})$, respectively. On the other hand, η_{retr} is a limiting point of the set of measures associated with tubes. Thus, the weak closure of the set of measures associated with R-hollows contains the following three measures.

(a) η_0. Recall that this measure has a density $\frac{1}{2}\cos\varphi\,\delta(\varphi + \varphi^+)$.
(b) η_{retr}. This measure has a density $\frac{1}{2}\cos\varphi\,\delta(\varphi - \varphi^+)$.
(c) $\frac{1}{2}(\eta_0 + \eta_{\text{retr}})$. This has a density $\frac{1}{4}\cos\varphi\,(\delta(\varphi + \varphi^+) + \delta(\varphi - \varphi^+))$.

We believe that each measure $\eta \in \mathcal{M}$ supported on the union of diagonals can be weakly approximated by measures associated with R-hollows. More precisely, let $\mathcal{M}_{\text{diag}}$ be the set of measures $\eta \in \mathcal{M}$ supported on the union of diagonals of $\square$. One easily sees that $\eta \in \mathcal{M}_{\text{diag}}$ if and only if the density of η has the form

$$g_1(\varphi)\delta(\varphi + \varphi^+) + g_2(\varphi)\delta(\varphi - \varphi^+),$$

where the measurable functions $g_1, g_2 : [-\pi/2, \pi/2] \to \mathbb{R}$ are even and nonnegative and $g_1(\varphi) + g_2(\varphi) = \frac{1}{2}\cos\varphi$.

Conjecture 9.1. The weak closure of the set of measures associated with R-hollows coincides with $\mathcal{M}_{\text{diag}}$.

Now consider hollows formed by broken lines, where the angle between any segment of the line and the inlet is a multiple of π/m. Such hollows will be called *Π_m-hollows*.

One easily sees that each measure associated with such a hollow is supported in the union of straight lines

$$\cup_{k \in \mathbb{Z}} \left(\{\varphi^+ = \varphi + 2\pi k/m\} \cup \{\varphi^+ = -\varphi + 2\pi k/m\} \right). \tag{9.29}$$

Conjecture 9.2. Each measure $\eta \in \mathcal{M}$ supported in the union of lines (9.29) can be weakly approximated by measures associated with Π_m-hollows.

If this conjecture is true, it will open the way to another proof of Theorem 4.1. We hope that the proof will provide a method of constructing hollows that have desirable properties (for instance, have nearly minimum or maximum resistance) and relatively simple shape. Note that the hollows constructed in the proof of that theorem have an extremely complicated shape that cannot by any means be reproduced in any practical application.

Bibliography

1. Aleksenko, A., Plakhov, A.: Bodies of zero resistance and bodies invisible in one direction. Nonlinearity **22**, 1247–1258 (2009)
2. Ambrosio, L.: Lecture notes on optimal transport problems. In: Mathematical Aspects of Evolving Interfaces (Funchal, Portugal, 2000), ed. P. Colli and J.F. Rodrigues, Lecture Notes in Math. **1812**, Springer, Berlin, 2003, 1–52
3. Bachurin, P., Khanin, K., Marklof, J., Plakhov, A.: Perfect retroreflectors and billiard dynamics. J. Mod. Dyn. **5**, 33–48 (2011)
4. Belloni, M., Kawohl, B.: A paper of Legendre revisited. Forum Math. **9**, 655–668 (1997)
5. Belloni, M., Wagner, A.: Newton's problem of minimal resistance in the class of bodies with prescribed volume. J. Convex Anal. **10**, 491–500 (2003)
6. Borg, K.I., Söderholm, L.H.: Orbital effects of the Magnus force on a spinning spherical satellite in a rarefied atmosphere. Eur. J. Mech. B/Fluids **27**, 623–631 (2008)
7. Borg, K.I., Söderholm, L.H., Essén, H.: Force on a spinning sphere moving in a rarefied gas. Phys. Fluids **15**, 736–741 (2003)
8. Brock, F., Ferone, V., Kawohl, B.: A symmetry problem in the calculus of variations. Calc. Var. **4**, 593–599 (1996)
9. Bucur, D., Buttazzo, G.: Variational Methods in Shape Optimization Problems. Progress in Nonlinear Differential Equations and Their Applications **65**, Birkhäuser, Boston, Inc., Boston, MA (2005)
10. Bunimovich. L.B.: Mushrooms and other billiards with divided phase space. Chaos **11**, 802–808 (2001)
11. Buttazzo, G., Ferone, V., Kawohl, B.: Minimum problems over sets of concave functions and related questions. Math. Nachr. **173**, 71–89 (1995)
12. Buttazzo, G., Kawohl, B.: On Newton's problem of minimal resistance. Math. Intell. **15**, 7–12 (1993)
13. Chechkin, G.A., Piatnitski, A.L., Shamaev, A.S.: Homogenization. Methods and Applications. Translations of Mathematical Monographs, vol. 234. American Mathematical Society, Providence (2007)
14. Chernov, N.: Entropy, Lyapunov exponents, and mean free path for billiards. J. Stat. Phys. **88**, 1–29 (1997)
15. Chow, T.S.: Wetting of rough surfaces. J. Phys.: Condens. Matter **10**(27), L445 (1998)
16. Comte, M., Lachand-Robert, T.: Newton's problem of the body of minimal resistance under a single-impact assumption. Calc. Var. Partial Differ. Eq. **12**, 173–211 (2001)
17. Comte, M., Lachand-Robert, T.: Existence of minimizers for Newton's problem of the body of minimal resistance under a single-impact assumption. J. Anal. Math. **83**, 313–335 (2001)

A. Plakhov, *Exterior Billiards: Systems with Impacts Outside Bounded Domains*, DOI 10.1007/978-1-4614-4481-7,

18. Comte, M., Lachand-Robert, T.: Functions and domains having minimal resistance under a single-impact assumption. SIAM J. Math. Anal. **34**, 101–120 (2002)
19. Eaton, J.E.: On spherically symmetric lenses. Trans. IRE Antennas Propag. **4**, 66–71 (1952)
20. Fieseler, P.D.: A method for solar sailing in a low Earth orbit. Acta Astron. **43**, 531–541 (1998)
21. Glutsyuk, A., Kudryashov, Yu.: On quadrilateral orbits in planar billiards. Dokl. Akad. Nauk **438**, 590–592 (2011) (in Russian)
22. Glutsyuk, A.A, Kudryashov, Yu.G: On quadrilateral orbits in planar billiards. Doklady Math. **83**, 371–373 (2011) [translated from Dokl. Akad. Nauk 438, 509–592 (2011) (Russian)]
23. Gouveia, P.D.F.: Computação de Simetrias Variacionais e Optimização da Resistência Aerodinâmica Newtoniana. Ph.D. Thesis, Universidade de Aveiro (2007)
24. Gouveia, P., Plakhov, A., Torres, D.: Two-dimensional body of maximum mean resistance. Appl. Math. Comput. **215**, 37–52 (2009). doi:10.1016/j.amc.2009.04.030
25. Gutkin, E.: Billiard dynamics: a survey with the emphasis on open problems. Regul. Chaotic Dyn. **8**, 1–13 (2003)
26. Harrison, I.K., Swinerd, G.G.: A free molecule aerodynamic investigation using multiple satellite analysis. Planet. Space Sci. **44**, 171–180 (1996)
27. Ivanov, S.G., Yanshin, A.M.: Forces and moments acting on bodies rotating around a symmetry axis in a free molecular flow. Fluid Dyn. **15**, 449 (1980)
28. Ivrii, V.Ya.: The second term of the spectral asymptotics for a laplace-beltrami operator on manifolds with boundary. Funct. Anal. Appl. **14**, 98–106 (1980)
29. Lachand-Robert, T., Oudet, E. Minimizing within convex bodies using a convex hull method. SIAM J. Optim. **16**, 368–379 (2006)
30. Lachand-Robert, T., Peletier, M.A.: Newton's problem of the body of minimal resistance in the class of convex developable functions. Math. Nachr. **226**, 153–176 (2001)
31. Lachand-Robert, T., Peletier, M.A.: An example of non-convex minimization and an application to Newton's problem of the body of least resistance. Ann. Inst. H. Poincaré Anal. Non Lin. **18**, 179–198 (2001)
32. Legendre, A.M.: Sur la maniére de distinguer les maxima des minima dans le calcul des variations. Memoires de L'Academie royale de Sciences annee 1786, pp. 7–37. Paris (1788)
33. Levin, V.L.: Solution of the Monge and the Monge–Kantorovich problems. Theory and applications. Dokl. Akad. Nauk **388**, 7–10 (2003) (in Russian)
34. Levin, V.L.: Optimality conditions and exact solutions of the two-dimensional Monge–Kantorovich problem. Zap. Nauchn. Sem. S.-Peterburg. Otdel. Mat. Inst. Steklov. (POMI), **312** (2004) [Teor. Predst. Din. Sist. Komb. i Algoritm. Metody. XI. Translation in J. Math. Sci. (N.Y.) **133**, 1456–1463 (2006) (in Russian)]
35. Levin, V.L.: Optimal solutions of the Monge problem. Adv. Math. Econ. **6**, 85–122 (2004)
36. Matheron, G.: Random Sets and Integral Geometry. Wiley, New York (1975)
37. McCann, R.J.: Exact solutions to the transportation problem on the line. Proc. R. Soc. Lond. A **455**, 1341–1380 (1999)
38. Mehta, R.D.: Aerodynamics of sport balls. Annu. Rev. Fluid Mech. **17**, 151–189 (1985)
39. Moe, K., Moe, M.M.: Gas–surface interactions and satellite drag coefficients. Planet. Space Sci. **53**, 793–801 (2005)
40. Newton, I.: Philosophiae naturalis principia mathematica. London: Streater (1687)
41. Ogilvy, J.A.: Theory of Wave Scattering from Random Rough Surfaces. A. Hilger in Bristol, England, Philadelphia (1991)
42. Persson, B.N.J.: Contact mechanics for randomly rough surfaces. Surf. Sci. Rep. **61**, 201–227 (2006)
43. Plakhov, A.: Newton's problem of a body of minimal aerodynamic resistance. Doklady Math. **390**, 314–317 (2003)
44. Plakhov, A.: Newton's problem of the body of minimal resistance with a bounded number of collisions. Russ. Math. Surv. **58**, 191–192 (2003)
45. Plakhov, A.: Newton's problem of the body of minimum mean resistance. Sbornik: Math. **195**, 1017–1037 (2004)

46. Plakhov, A.: Precise solutions of the one-dimensional Monge–Kantorovich problem. Sbornik: Math. **195**, 1291–1307 (2004)
47. Plakhov, A.: Newton's problem of minimal resistance for bodies containing a half-space. J. Dyn. Contr. Syst. **10**, 247–251 (2004)
48. Plakhov, A., Torres, D.: Newton's aerodynamic problem in media of chaotically moving particles. Sbornik: Math. **196**, 885–933 (2005). doi:10.1070/SM2005v196n06ABEH000904
49. Plakhov, A.: Billiards in unbounded domains reversing the direction of motion of a particle. Russ. Math. Surv. **61**, 179–180 (2006)
50. Plakhov, A.: Billiards and two-dimensional problems of optimal resistance. Arch. Ration. Mech. Anal. **194**, 349–382 (2009)
51. Plakhov, A.: Billiard scattering on rough sets: two-dimensional case. SIAM J. Math. Anal. **40**, 2155–2178 (2009)
52. Plakhov, A.: Scattering in billiards and problems of Newtonian aerodynamics. Russ. Math. Surv. **64**, 873–938 (2009)
53. Plakhov, A.: Comment on "Functions and domains having minimal resistance under a single-impact assumption" [SIAM J. Math. Anal. 34 (2002), pp.101–120]. SIAM J. Math. Anal. **41**, 1721–1724 (2009). doi:10.1137/09075439X
54. Plakhov, A.: Mathematical retroreflectors. Discr. Contin. Dyn. Syst. A **30**, 1211–1235 (2011)
55. Plakhov, A.: Optimal roughening of convex bodies. Can. J. Math. **64**, 1058–1074 (2012)
56. Plakhov, A., Gouveia, P.: Problems of maximal mean resistance on the plane. Nonlinearity **20**, 2271–2287 (2007)
57. Plakhov, A., Roshchina, V.: Invisibility in billiards. Nonlinearity **24**, 847–854 (2011)
58. Plakhov, A., Tchemisova, T.: Force acting on a spinning rough disk in a flow of non-interacting particles. Doklady Math. **79**, 132–135 (2009)
59. Plakhov, A., Tchemisova, T., Gouveia, P.: Spinning rough disk moving in a rarefied medium. Proc. R. Soc. A. **466**, 2033–2055 (2010)
60. Plakhov, A., Torres, D.: Newton's aerodynamic problem in media of chaotically moving particles. Sbornik: Math. **196**, 885–933 (2005)
61. Pontryagin, L.S., Boltyanskii, V.G., Gamkrelidze, R.V., Mishchenko, E.F.: The Mathematical Theory of Optimal Processes. Wiley (Interscience), New York (1962)
62. Prandtl, L.: Application of the "Magnus Effect" to the wind propulsion of ships. Die. Naturwiss. **13**, 93–108 (1926) [Transl. NACA-TM-367]
63. Rachev, S.T., Rüschendorf, L.: Mass Transportation Problems, vol. 1: Theory. Probability and its Applications (New York). Springer, New York (1998)
64. Rubinov, S.I., Keller, J.B.: The transverse force on a spinning sphere moving in a viscous fluid. J. Fluid Mech. **11**, 447–459 (1961)
65. Rychlik, M.R.: Periodic points of the billiard ball map in a convex domain. J. Diff. Geom. **30**, 191–205 (1989)
66. Santaló, L.A.: Integral Geometry and Geometric Probability. Addison-Wesley, Reading (1976)
67. Stojanov, L.: Note on the periodic points of the billiard. J. Diff. Geom. **34**, 835–837 (1991)
68. Tabachnikov, S.: Billiards. Société Mathématique de France, Paris (1995)
69. Tikhomirov, V.M.: Newton's aerodynamic problem. Kvant (5), 11–18 (1982) (in Russian)
70. Tikhomirov, V.M.: Stories about maxima and minima. In: Mathematical World, vol. 1. AMS, Providence (1990) [Translated from: V.M. Tikhomirov. Rasskazy o maksimumakh i minimumakh. Nauka, Moscow, 1986 (in Russian)]
71. Tyc, T., Leonhardt, U.: Transmutation of singularities in optical instruments. New J. Phys. **10**, 115038 (8 pp.) (2008)
72. Uckelmann, L.: Optimal couplings between one-dimensional distributions. In: Benes, V., Stepan, J. (eds.) Distributions with Given Marginals and Moment Problems, pp. 275–281. Kluwer, Dordrecht (1997)
73. Vorobets, Ya.B.: On the measure of the set of periodic points of a billiard. Math. Notes **55**, 455–460 (1994)
74. Wang, C.T.: Free molecular flow over a rotating sphere. AIAA J. **10**, 713 (1972)

75. Weidman, P.D., Herczynski, A.: On the inverse Magnus effect in free molecular flow. Phys. Fluids **16**, L9–L12 (2004)
76. Wojtkowski, M.P.: Principles for the design of billiards with nonvanishing Lyapunov exponents. Commun. Math. Phys. **105**, 391–414 (1986)
77. Wojtkowski, M.P.: Two applications of Jacobi fields to the billiard ball problem. J. Diff. Geom. **40**, 155–164 (1994)

Index

A. Plakhov, *Exterior Billiards: Systems with Impacts Outside Bounded Domains*, DOI 10.1007/978-1-4614-4481-7,

MIX
Papier aus verantwortungsvollen Quellen
Paper from responsible sources
FSC® C105338

If you have any concerns about our products,
you can contact us on
ProductSafety@springernature.com

In case Publisher is established outside the EU,
the EU authorized representative is:
Springer Nature Customer Service Center GmbH
Europaplatz 3, 69115 Heidelberg, Germany

Printed by Libri Plureos GmbH
in Hamburg, Germany